英文E-mail实用大全
修订本

张慈庭英语研发团队　著

江苏凤凰科学技术出版社

图书在版编目（CIP）数据

英文E-mail实用大全 / 张慈庭英语研发团队著. -- 修订本. -- 南京：江苏凤凰科学技术出版社, 2020.5
（易人外语）
ISBN 978-7-5713-0722-6

Ⅰ. ①英… Ⅱ. ①张… Ⅲ. ①电子邮件 - 英语 - 写作 - 自学参考资料 Ⅳ. ①TP393. 098

中国版本图书馆CIP数据核字(2020)第007173号

英文 E-mail 实用大全 修订本

著　　者	张慈庭英语研发团队
责任编辑	葛　昀
责任校对	杜秋宁
责任监制	方　晨
出版发行	江苏凤凰科学技术出版社
出版社地址	南京市湖南路 1 号 A 楼，邮编：210009
出版社网址	http://www.pspress.cn
印　　刷	天津旭丰源印刷有限公司
开　　本	718 mm × 1000 mm　1/16
印　　张	30.5
字　　数	460 000
版　　次	2020 年 5 月第 1 版
印　　次	2020 年 5 月第 1 次印刷
标准书号	ISBN 978-7-5713-0722-6
定　　价	56.00 元

图书如有印装质量问题，可随时向我社出版科调换。

序 | PREFACE

在职场上，尤其处于外贸行业，不懂得如何使用E-mail，就会处于劣势。许多英语学习者往往背了很多单词，也学了一些日常生活的会话，但在国际贸易等外贸场合却完全不知道该如何用英文E-mail表达自己的想法或传递信息，甚至就连学了多年英文，已经能用英文流利聊天的人，一想到要写一封正式的英文E-mail，也会觉得无从下手。不少人可能都会希望有一本能直接在各个情境中套用的英文E-mail范本大全，只要替换填入适当的名字、日期等内容，就能轻松传达信息。

本书就是能满足读者需求的英文E-mail范本大全，让所有害怕写英文E-mail的读者能够随抄随用，省时省力，在职场中游刃有余。除了200篇各种情境的英文E-mail范例，更收录了“英文E-mail写作指南”，让读者了解E-mail书写的基本规则与格式。还有“英文E-mail相关词汇”，让读者能够在关键时刻轻松套用专业单词，保证不出丑。我们的期望就是无论是职场菜鸟还是谈判老手，都不再畏惧书写英文E-mail，在职场上尽情发挥真正的实力。

书中最后还特别收录了“商用书信必抄200惯用句”，让读者在遇到某些不知道如何表达才最礼貌的情况时，能轻松使用这些规范惯用句，不必再有“我这样写，会不会好像在责怪对方？对方会不会生我的气？”的顾虑，更不用担心会因为不清楚国外风土民情而不小心触犯禁忌。所以请大家抛开对英文E-mail的恐惧，自信且自在地迈向职场生涯的更高点吧！

张慈庭英语研发团队

張慈庭

目录 | CONTENTS

Part 1 英文E-mail 写作指南

Part 2 英文E-mail 实例集

Unit 1 交际 Socializing

Unit 2 申请 Application

Unit 3 求职 Applying for a Job

Unit 4 感谢 Gratitude

Unit 5 邀请 Invitation

Unit 6 通知 Notice

Unit 7 业务开发维护 Business Establishing & Maintaining

Unit 8 询问 Inquiry

Unit 9 | 请求 Request

Unit 10 | 催促 Urging

Unit 11 | 投诉 Complaint

Unit 12 拒绝 Refusal

Unit 13 道歉 Apology

Unit 14 恭贺 Congratulations

Unit 15 慰问吊唁 Consolations and Condolences

Part 3 英文E-mail 词汇篇

Part 4 英文E-mail 超值附赠

使用说明 USER'S GUIDE

Part 1 英文E-mail 写作指南

01 英文E-mail写作指南，建立完整写作基础：

Part 1中收录了英文E-mail写作的7C原则、图解英文E-mail 八大组成元素以及英文E-mail电脑相关实用技巧三个单元。

这三个单元为学习者提供了基础写作知识及电脑实用技巧，不但能够让E-mail写作更具有逻辑性及完整性，还能提高写作效率！

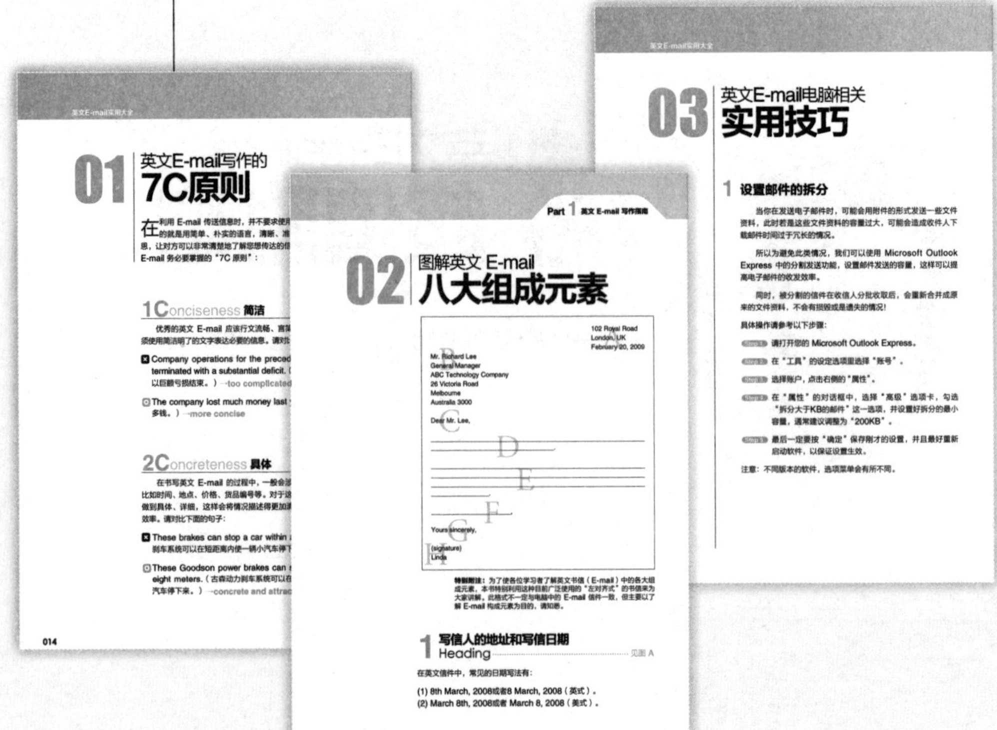
英文E-mail实用大全

01 英文E-mail写作的 7C原则

在利用 E-mail 传送信息时，并不要求使用
的就是用简单、朴实的语言，清晰、准
息，让对方可以非常清楚地了解您想传达的信
E-mail 务必要掌握的"7C 原则"：

1Conciseness 简洁

优秀的英文 E-mail 应该行文流畅、言简
须使用简洁明了的文字表达必要的信息。请对比

Company operations for the preced
terminated with a substantial deficit. (
以巨额亏损结束。) →too complicated

The company lost much money last
多钱。) →more concise

2Concreteness 具体

在书写英文 E-mail 的过程中，一般会涉
比如时间、地点、价格、货品编号等。对于这
做到具体、详细，这样会将情况描述得更加清
效率。请对比下面的句子：

These brakes can stop a car within
刹车系统可以在短距离内使一辆小汽车停下

These Goodson power brakes can
eight meters.（古森动力刹车系统可以在
汽车停下来。）→concrete and attrac

014

Part 1 英文 E-mail 写作指南

02 图解英文 E-mail 八大组成元素

102 Royal Road
London, UK
February 20, 2009

Mr. Richard Lee
General Manager
ABC Technology Company
26 Victoria Road
Melbourne
Australia 3000

Dear Mr. Lee,

Yours sincerely,

(signature)
Linda

A B C D E F G H

特别附注：为了使各位学习者了解英文书信（E-mail）中的各大组成元素，本书特别利用这种目前广泛使用的"左对齐式"的书信来为大家讲解。此格式不一定与电脑中的 E-mail 信件一致，但主要以了解 E-mail 构成元素为目的，请知悉。

1 写信人的地址和写信日期 Heading ……见图 A

在英文信件中，常见的日期写法有：

(1) 8th March, 2008或者8 March, 2008（英式）。
(2) March 8th, 2008或者 March 8, 2008（美式）。

017

英文E-mail实用大全

03 英文E-mail电脑相关 实用技巧

1 设置邮件的拆分

当你在发送电子邮件时，可能会用附件的形式发送一些文件资料，此时若是这些文件资料的容量过大，可能会造成收件人下载邮件时间过于冗长的情况。

所以为避免此类情况，我们可以使用 Microsoft Outlook Express 中的分割发送功能，设置邮件发送的容量，这样可以提高电子邮件的收发效率。

同时，被分割的信件在收信人分批收取后，会重新合并成原来的文件资料，不会有损毁或是遗失的情况！

具体操作请参考以下步骤：

Step 1 请打开您的 Microsoft Outlook Express。

Step 2 在"工具"的设定选项里选择"账号"。

Step 3 选择账户，点击右侧的"属性"。

Step 4 在"属性"的对话框中，选择"高级"选项卡，勾选"拆分大于KB的邮件"这一选项，并设置好拆分的最小容量，通常建议调整为"200KB"。

Step 5 最后一定要按"确定"保存刚才的设置，并且最好重新启动软件，以保证设置生效。

注意：不同版本的软件，选项菜单会有所不同。

Part 2

英文E-mail实例集

本单元收录多达200篇的E-mail实例，让学习者可以灵活应对各种情况。本单元版面模拟电子邮箱界面，让学习者可以迅速了解各个栏位的使用方式，搭配信件下方的中文翻译，E-mail学习绝对简单到不行！遇到紧急状况的时候，学习者可以直接套用，只需几分钟就能完成一封信件，工作效率高到惊人！

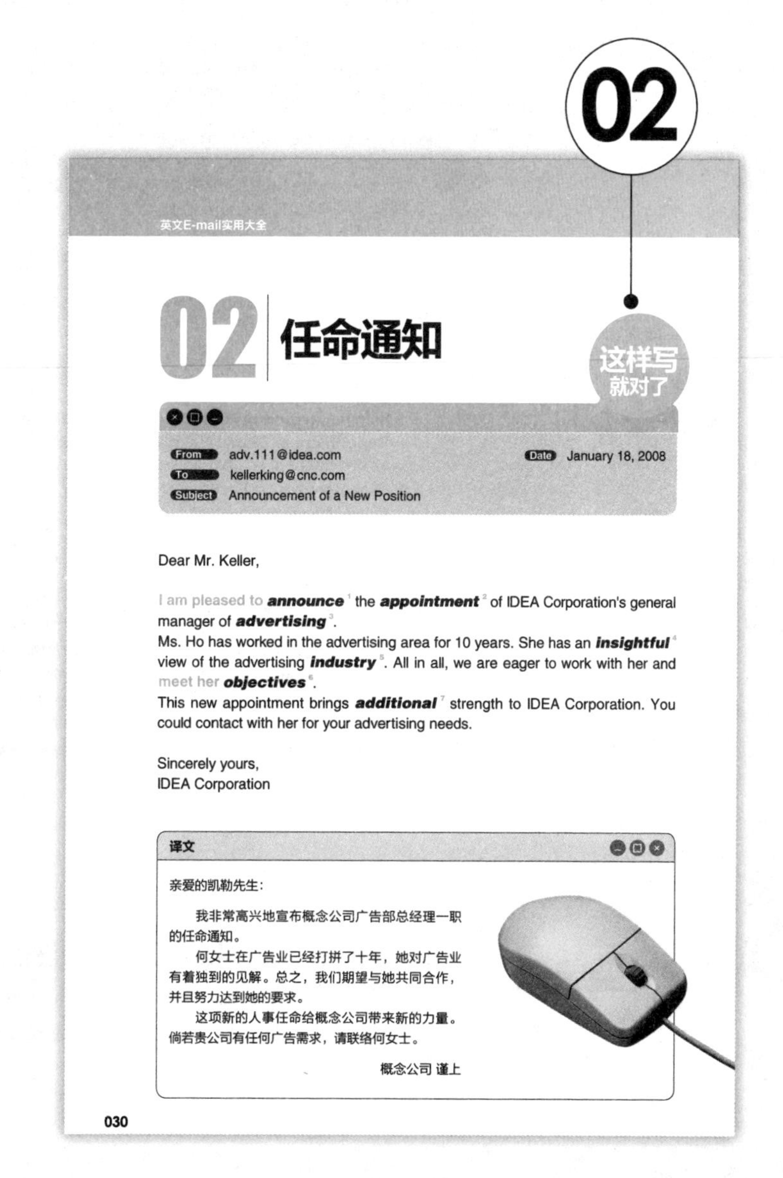

02
英文E-mail实用大全
02 任命通知
这样写就对了
From adv.111@idea.com
Date January 18, 2008
To kellerking@cnc.com
Subject Announcement of a New Position
Dear Mr. Keller,
I am pleased to announce the appointment of IDEA Corporation's general manager of advertising.
Ms. Ho has worked in the advertising area for 10 years. She has an insightful view of the advertising industry. All in all, we are eager to work with her and meet her objectives.
This new appointment brings additional strength to IDEA Corporation. You could contact with her for your advertising needs.
Sincerely yours,
IDEA Corporation
译文
亲爱的凯勒先生：
我非常高兴地宣布概念公司广告部总经理一职的任命通知。
何女士在广告业已经打拼了十年，她对广告业有着独到的见解。总之，我们期望与她共同合作，并且努力达到她的要求。
这项新的人事任命给概念公司带来新的力量。倘若贵公司有任何广告需求，请联络何女士。
概念公司 谨上
030

本书中变色或粗体标示之处，即为大部分学习者容易犯错或是混淆的语法，本书作者特别针对此部分进行了完整的解析，让学习者不再写出错误百出的信件。

Part 2 英文 E-mail 实例集　Unit 1 交际

语法重点解析

1 解析重点1 **I am pleased to...**

I am pleased to...意为“我非常高兴……”英语书信中，新产品开发、就职祝贺等充满喜悦气氛的场合，都可以用be pleased to，be happy to，be glad to 等表达方式。

例如：

I am pleased to hear of your success.（听到了你成功的消息，我很高兴。）

I am happy to accept the position of flight attendant with your company.（我很高兴接受贵公司空乘员的职位。）

2 解析重点2 **meet one's objective**

meet one's objective 意为“达到某人的要求”，meet 意为“遇见；满足”。objective 本来有形容词和名词两种词性，但在此作名词用，意为“目标”。与它同义的词还有 target，goal 等。所以要表达“实现或达到某人的目标”还可以说reach one's target，reach one's goal。例如：

The young people tried their best to reach their targets.（那些年轻人尽全力去达成他们的目标。）

I will keep on striving until I reach my goal!（我会继续奋斗直到实现我的目标！）

04

高频例句

1. **I am being promoted to Manager of the Marketing Department at Global Links as of March 21.**
   3月21日起我将正式升职为Global Links的市场部经理。
2. **Miss Bree will be taking over my territory as your sales representative.**
   布里小姐将接手我的工作，担任你们的销售代表。
3. **Thank you so much for your support *during*[8] the past years.**
   非常感谢您这些年来的支持。
4. **I am your new Diamond International sales representative.**
   我现在担任钻石国际新的销售代表。

必背关键单词

1. ***announce*** [əˈnaʊns] v. 宣布；公告
2. ***appointment*** [əˈpɔɪntmənt] n. 任命
3. ***advertising*** [ˈædvətaɪzɪŋ] n. 广告
4. ***insightful*** [ˈɪnsaɪtful] adj. 有洞察力的
5. ***industry*** [ˈɪndəstrɪ] n. 工业；行业
6. ***objective*** [əbˈdʒektɪv] n. 目标
7. ***additional*** [əˈdɪʃənl] adj. 额外的；另外的
8. ***during*** [ˈdjʊərɪŋ] prep. 在……期间

031

本书特别收录了各情境中使用频率最高的例句，让已熟悉E-mail写作的人进一步学习，根据自身需求写出语言更漂亮、表达更精确的信件。

必备关键单词： 此处单词对应了“这样写就对了”和“高频例句”中以变色编号标示的单词，每个单词都清楚标示国际上通用的英式音标、词性以及中文释义，省去学习者查找单词的时间。

Part 3

英文E-mail 词汇篇

06 英文E-mail相关词汇：

为了满足求职以及商业往来等各种E-mail需求，本书特别收录了五大类使用频率最高的英文E-mail相关词汇，让学习者能够在最短的时间内快速套用，想写什么主题都没问题，真正做到“遇到各种情境”都不怕！

英文E-mail实用大全

01 公司部门名称

Accounting Department	财务部
Advertising Department	广告部
Branch Office	分公司
Business Office	营业部
Export Department	出口部
General Affairs Department	总务部
Head Office	总公司
Human Resources Department	人力资源部
Import Department	进口部
International Department	国际部
Management Department	管理部
Market Department	市场部
Personnel Department	人事部
Planning Department	企划部
Product Development Department	产品开发部
Public Relations Department	公关部
Real Estate Development Department	地产开发部
Research and Development Department	研发部
Sales Department	销售部
Sales Promotion Department	销售推广部
Secretarial Pool	秘书室

444

Part 4

英文E-mail 超值附赠

07 商用书信必抄200惯用句：

商用书信和写给朋友的一般书信不太一样，要礼貌、诚恳，收信者才会更愿意与你进行商业上的往来。本书特别整理了8大常用情境中的200句惯用例句，适时插入到你的E-mail中，起画龙点睛之效！

英文E-mail实用大全

商用书信必抄200惯用句

因为商用书信有许多惯用表达，所以和我们平常写给朋友、家人的信件并不相同。像是要委婉地表达要求、不满、拒绝或想要真诚地表示谢意、歉意，都有一些固定的句型可以套用。套上外国人惯用的句型，不但能精准地表达意思，更能让国外客户惊呼“你的英文好地道！”以下为大家整理出了200句商用书信惯用必抄句，找到你需要的情境，大胆地用吧！

委婉表达希望对方快点回复的心情

1. Your prompt reply will be very much appreciated.
如果您能尽快回复，我将感激不尽。
prompt adj. 迅速的；及时的　appreciate v. 感激

2. I look forward to hearing your opinions on this matter.
我很期待听到您对这件事的看法。
look forward to phr. 期待　opinion n. 意见

3. Feel free to communicate with me through e-mail any time.
欢迎随时通过e-mail与我联系。
feel free to... phr. 随意……；想要……就请自便
communicate v. 沟通

4. I'm anxiously awaiting your response.
我急切地等待您的回复。
anxiously adv. 焦急地；担忧地　await v. 等待

458

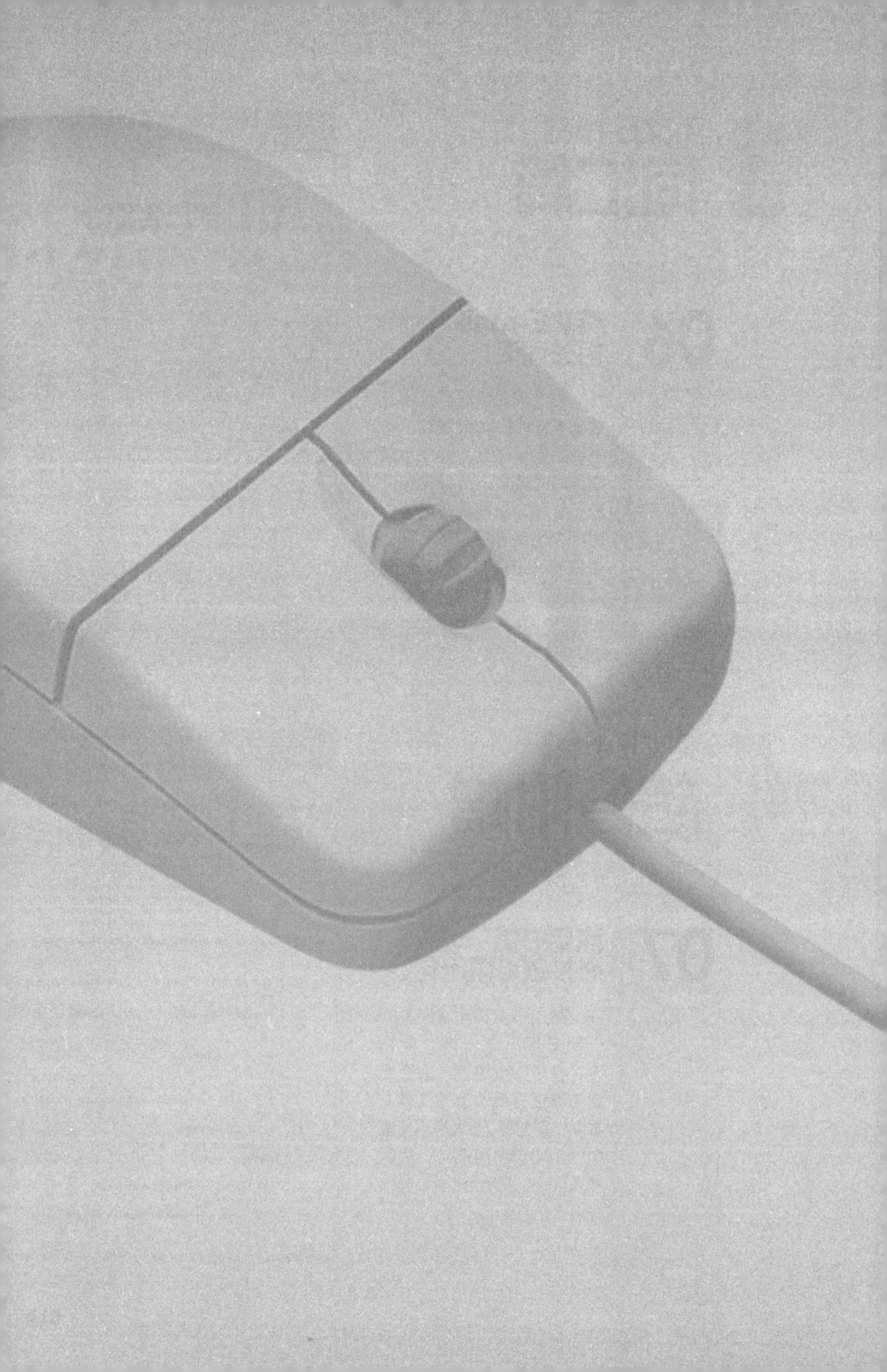

Part 1

英文E-mail 写作指南

01 英文E-mail写作的 7C原则

在利用E-mail传送信息时，并不要求使用华丽优美的语句，需要做的就是用简单、朴实的语言，清晰、准确、完整地表达自己的意思，让对方可以非常清楚地了解您想传达的信息。以下介绍了写好英文E-mail务必要掌握的“7C原则”：

1 Conciseness 简洁

优秀的英文E-mail应该行文流畅、言简意赅。也就是说写信者必须使用简洁明了的文字表达必要的信息。请对比下面的句子：

✖ Company operations for the preceding accounting period terminated with a substantial deficit.（上一会计年度，公司运营以巨额亏损结束。）→**too complicated**

⊙ The company lost much money last year.（公司去年损失了许多钱。）→**more concise**

2 Concreteness 具体

在书写英文E-mail的过程中，一般会涉及一些具体情况的描述，比如时间、地点、价格、货品编号等。对于这类信息，发邮件者应尽可能做到具体、详细，这样会将情况描述得更加清楚，并且有助于提高办事效率。请对比下面的句子：

✖ These brakes can stop a car within a short distance.（这些刹车系统可以在短距离内使一辆小汽车停下来。）→**general**

⊙ These Goodson power brakes can stop a 2-ton car within eight meters.（古德森动力刹车系统可以在八米内使一辆两吨重的小汽车停下来。）→**concrete and attractive**

3Clearness 清晰

“清晰”是英文书信写作最重要的原则。一封含糊不清、辞不达意的商业书信会引起误会与意见分歧，甚至会造成经济损失。想做到清楚表达应选择正确、简练的词汇以及恰当的句子结构。商业书信的撰写者必须将自己的意思清晰地表达出来，以便对方准确理解。请对比下面的句子：

☒ We can supply 50 tons of the item only.（我们只能提供五十吨的商品。）→**focus on the "item"**

◎ We can supply only 50 tons of the item.（这个商品，我们只能提供五十吨。）→**focus on the "quantity"**

4Courtesy 礼貌

即便是用E-mail联系，也需要注意礼仪。因而在写信过程中，要多选用礼貌、委婉的词语，像would, could, may, please, thank you等。礼貌并不意味着低三下四。请对比下面的句子：

☒ We are sorry that you misunderstood me.（我们很遗憾您对我们有所误解。）→**put the blame on "you"**

◎ We are sorry that we didn't make ourselves clear.（我们很抱歉没把我们的意思表达清楚。）→**put the blame on "ourselves"**

5Consideration 体贴

在英文E-mail写作过程中，发文者应设身处地地为对方着想，尊重对方的风俗习惯，即采取所谓的“You-Attitude”（对方态度），尽可能地避免使用“I-Attitude”或“We-Attitude”（我方态度）。另外，还应该考虑收信者的文化程度、性别等多方面的因素。请对比下面的句子：

☒ I received your letter of June 23, 2014 this morning.（我今天早上收到了您于2014年6月23日给我的信件。）→**I-Attitude**

◎ Your letter of June 23, 2014 arrived this morning.（您2014年6月23日写的信件已于今天早上送达。）→You-Attitude

6Correctness 准确

在英文E-mail写作中，除了避免语法、拼写及标点错误外，其所引用的史料、资料等也应准确无误。尤其是在商务信函中提到具体日期、资料等内容时更要准确表达以免发生歧义。请对比下面的句子：

☒ This contract will come into effect from Oct.1.（这份合约将于10月1日起生效。）→ambiguous

◎ This contract will come into effect from and including October 1, 2014.（这份合约将于2014年10月1日当日开始生效。）→accurate

7Completeness 完整

在英文E-mail写作中，信息的完整性很关键，所以商务信函中应包括所有必需的信息。例如下述通知，短短的几句话就包含了应有的全部信息：

Notice

All the staffs of Accounts Department are requested to be ready to attend the meeting in the conference room on Tuesday, at 3:00 p.m., Jan. 6, 2015, to discuss the financial statement of last year.

Accounts Department

02 图解英文E-mail 八大组成元素

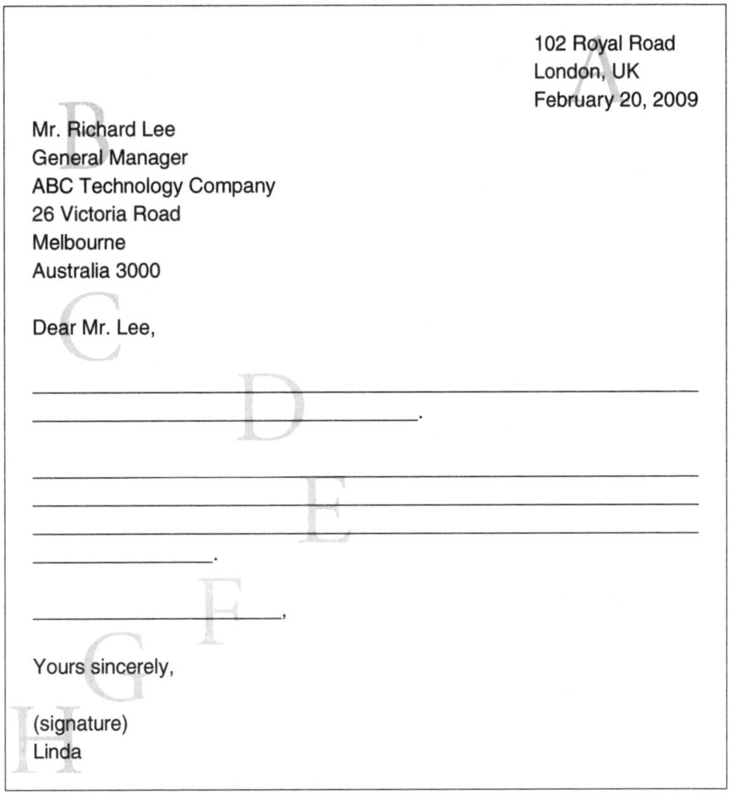

102 Royal Road
London, UK
February 20, 2009 A

B
Mr. Richard Lee
General Manager
ABC Technology Company
26 Victoria Road
Melbourne
Australia 3000

C
Dear Mr. Lee,

D
____
__________________________.

E
____
____
____
____________.

F
____________________,

G
Yours sincerely,

H
(signature)
Linda

特别附注：为了使各位学习者了解英文书信（E-mail）中的各大组成元素，本书特别利用这种目前广泛使用的“左对齐式”的书信来为大家讲解。此格式不一定与电脑中的E-mail信件一致，但主要以了解E-mail构成元素为目的，请知悉。

1 写信人的地址和写信日期 Heading

见图 A

在英文信件中，常见的日期写法有：

(1) 8th March, 2008或者8 March, 2008（英式）。
(2) March 8th, 2008或者March 8, 2008（美式）。

日期写法宜遵循下列规则：

(1) 年份宜书写完整，尽量避免使用“08”代替“2008”。
(2) 月份须采用规范的简写，详见下表：

January（Jan.）	February（Feb.）	March（Mar.）	April（Apr.）
May	June	July	August（Aug.）
September（Sep.）	October（Oct.）	November（Nov.）	December（Dec.）

(3) 日期可用序数词（Ordinal Numbers），例如：1st, 2nd, 3rd, 4th 等。也可用基数词（Cardinal Numbers），例如：1，2，3，4等，但美式书信大多采用后者。
(4) 在年份与月日之间必须用逗号隔开。
(5) 日期不可全部使用如7.12.2008或7／12／2008这样的阿拉伯数字书写，否则会引起误解。因为英美国家在这方面的书写不同。按照美国人的习惯，上述日期为2008年7月12日，而按照英国人的习惯则是2008年12月7日。

2 收信人的全名和地址 Inside Address

见图 B

这部分应置放在信件的左上方。

3 称呼 Salutation

见图 C

(1) 一般对不熟的人，常用称呼如表所示，但要注意的是，这些称呼不能单独使用，后面一定要接具体的姓氏或人名。需要说明的是，这里的Dear只是一个客气的称谓，并非完全表示“亲爱的”的意思。

Dear Sir	亲爱的先生
Dear Madam	亲爱的女士
Mr.	先生
Mrs.	夫人（已婚），后接夫姓
Miss	小姐（未婚），后接原姓
Ms. = Miss	夫人、小姐（统称），后接原姓
Mr. and Mrs.	夫妇

(2) 对熟人则应写明姓名，并冠以先生或女士的称谓，如：Dear Mr. Bill Smith（亲爱的比尔·史密斯先生），Dear Mrs. Sally Smith（亲爱的莎莉·史密斯夫人）。
(3) 对关系亲密者，则不用先生或女士等称谓，也不用姓，只用其名，如：Dear Bill（亲爱的比尔），Dear Sally（亲爱的莎莉）。
(4) 对亲人，在其名的基础上，需加上Darling（亲爱的）或Dearest（最亲爱的）等词语，如：My Darling Mary（我亲爱的玛莉）或My Dearest Mary（我最亲爱的玛丽）。
(5) 称呼之后的标点符号用逗号表示。

4 开头语 Opening Sentence

见图 D

邮件的开头语在书写时可参考p020“常见开头语”。

5 信件正文 Body of the Letter

见图 E

正文可根据内容决定长短。

6 祝福 / 结束语 Concluding Sentence

见图 F

如果想在信件结束后表达对收件人的祝福或是提醒，可参考p021“常见祝福 / 结束语”。

7 结尾敬语 Complimentary Close

见图 G

(1) 对一般人，可用Sincerely yours,（你诚挚的）或Faithfully yours,（你真诚的）。
(2) 对熟悉的人，可用Truly yours,（你忠实的）。
(3) 对亲近的人，可用Affectionately yours,（挚爱你的）或Lovingly yours,（深爱你的）。
(4) 对上级、长辈，可用Respectfully yours,（敬重您的）。

注意：以上结束语以及表示敬意的结束称呼中，第一个字母要大写，结尾要有逗号。另外，结束称呼中的两个单词也可以倒过来使用，如：Yours sincerely，Yours lovingly，Yours respectfully 等。

8 签名 / 署名 Signature

见图 H

如在较正式的信件中，请务必标示全名。

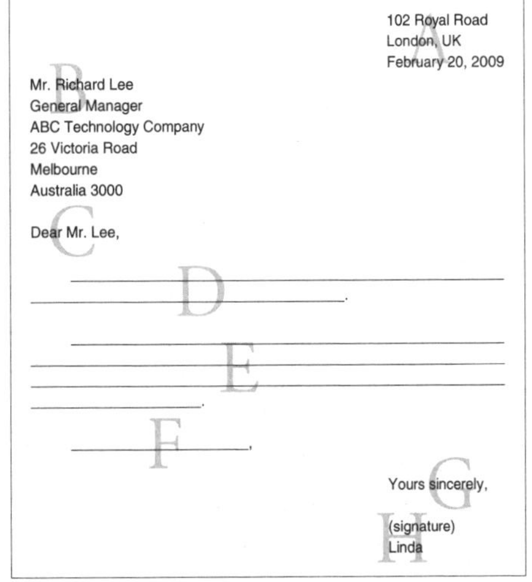

附注：传统的书信形式多采用缩进式。

常见开头语

01 Thank you for your letter dated June 20, 2013.
02 I am writing to you to ask about the conference to be held in New York next week.
03 We learn from your e-mail that you are interested in our products and would like to establish business relationship with us.
04 In reply to your letter of May 16th, I want to say...
05 We are pleased / glad to inform you that...
06 We are pleased to send you our catalog.
07 I must apologize for being reply to you so late.
08 May I take the liberty of mailing you and confirm some points?
09 I regret being unable to attend your banquet on Friday.
10 I am very excited and delighted at your good news.

11 Many thanks for your letter of September 4, 2008.
12 A thousand thanks for your kind letter of November 24, 2013.
13 Your letter that arrived today gave me great comfort.
14 Thank you very much for your letter of August 1 and the gift you sent me on Christmas Eve.
15 What a treat to receive your kind letter of May 5th!
16 It is always a thrill to receive your e-mail.
17 First of all, I must thank you for your kind assistance and attention to me.
18 With great delight, I learned from your letter of this Sunday that...
19 I was so glad to receive your letter of March 23rd.
20 I am very much pleased to inform you that my visit to your country has been approved.

21 I wish to apply for the teaching position you are offering.
22 I am very obliged to you for your warm congratulations.
23 My wife joins me in thanking you for the dinner party you gave in our honor last Monday.
24 We acknowledge with thanks the receipt of your letter dated Feb 5.
25 I regret being unable to reply to your letter earlier due to pressure of work.
26 I hope that you will excuse me for this late reply to your kind letter.
27 I must apologize for not being able to reply to your kind letter until today.
28 May I take the liberty of writing to you and appeal for your kind attention to...?
29 Owing to busy work, I have not been able to reply to your letter earlier, for which I must apologize.
30 With great delight, I learned that...

常见祝福 / 结束语

01 We look forward to your reply at your earliest convenience.
02 Your early reply will be highly appreciated.
03 Please let me know if you want more information.
04 Any other particulars wanted, we shall be pleased to send you.
05 I wish you every success in the coming year.
06 I look forward to our next meeting in Los Angeles.
07 Hoping to receive your early reply.
08 The help you gave me is sincerely valued.
09 With best regards to your family.
10 I hope everything will be well with you.

11 Awaiting your good news.
12 Looking forward to your early reply.
13 Hoping to hear from you very soon.
14 We await your good news.
15 I hope to hear from you very soon.
16 Expecting your immediate response.
17 Please remember me to your family.
18 Thank you very much for your consideration.
19 With love and good wishes.
20 Best wishes for all of you.

21 I expect your early reply soon.
22 I wish you all the best.
23 I appreciate your immediate reply. Thanks once more!
24 If you need any assistance, I am available any time.
25 Thank you once again for your kind letter.
26 Please let me know if you require further information.
27 I am always glad to be of serving to you.
28 Please accept my sincere thanks for your kind attention to this matter.
29 With thanks and regards.
30 Please do not hesitate to contact me if you...

03 英文E-mail电脑相关 实用技巧

1 设置邮件的拆分

当你在发送电子邮件时，可能会用附件的形式发送一些文件资料，此时若是这些文件资料的容量过大，可能会造成收件人下载邮件时间过长的情况。

所以为避免此类情况，我们可以使用Microsoft Outlook Express中的分割发送功能，设置邮件发送的容量，这样可以提高电子邮件的收发效率。

同时，被分割的信件在收信人分批收取后，会重新合并成原来的文件资料，不会有损毁或是遗失的情况！

具体操作请参考以下步骤：

Step 1. 请打开您的Microsoft Outlook Express。

Step 2. 在“工具”的设定选项里选择“账号”。

Step 3. 选择账户，点击右侧的“属性”。

Step 4. 在“属性”的对话框中，选择“高级”选项卡，勾选“拆分大于KB的邮件”这一选项，并设置好拆分的最小容量，通常建议调整为“200KB”。

Step 5. 最后一定要按“确定”保存刚才的设置，并且最好重新启动软件，以保证设置生效。

注意：不同版本的软件，选项菜单会有所不同。

2 避免Outlook Express自动将病毒邮件寄出

许多人在使用Outlook Express时发现，它容易自动将病毒邮件寄出。其实绝大部分的病毒都是通过Outlook Express的通讯录自行寄出，只要手动关闭下列功能，就可以减少一些风险：

Step 1. 启动Outlook Express。

Step 2. 选择“工具”→“选项”→“发送”，将“立即发送邮件”前的“√”取消。

3 避免收到广告信件：保护自己与朋友的隐私

生活在互联网时代，您和您的朋友一定经常用邮件传递信息。

而在朋友之间不断转发的信件中，有相当一部分是由广告商最先发送的。利用一些吸引人的文章和图片，让大家不自觉地传来传去，没多久，信件中的收件人名单就会越来越多；接着等这些信在各地传播后再绕回来，广告商手里就有很可观的名单。那该怎么办呢？事实上，您得靠您的朋友来保护您的隐私，反过来说，您的朋友也得靠您来免除垃圾邮件的骚扰。想避免收到这些骚扰信函，请参考以下的建议：

Step 1. 转寄信件前，注意把原来的信头去掉，避免泄露其他人的信息。

Step 2. 如果您一次要把邮件转发给很多人，可以选择密送的方式，这样一来可以保护各个收件人的地址不被其他人轻易获取，二来可以使收件人节省下收取大量抄送地址的时间。

4 避免收到广告信件：邮箱使用技巧

邮箱要按功能来分类才更高效！像新闻栏目主编的邮箱可能分为：发报邮箱、订报邮箱、投稿专用邮箱、个人亲朋好友邮箱及特殊邮箱。当然，您可能没有那么多的邮箱，不过现在您就可以在网上再申请一个免费的邮箱。也就说最好注册两个邮箱，一个（最好是您现在用的这一个，因为一些信息可能已经被泄漏）专门用来订阅电子报、填写网络资料、参加抽奖等，也就是可能会有广告信来源的都用这一个邮箱，而另一个则完全用来和朋友交流。为什么会有这种分法呢？因为不管是订电子报还是参加网上抽奖游戏等，都有可能成为别人的网络受众样本，所以建议您用一个邮箱专门来进行此类行为。虽然申请免费邮箱也有可能会收到广告信，但概率要小很多。

5 让邮件自动回到自己的家

不知道大家有没有这种经历，每次收到的邮件总是杂乱地放在同一个文件夹内，还要一封一封整理归类。以下利用“邮件规则”这一功能，教大家如何让邮件自动发送到指定的文件夹：

Step 1. 让好友来信跑到自己的位置：首先假设您的好友叫小明，那我们就订一个规则，选择“寄件者包含人员”后再选“移到指定的文件夹”。另外在规则说明中点击“包含人员”，新增小明的E-mail地址并按“确定”（也可以用通讯录一次同时指定多人），接下来就是指定收件的文件夹。最后，记得替这个规则取个名字吧！

Step 2. 让喜欢的新闻报纸跑到自己的位置：首先假设您要收集的报纸叫“时事前沿电子报”，那我们就制订一个规则，选择“主题包含特定文字”后再选“移到指定的文件夹”。另外在规则说明中点击“特定文字”，新增“时事前沿电子报”并按“确定”，接下来就是指定您收集报纸的文件夹（如果没有适合的文件夹，请点击新建文件夹），最后替这个规则命名。

Step 3. 让某个账号的信件有自己的家：首先假设您要设定的账号是MS1，那我们就订一个规则，选择“指定特定的账号”再选“移到指定的文件夹”。另外要打勾的就是“停止处理更多的规则”，再在规则说明中点选“特定”，选择 MS1并按“确定”，然后指定存放文件夹，并为这个规则取名。

Note

如果你还发现了更多关于E-mail使用的小技巧，不妨记录在这里，这样可以大大地提升你的工作效率哦！

Part 2

英文E-mail
实例集

Unit 1 交际 Socializing

01 欢迎来访

From mike4321@tom.com　　Date January 2, 2008
To whitney@cnd.com
Subject Welcome Letter

Dear Mr. Whitney,

I am glad to learn that you will be ***visiting***[1] Toronto next month. I'll be in Montreal between February 21 and 25, so if it's ***convenient***[2] for you, shall we ***arrange***[3] to meet on the 28th? Please let me know ***whether***[4] I can arrange hotel accommodations and ***transportation***[5] from the ***airport***[6] for you.

Expecting your ***arrival***[7] soon!

Truly yours,
Mike Smith

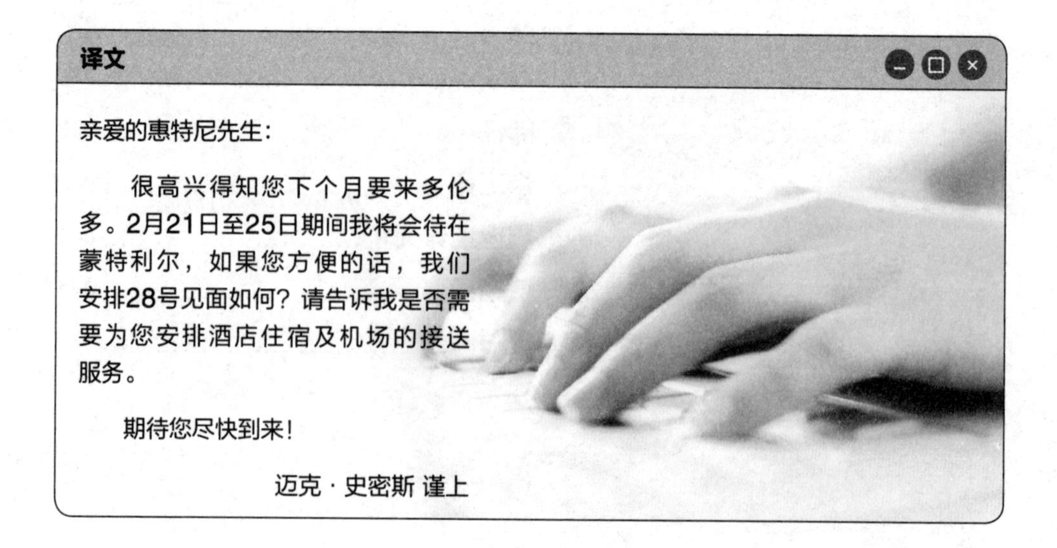

译文

亲爱的惠特尼先生：

很高兴得知您下个月要来多伦多。2月21日至25日期间我将会待在蒙特利尔，如果您方便的话，我们安排28号见面如何？请告诉我是否需要为您安排酒店住宿及机场的接送服务。

期待您尽快到来！

迈克 · 史密斯 谨上

语法重点解析

1 解析重点1 learn

learn 的意思是“学习；学会；得知；获悉”，一般作“学习；学会”解释的情况比较多。而在表达知道某事的时候，不光可以用 know，也可以用 learn，这样句子表达更加灵活多样。

请对照以下例句：

I don't know if he will make it.（我不知道他能否成功。）

I learned this news from the newspaper.（我从报纸上得知了这个消息。）

2 解析重点2 Expecting your arrival soon!

实际上这是个省略句，完整的句子应该是 I am expecting your arrival soon! expecting 是 expect 的现在分词，是“期望”的意思。要注意的是 expecting 还有“怀孕的”的意思。请对照以下例句：

We are expecting you in London on Tuesday.（我们星期二在伦敦等你到来。）

She is expecting another baby.（她又怀孕了。）

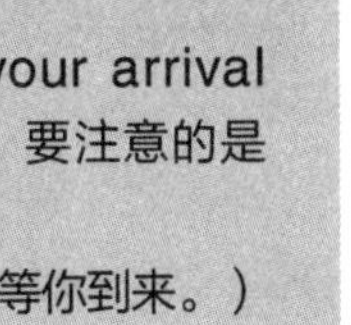

1. I am delighted to learn that you will be visiting Shanghai next month.

得知您将于下个月抵达上海的消息，我很高兴。

2. We will be glad to arrange a tour of our factory for July 3.

我们乐意安排7月3日参观我们工厂的相关事宜。

3. Please let me know once you have confirmed the exact date of your arrival.

一旦您确定了确切的抵达日期，请通知我。

4. I look forward to meeting with you soon.

我期待能尽快与您会面。

5. Please inform us if we can arrange *hotel*[8] *accommodations*[9] for you.

如果需要我们为您安排酒店食宿，请通知我们。

6. We are looking forward to meeting with you soon.

我们期待能尽快与您见面。

必背关键单词

1. ***visit*** [ˈvɪzɪt] n. & v. 访问
2. ***convenient*** [kənˈviːnɪənt] adj. 方便的
3. ***arrange*** [əˈreɪndʒ] v. 安排
4. ***whether*** [ˈweðə(r)] conj. 是否
5. ***transportation*** [ˌtrænspɔːˈteɪʃn] n. 运输
6. ***airport*** [ˈeəpɔːt] n. 机场
7. ***arrival*** [əˈraɪvl] n. 到达
8. ***hotel*** [həʊˈtel] n. 旅馆
9. ***accommodation*** [əˌkɒməˈdeɪʃn] n. 住宿；膳宿（供应）

02 任命通知

这样写就对了

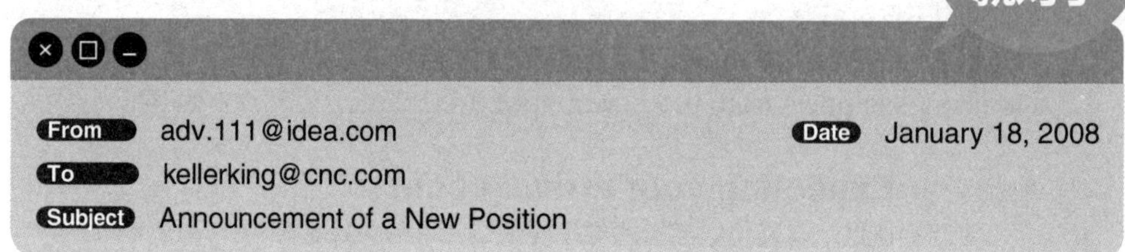

From adv.111@idea.com
Date January 18, 2008
To kellerking@cnc.com
Subject Announcement of a New Position

Dear Mr. Keller,

I am pleased to ***announce***[1] the ***appointment***[2] of IDEA Corporation's general manager of ***advertising***[3].
Ms. Ho has worked in the advertising area for 10 years. She has an ***insightful***[4] view of the advertising ***industry***[5]. All in all, we are eager to work with her and meet her ***objectives***[6].
This new appointment brings ***additional***[7] strength to IDEA Corporation. You could contact with her for your advertising needs.

Sincerely yours,
IDEA Corporation

译文

亲爱的凯勒先生:

我非常高兴地宣布对概念公司广告总经理一职的任命通知。

何女士在广告业已经打拼了十年，她对广告业有着独到的见解。总之，我们期望与她共同合作，并且努力实现她的目标。

这项新的人事任命给概念公司带来了新的力量。倘若贵公司有任何广告需求，请联络何女士。

概念公司 谨上

语法重点解析

1 解析重点1 **I am pleased to...**

I am pleased to...意为“我非常高兴……”英语书信中，新产品开发、就职祝贺等充满喜悦气氛的场合，都可以用be pleased to，be happy to，be glad to 等表达方式。例如：

I am pleased to hear of your success.（听到了你成功的消息，我很高兴。）

I am happy to accept the position of flight attendant with your company.（我很高兴接受贵公司空乘员的职位。）

2 解析重点2 **meet one's objective**

meet one's objective 意为“达到某人的要求”，meet 意为“遇见；满足”。objective 本来有形容词和名词两种词性，但在此做名词用，意为“目标”。与它同义的词还有 target，goal 等。所以要表达“实现或达到某人的目标”还可以说reach one's target，reach one's goal。例如：

The young people tried their best to reach their targets.（那些年轻人尽全力去达成他们的目标。）

I will keep on striving until I reach my goal!（我会继续奋斗直到实现我的目标！）

1. **I am being promoted to Manager of the Marketing Department at Global Links as of March 21.**
   3月21日起我将正式升职为Global Links公司的市场部经理。
2. **Miss Bree will be taking over my territory as your sales representative.**
   布里小姐将接手我的工作，担任你们的销售代表。
3. **Thank you so much for your support *during*[8] the past years.**
   非常感谢您这些年来的支持。
4. **I am your new Diamond International sales representative.**
   我现在担任钻石国际新的销售代表。

必背关键单词 

1. ***announce*** [əˈnaʊns] v. 宣布；公告
2. ***appointment*** [əˈpɔɪntmənt] n. 任命
3. ***advertising*** [ˈædvətaɪzɪŋ] n. 广告
4. ***insightful*** [ˈɪnsaɪtfʊl] adj. 有洞察力的
5. ***industry*** [ˈɪndəstrɪ] n. 工业；行业
6. ***objective*** [əbˈdʒektɪv] n. 目标
7. ***additional*** [əˈdɪʃənl] adj. 额外的；另外的
8. ***during*** [ˈdjʊərɪŋ] prep. 在……期间

03 新职员到职通知

这样写就对了

From personnel@abc.com
To allstaff@abc.com
Subject Introducing New Colleagues
Date December 1, 2008

Dear members,

All attention, please!
I hereby make an ***announcement***[1] that there will be three new ***colleagues***[2] coming into our ***company***[3]. They are Cindy Wang, Nancy Li and May Chou, who ***graduated***[4] from HKU, UC and NTU respectively. They will serve in the ***Marketing***[5] Department from tomorrow on, namely, December 2, 2008.
Please get ***along***[6] well with new colleagues and ***pursue***[7] better ***development***[8] of our company.
Thanks in advance for your cooperation!

Sincerely yours,
Personnel Department

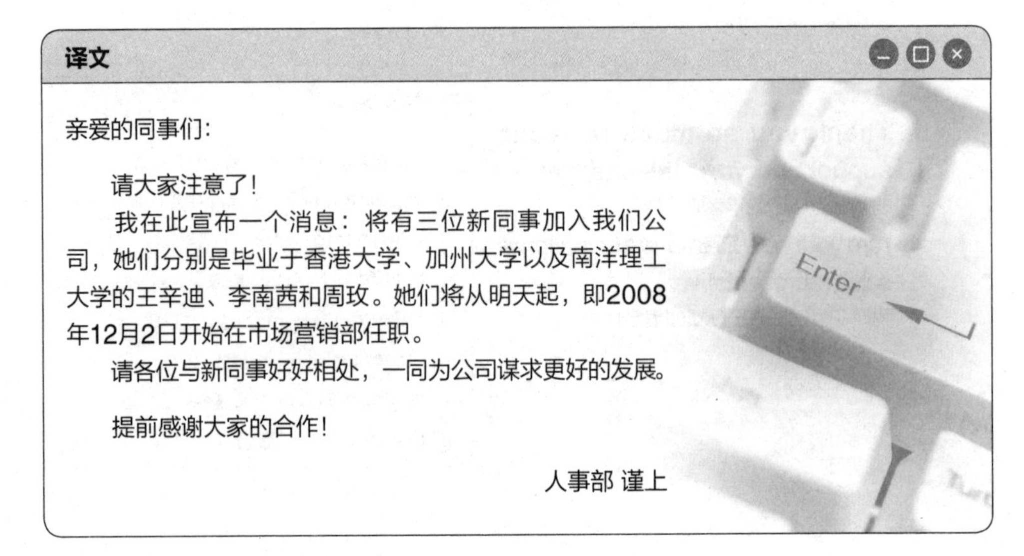

译文

亲爱的同事们：

请大家注意了！

我在此宣布一个消息：将有三位新同事加入我们公司，她们分别是毕业于香港大学、加州大学以及南洋理工大学的王辛迪、李南茜和周玫。她们将从明天起，即2008年12月2日开始在市场营销部任职。

请各位与新同事好好相处，一同为公司谋求更好的发展。

提前感谢大家的合作！

人事部 谨上

语法重点解析

1 解析重点1 **They will serve in the Marketing Department from tomorrow on, namely, December 2, 2008.**

此句的意思是“他们将从明天起，即2008年12月2日开始在市场营销部任职”。可能有人会认为此句到 from tomorrow on 就已经表达完整了，但是依据英文书信写作7C原则中的 Correctness（准确）原则，涉及具体日期、时间等一定要描述清楚。所以，此句后才会有 namely, December 2, 2008，这个绝对不是多余的。

2 解析重点2 **in advance**

这个短语是“提前；预先”的意思。advance 在这里作为形容词“预先的；先前的”使用。表达“提前；事先”还可以用 ahead of time，ahead of schedule，beforehand 等表达。例如：

A punctual person always finishes everything ahead of time.（一个守时的人总是提前把事情做好。）

You should tell me in advance.（你应该事先告诉我。）

We have completed our work ahead of schedule.（我们提前完成了工作。）

He arrived at the meeting place beforehand.（他提前到达了会面地点。）

1. **May I have your attention, please?**
   请注意听好吗?
2. **I would like to introduce you a new colleague.**
   我要为大家介绍一位新同事。
3. **Thank you so much for your support and encouragement.**
   感谢大家的支持和鼓励。
4. **I hope all *members*[9] could get along well with each other.**
   我希望大家都能和睦相处。
5. **The new colleague will serve as assistant to the general manager.**
   这位新同事将担任总经理助理一职。
6. **He is a top student of MIT.**
   他是麻省理工学院的顶尖学生。

必背关键单词

1. ***announcement*** [əˈnaʊnsmənt] n. 宣布
2. ***colleague*** [ˈkɒli:g] n. 同事
3. ***company*** [ˈkʌmpənɪ] n. 公司
4. ***graduate*** [ˈgrædʒʊət] v. 毕业
5. ***marketing*** [ˈmɑ:kɪtɪŋ] n. 市场营销
6. ***along*** [əˈlɒŋ] adv. 一起
7. ***pursue*** [pəˈsju:] v. 追求
8. ***development*** [dɪˈveləpmənt] n. 发展
9. ***member*** [ˈmembə(r)] n. 成员

04 辞职

这样写就对了

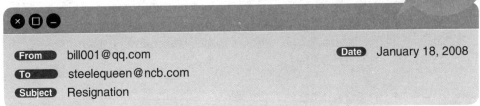

Dear Ms. Steele,

I am writing to let you know a ***decision***[1] of mine.
I want to ***resign***[2] from the NCB Company for some ***personal***[3] reasons. ***Moreover***[4], I will be starting a new ***position***[5] with the Government Bookstore as an office manager. **I think perhaps that job would be more *suitable*[6] for me**. Anyway, I am still going to ***express***[7] many thanks for your support while I was here.

I hope everything goes well!

Faithfully yours,
Bill

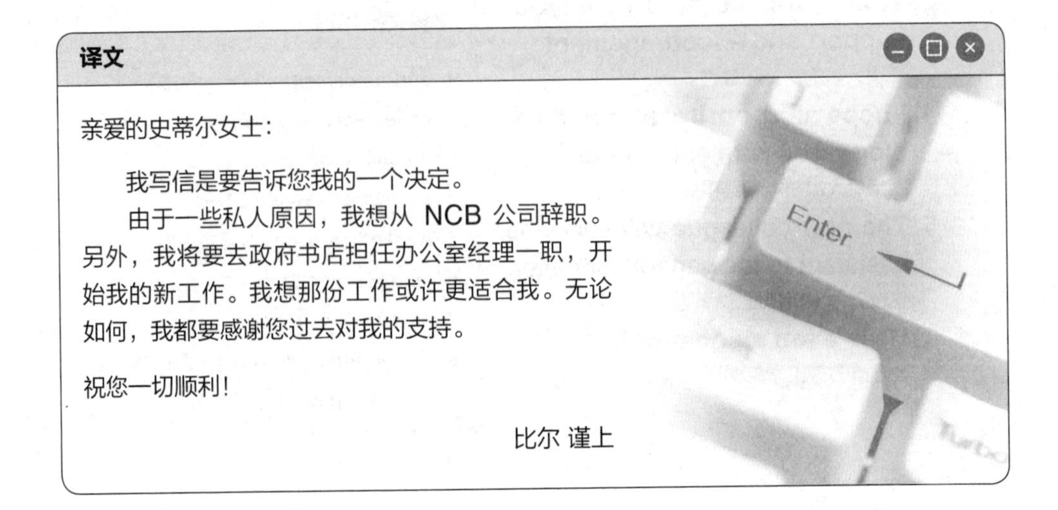

译文

亲爱的史蒂尔女士:

我写信是要告诉您我的一个决定。

由于一些私人原因，我想从 NCB 公司辞职。另外，我将要去政府书店担任办公室经理一职，开始我的新工作。我想那份工作或许更适合我。无论如何，我都要感谢您过去对我的支持。

祝您一切顺利!

比尔 谨上

语法重点解析

1 解析重点1 resign

表达“辞职；离职”一般用 resign。要注意的是 retire 虽然也有“离职”的意思，但是通常适用于因年老或其他原因而“退休”或“退职”的场合。请对照以下例句：

Do you want me to resign?（你想要我辞职吗？）

Some of the older workers were retired early.（有些老工人提前退休了。）

2 解析重点2 I think perhaps that job would be more suitable for me.

此句的意思是“我想那份工作或许更适合我”。其中的短语 be suitable for 是“适合的；恰当的”的意思。fit 指大小、形状的合适，可引申为“吻合；协调”。match 多指大小、色调、形状、性质等方面的搭配。请对比下面的例句：

What time is suitable for us to meet?（我们什么时候会面合适呢？）

This new jacket fits her well.（这件新夹克很合她的身。）

The tie does not match my suit.（这条领带和我的西装不搭。）

1. I'd like to say that I've really enjoyed working with you. However, I think it's time for me to leave.

我真的很高兴能与你共事。但是，我觉得现在是我离开的时候了。

2. I want to expand my *horizons*[8].

我想拓展我的视野。

3. I've made a tough decision; here is my resignation.

我做了一个很困难的决定，这是我的辞呈。

4. I've been trying, but I don't think I'm up to this job.

我一直很努力，但我觉得我无法胜任这份工作。

5. I've been here for too long. I want to change my environment.

我在这里待太久了，我想换一下环境。

6. I'm sorry to bring up my resignation at this moment, but I've decided to study abroad.

我很抱歉在这个时候递交辞呈，但我已经决定要出国读书了。

必背关键单词

1. ***decision*** [dɪˈsɪʒn] n. 决定
2. ***resign*** [rɪˈzaɪn] v. 辞职
3. ***personal*** [ˈpɜːsənl] adj. 私人的
4. ***moreover*** [mɔːrˈəʊvə(r)] adv. 此外；而且
5. ***position*** [pəˈzɪʃn] n. 方位；位置；职位
6. ***suitable*** [ˈsuːtəbl] adj. 合适的
7. ***express*** [ɪkˈspres] v. 表达
8. ***horizon*** [həˈraɪzn] n. 视野；地平线

05 调职

这样写就对了

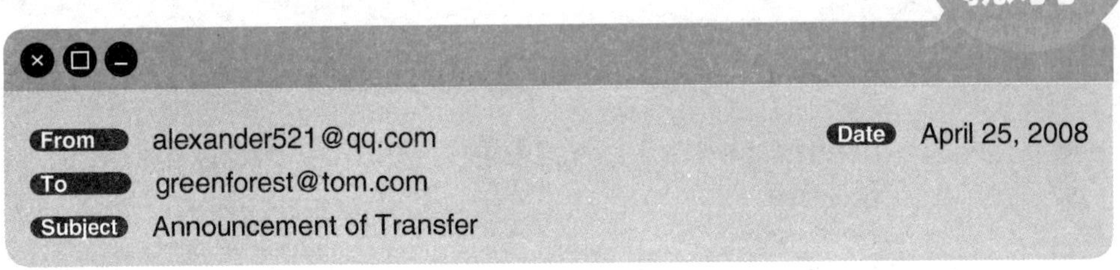
From alexander521@qq.com
To greenforest@tom.com
Subject Announcement of Transfer
Date April 25, 2008

Dear Mr. Green,

This is to let you know that I will be **transferred** to the Beijing ***office***[1] of our company as of May 11, 2008.
I would like to thank you for all your support during the ***past***[2] years and hope that you will **continue to** ***extend***[3] the same to my ***replacement***[4], Miss Gao.

With thanks and ***regards***[5],

Sincerely yours,
Alex

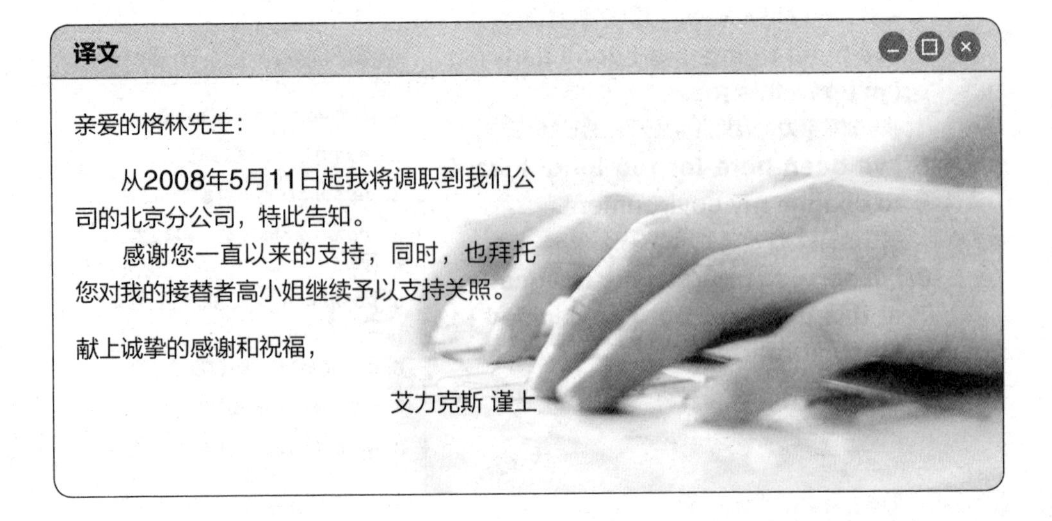
译文

亲爱的格林先生：

从2008年5月11日起我将调职到我们公司的北京分公司，特此告知。
感谢您一直以来的支持，同时，也拜托您对我的接替者高小姐继续予以支持关照。

献上诚挚的感谢和祝福，

艾力克斯 谨上

1 解析重点1 transfer

“调职”一般用 transfer 表达。也可以用 will be transferred to, will be posted at 等被动语态表达。但是，使用 will be moved to 时，有“被强行调动”的意思。请对照以下例句：

He put in for a transfer to another position.（他申请调职。）

Unfortunately, I will be moved to the suburb.（不幸的是，我要被调到郊区工作了。）

2 解析重点2 continue to

要表达“继续做某事”时可以用这个短语。表达同样的意思还可以说 keep on。但是请注意，continue to 后面要接动词原形，而 keep on 后面要接 V-ing 形式。请对照以下例句：

Please continue to support me!（请继续支持我吧！）

She kept on working although she was tired.（尽管很疲劳，但她仍继续工作。）

1. **I am writing to tell you something about my transfer.**
   我写这封信是想告诉您关于我调职的事。
2. **I've really enjoyed working with you.**
   我真的很高兴与你共事。
3. **I would like to thank you for all your assistance.**
   感谢您所有的协助。
4. **Mr. Wang will be my replacement.**
   王先生将接替我原来的工作。
5. **I hope you will get along well with each other.**
   我希望你们能愉快相处。
6. **Please continue to *support*[6] our work.**
   请继续支持我们的工作。
7. **I would like to *express*[7] my great thanks during the past years.**
   非常感谢您一直以来的支持和关照。
8. **Expecting all of you will *succeed*[8].**
   期待你们都能成功。

必背关键单词 

1. ***office*** [ˈɒfɪs] n. 办公室；办公处；事务所
2. ***past*** [pɑːst] adj. 过去的
3. ***extend*** [ɪkˈstend] v. 延伸；给予
4. ***replacement*** [rɪˈpleɪsmənt] n. 代替；更换
5. ***regard*** [rɪˈgɑːd] n. 关心；问候
6. ***support*** [səˈpɔːt] v. 支持；支援
7. ***express*** [ɪkˈspres] v. 表达
8. ***succeed*** [səkˈsiːd] v. 成功

06 卸任

From mikepan400@163.com
To abc@bbs.com
Subject Announcement of My Resignation
Date December 18, 2008

Dear all,

All good things must come to an end. I am leaving this company.
When I look back on the past years, all the memories I have of working with you are **invaluable** to me. I would like to ***deliver***[1] my ***heartfelt***[2] thanks to all who have ***shown***[3] me your ***guidance***[4], support and assistance.
Attached[5] is my personal contact information.
Keep in touch and I wish all of you a ***promising***[6] ***future***[7]!

Warmly regards,
Mike Pan

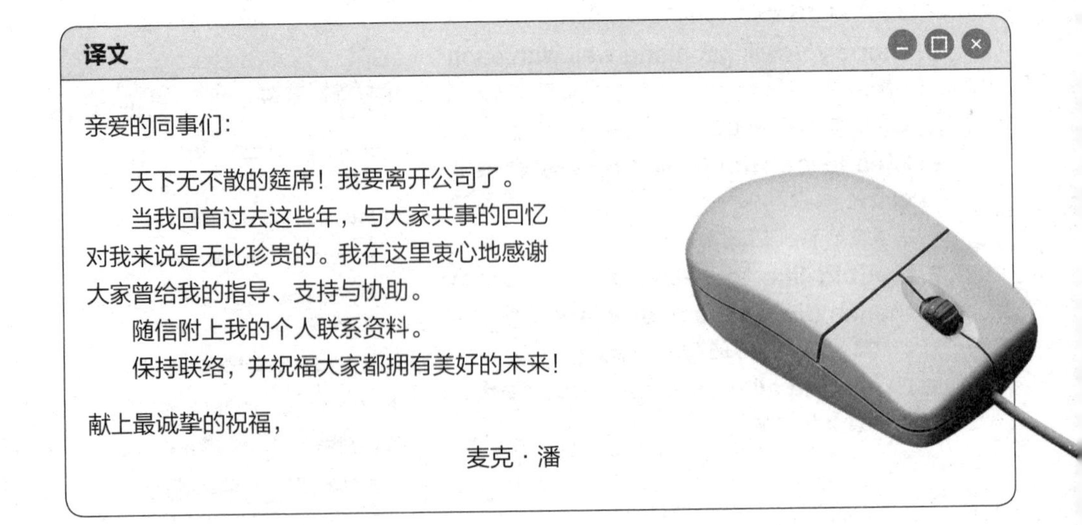

译文

亲爱的同事们:

天下无不散的筵席！我要离开公司了。

当我回首过去这些年，与大家共事的回忆对我来说是无比珍贵的。我在这里衷心地感谢大家曾给我的指导、支持与协助。

随信附上我的个人联系资料。

保持联络，并祝福大家都拥有美好的未来！

献上最诚挚的祝福，

麦克·潘

语法重点解析

1 解析重点1 All good things must come to an end.

这句话可直译为“所有美好的事情都有结束的时候”，但是我们可以灵活地翻译成一句中文常用语：天下无不散之筵席。又如英文中的 How time flies 可以翻译为“光阴似箭，岁月如梭”。对于这些与中文成语、谚语或俗语等意思很相近的英文，我们在翻译时可以直接套用，使译文在增加知识性的同时更加地道。

2 解析重点2 invaluable

以下是invaluable（无价的）、valuable（贵重的）、valued（重要的；宝贵的）的用法：

Your support is invaluable to me.（你的支持对我来说是无价的。）
He bought me a valuable diamond.（他买了一颗贵重的钻石给我。）
She is one of our valued customers.（她是我们重要的客户之一。）

高频例句

1. **I am leaving the company that I have *served*[8] for nearly 20 years.**
   我将要离开这个我服务了近20年的公司了。
2. **It was wonderful working with you.**
   跟大家一起工作非常美好。
3. **It was, in retrospect, the happiest day of my life.**
   回想起来，那是我最幸福的日子。
4. **I hope you always have a wonderful time.**
   希望你们一直拥有美好的时光。
5. **I would like to deliver my heartfelt thanks to your support.**
   我要衷心地感谢大家对我的支持。
6. **Attached is my new contact information.**
   随信附上我新的联系方式。
7. **I will remember all of you forever.**
   我会永远记住大家的。

必背关键单词

1. ***deliver*** [dɪˈlɪvə(r)] v. 发出；提出
2. ***heartfelt*** [ˈhɑ:tfelt] adj. 衷心的；真诚的
3. ***show*** [ʃəʊ] v. 表现；展示
4. ***guidance*** [ˈgaɪdns] n. 指导
5. ***attach*** [əˈtætʃ] v. 附上
6. ***promising*** [ˈprɒmɪsɪŋ] adj. 有希望的；有前途的
7. ***future*** [ˈfju:tʃə(r)] n. 未来
8. ***serve*** [sɜ:v] v. 服务

07 换工作

这样写就对了

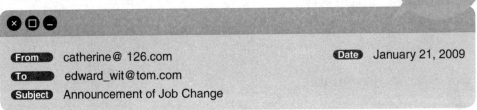

From catherine@ 126.com　　Date January 21, 2009
To edward_wit@tom.com
Subject Announcement of Job Change

Dear Mr. Edward,

I have ***recently***[1] ***changed***[2] my job and become a ***consultant***[3] in Milestone Consultation International Co., so I have also moved to a new place near our company.
My new ***address***[4] and ***contact***[5] number are as ***follows***[6],
Address: **8F, No.130, Sec. 1, Fu Hsin Road, Shanghai**
Telephone number: 2730-8888
I hope to **keep in touch with** you.

With my best wishes,
Catherine

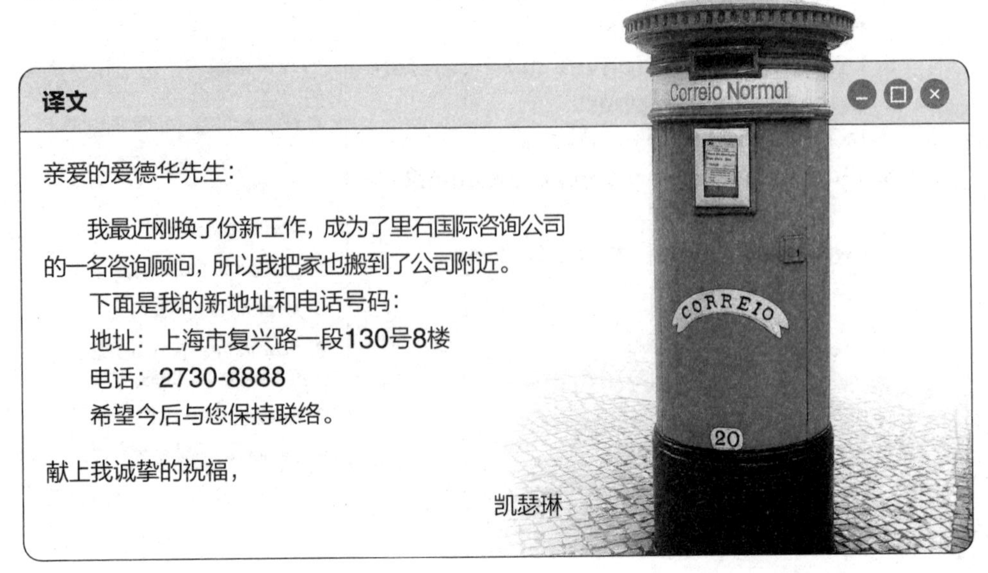

译文

亲爱的爱德华先生：

我最近刚换了份新工作，成为了里石国际咨询公司的一名咨询顾问，所以我把家也搬到了公司附近。
下面是我的新地址和电话号码：
地址：上海市复兴路一段130号8楼
电话：2730-8888
希望今后与您保持联络。

献上我诚挚的祝福，
凯瑟琳

语法重点解析

1 解析重点1 8F, 130, Sec 1, Fu Hsin Road, Shanghai

住址的写法是英文书信与中文书信中的一个重要区别。如8F, 130, Sec 1, Fu Hsin Road, Shanghai 翻译成中文就是“上海市复兴路一段130号8楼”。中文描述地址是从大到小，而英文正好相反，是从小到大。如下面的例子：

7F, No.130 , Sec. 3 , Wenhua Road , Banqiao Dist. ,Shanghai City

翻译成中文就是“上海市板桥区文化路三段130号7楼”。

2 解析重点2 keep in touch with

keep in touch with 是“保持联络”的意思，跟它同义的短语还有 stay in touch with。这里接在 I hope to... 的句型之后是“希望能保持联络”意思，但是日常比较常用的是 please keep / stay in touch。contact 虽然也有“联络”的意思，但在这里使用会略显死板。请对照以下例句：

They keep in touch with each other by mail.（他们通过写信保持联系。）

We agreed to contact each other again as soon as possible.（我们同意尽快再次联络。）

高频例句

1. **I plan to change my job.**
   我打算换份工作。
2. **I am going to work for the *Business Daily*.**
   我将要去《商务日报》工作。
3. **I *quitted*[7] my job and found another job that I like better.**
   我辞职了，换了一份我更喜欢的工作。
4. **It's time to change my job!**
   是换份工作的时候了！
5. **I intend to job-hop to that famous computer company.**
   我想跳槽去那家知名电脑公司。
6. **I finally found my *niche*[8] after several job-hopping.**
   在几次跳槽之后，我终于找到了最适合我的职位。
7. **Please remember my new address and contact method.**
   请记下我的新地址和联系方式。

必背关键单词

1. ***recently*** ['ri:sntlɪ] adv. 最近；近来
2. ***change*** [tʃeɪndʒ] v. 改变
3. ***consultant*** [kən'sʌltənt] n. 顾问
4. ***address*** [ə'dres] n. 地址
5. ***contact*** ['kɒntækt] n. 联系；联络
6. ***follow*** ['fɒləʊ] v. 跟随；接着
7. ***quit*** [kwɪt] v. 放弃；辞职
8. ***niche*** [nɪtʃ] n. 合适的职位

08 重返工作岗位

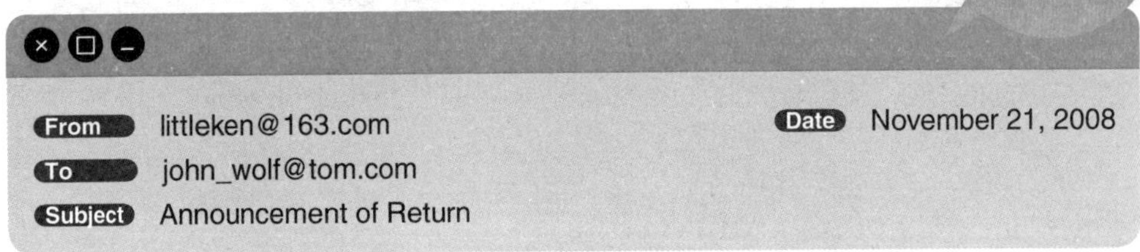

From littleken@163.com
Date November 21, 2008
To john_wolf@tom.com
Subject Announcement of Return

Dear John,

I hope this letter finds you well.
I just want to let you know that I have ***recovered***[1] from my ***recent***[2] ***appendicitis***[3]. Now I have come back to work ***again***[4].
I look ***forward***[5] to working with you again and hearing from you soon.
Thank you very much for your ***consolation***[6].

Yours ***truly***[7],
Ken

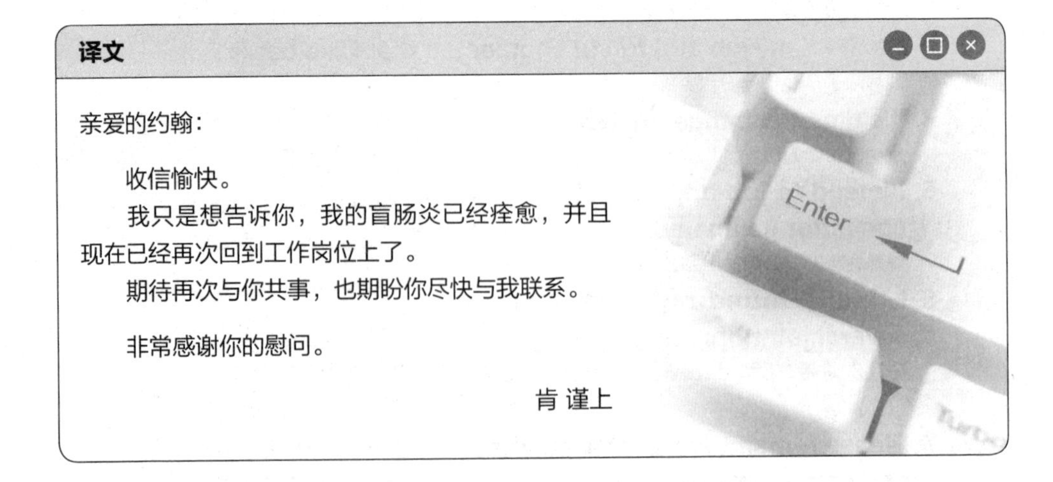

译文

亲爱的约翰：

收信愉快。

我只是想告诉你，我的盲肠炎已经痊愈，并且现在已经再次回到工作岗位上了。

期待再次与你共事，也期盼你尽快与我联系。

非常感谢你的慰问。

肯 谨上

1 解析重点1 **I hope this letter finds you well.**

I hope this letter finds you well.是在英文书信中常用到的开头语。与它类似的句子还有I hope you are doing well.等。这些句子有一些也可以应用于对话中或作为结尾句使用。

2 解析重点2 **come back**

"回归；回来"可以用 come back，也可以用另一个单词 return。虽然它们都有"回归"的意思，但是 return 还有"归还；返还"的意思。请对照以下例句：
Will you wait here until I come back?（您能在这里等到我回来吗？）
Please return the book to me.（请把书还给我。）

1. I hope this letter finds you well.
收信愉快。

2. I have recovered from the *pneumonia*[8].
我的肺炎已经痊愈了。

3. I come back to work with vigor again.
我又精神饱满地回到了工作岗位。

4. It's time for me to come back!
是我回来的时候了！

5. I just want to tell you the good news.
我只是想告诉你这个好消息。

6. I have always looked forward to working with you again.
我一直期待着再次与您共事。

7. I expect to hear from you as soon as possible.
期待你尽快回信。

8. Thank you very much for your sincere consolation.
非常感谢你诚挚的慰问。

9. I can *concentrate*[9] on my work again.
我可以再次投入工作了。

必背关键单词

1. ***recover*** [rɪˈkʌvə(r)] v. 恢复
2. ***recent*** [ˈriːsnt] adj. 不久前的；近来的
3. ***appendicitis*** [əˌpendəˈsaɪtɪs] n. 盲肠炎
4. ***again*** [əˈgen] adv. 再次
5. ***forward*** [ˈfɔːwəd] adv. 向前
6. ***consolation*** [ˌkɒnsəˈleɪʃn] n. 安慰；慰问
7. ***truly*** [ˈtruːlɪ] adv. 真诚地
8. ***pneumonia*** [njuːˈməʊnɪə] n. 肺炎
9. ***concentrate*** [ˈkɒnsntreɪt] v. 集中；专心于

09 赞扬同事

这样写就对了

From johndiamond@SIS.com
To kevin09@yahoo.com
Subject Great Performance!
Date November 2, 2008

Dear Kevin,

First of all, ***congratulations***[1] on meeting and ***exceeding***[2] our goals for school ***instrument***[3] ***sales***[4] in October!
You worked on ***arranging***[5] for a ***trade-in***[6] for a completely new set of instruments and helped make October a month to remember.
I hope you will put the ***bonus***[7] check to good use, and continue to bring new ideas to the sales ***department***[8].

Sincerely yours,
John Diamond

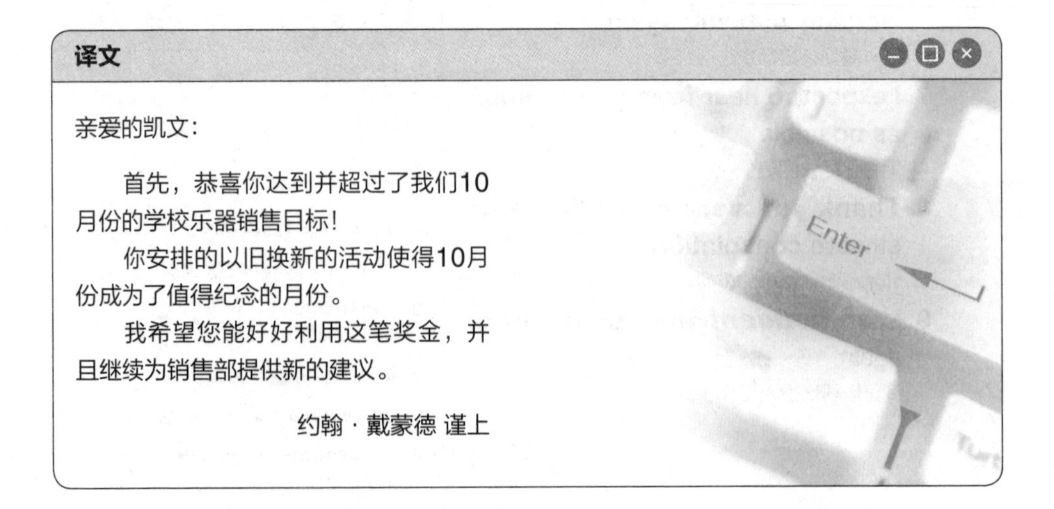

译文

亲爱的凯文:

首先，恭喜你达到并超过了我们10月份的学校乐器销售目标!

你安排的以旧换新的活动使得10月份成为了值得纪念的月份。

我希望您能好好利用这笔奖金，并且继续为销售部提供新的建议。

约翰·戴蒙德 谨上

1 解析重点1 **congratulations on**

恭喜别人的时候一定会用到这个句型。注意介词 on 后面可以直接接名词或者 V-ing。例如：

Congratulations on your engagement!（恭喜你们订婚！）

Congratulations on fulfilling your dream!（恭喜你实现梦想！）

2 解析重点2 **put the bonus check to good use**

这句话的意思是“好好利用奖金”。put...to use 是“利用；使用”的意思，make use of 也有“利用；使用”的意思，例如：

We must put everything to its best use.（我们要充分利用好一切。）

We must make good use of our spare time.（我们必须好好利用我们的空闲时间。）

1. **Congratulations on reaching your target!**
   恭喜你达成目标！
2. **You have exceeded the sales goal this month.**
   你这个月已经超过销售目标了。
3. **You are a creative staff member.**
   你是一名富有创造力的员工。
4. **It is really a wonderful idea.**
   这真是一个不错的点子。
5. **Your experience is being put to good use there.**
   在那里可以好好利用你的经验。
6. **Please continue to bring new ideas to the sales department.**
   请继续为销售部提供好点子。
7. **Thank you for all your effort.**
   感谢你做出的所有的努力。
8. **We must make good use of the available resources.**
   我们必须充分利用现有资源。

必背关键单词 

1. ***congratulation*** [kənˌɡrætjʊˈleɪʃən] n. 祝贺
2. ***exceed*** [ɪkˈsiːd] v. 超过
3. ***instrument*** [ˈɪnstrəmənt] n. 器具；乐器
4. ***sale*** [seɪl] n. 销售
5. ***arrange*** [əˈreɪndʒ] v. 安排；准备
6. ***trade-in*** [ˈtreɪdˌɪn] n. 折价物
7. ***bonus*** [ˈbəʊnəs] n. 奖金
8. ***department*** [dɪˈpɑːtmənt] n. 部门

10 佳节问候

Dear Mr. Smith,

Holiday[1] ***greetings***[2] and best wishes for the New Year!
May[3] you and all your family members have a ***joyous***[4] holiday ***season***[5]. Thank you for your ***patronage***[6] over the past few years and I hope we will enjoy more years of business cooperation together.

Best regards,

Yours faithfully,
Walt Lin

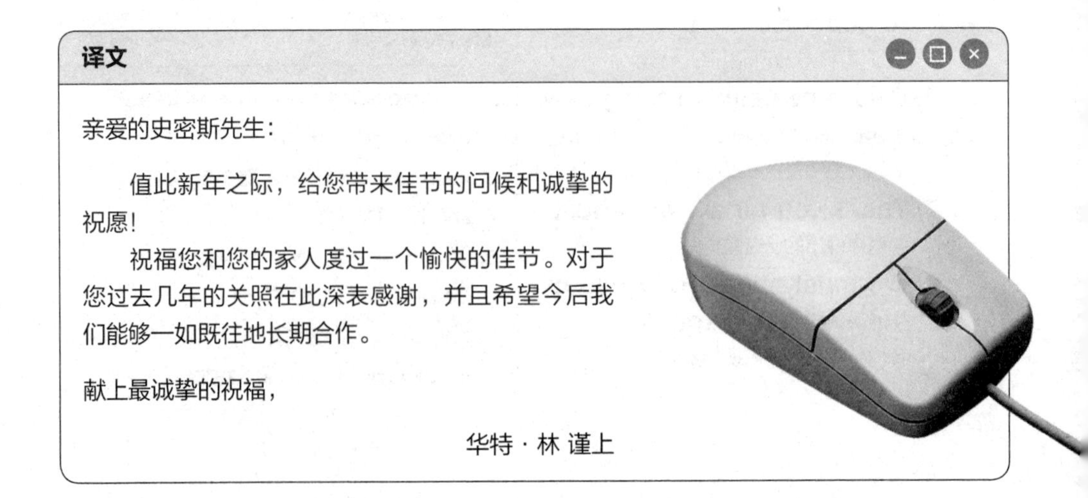

译文

亲爱的史密斯先生：

值此新年之际，给您带来佳节的问候和诚挚的祝愿！

祝福您和您的家人度过一个愉快的佳节。对于您过去几年的关照在此深表感谢，并且希望今后我们能够一如既往地长期合作。

献上最诚挚的祝福，

华特·林 谨上

语法重点解析

1 解析重点1 May

May 不但有名词“5月”和助动词“可以”的意思，还可表示“祝；愿（亦为助动词）”。注意，表示“祝；愿”时，一般要放于句首，且句中用动词原形表示。请对照以下例句：

I graduated from the college in May last year.（我去年5月大学毕业。）

May I make an appointment now?（我现在可以预约会面吗？）

May you have a happy journey.（祝你旅途愉快。）

2 解析重点2 holiday seasons

在此 season（季节）通常用于表示“时节”，而一般不使用在圣诞和新年假期以外的时节。holiday 广义上意为“假期；休假”，要注意的是当使用 holiday seasons 或者 holiday greetings 时，一般指圣诞节和新年的休假。所以如果职场中有人问 What are you doing for the holiday? 其实是在问年假状况。

1. I wish you a Happy New Year!
祝你新年快乐！

2. I would like to wish you all the best for a wonderful holiday season.
诚挚地祝你假期快乐。

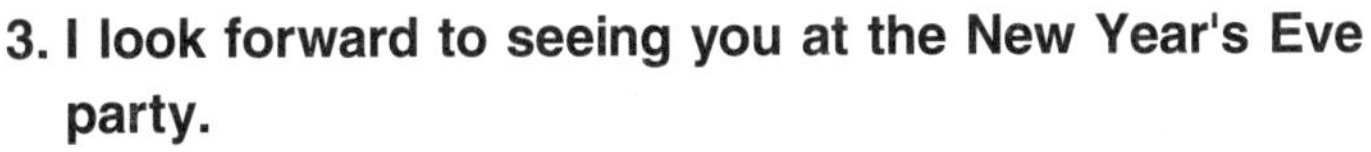

3. I look forward to seeing you at the New Year's Eve party.
我期盼在新年前夜的晚会上见到你。

4. May your new year be filled with health and *happiness*[7].
祝您在新的一年里身体健康、生活愉快。

5. We send you our best *wishes*[8] for the holidays.
我们送出最诚挚的祝福，祝您假期愉快。

6. May all the joys of Christmas be yours!
祝你圣诞快乐！

必背关键单词 

1. ***holiday*** [ˈhɒlədeɪ] n. 假期
2. ***greeting*** [ˈgriːtɪŋ] n. 问候
3. ***may*** [meɪ] v. 祝；愿
4. ***joyous*** [ˈdʒɔɪəs] adj. 快乐的
5. ***season*** [ˈsiːzn] n. 季节
6. ***patronage*** [ˈpætrənɪdʒ] n. 支持；赞助；惠顾
7. ***happiness*** [ˈhæpɪnəs] n. 幸福；快乐
8. ***wish*** [wɪʃ] n. 祝福

Unit 2 申请 Application

01 申请留学

这样写就对了

From davidw1985@163.com
Date February 22, 2008
To gradadmission@upenn.edu
Subject Application for Studying Abroad

Dear Sir or Madam,

I would like to apply for admission to your ***university***[1] as a Master's student in Applied Economics next September.
I am in my fourth year of ***undergraduate***[2] studies at Peking University at present, and I will receive a ***Bachelor***[3] of Arts in ***Economics***[4] in July. It has been my dream to ***pursue***[5] graduate studies at the University of Pennsylvania, an ***institution***[6] well-known for its excellent ***faculty***[7] and students as well as the strong leadership in the field of economics. I am ***confident***[8] that I would benefit a lot from the rich academic and cultural community of your university.
As requested, I have sent two letters of ***recommendation***[9], an original copy of my university transcript, and a copy of my TOEFL certificate. Please also find attached my completed application form.

Thank you very much for your consideration. I look forward to hearing from you soon.

Sincerely yours,
David Wang

译文

亲爱的女士或先生：

我想申请明年九月到贵校攻读应用经济学硕士学位。
我现在是一名北京大学的大四学生，将于今年七月获得经济学文科学士学位。能去宾夕法尼亚大学读研究生一直是我的梦想。贵校师资水平和学生都相当优秀，而且贵校在经济学领域里享有极高的声誉。我相信我将在这个富有学术和文化气息的校园里获益匪浅。
我已经按照贵校的要求，将两份推荐信、一份大学成绩单原件，以及一份托福成绩证明复印件寄出。入学申请表请参见附件。

非常感谢您对我的申请予以考虑。期待能尽快收到您的回复。

大卫 · 王 敬上

语法重点解析

1 解析重点1 admission

admission 表示“进入；许可；承认；录用”之意。要注意的是，获准进入大学要用 admission 而不能用 entry。虽然 entry 也有“进入”的意思，但是它主要强调进入的动作或状态。请对照以下例句：

Tom gained admission to that famous university.（汤姆获准进入那所著名的大学。）

Entry to the museum is free.（进入博物馆不需要门票。）

2 解析重点2 Sincerely yours

Sincerely yours 属于英文书信中的结尾敬语（Complimentary Close），相当于中文书信中的“谨上”“敬上”“敬启”等。结尾敬语的第一个字母必须大写。这里要注意的是，英国人习惯将 yours 放在前面，而美国人习惯将 yours 放在后面，这封 E-mail 是要申请美国的大学，所以最好要把 yours 放在后面。

1. **I would like to apply for the two-year intensive English course at your school.**
   我想申请贵校两年制的英语强化课程。
2. **I am writing to apply for the English immersion program at the University of Birmingham.**
   我想申请伯明翰大学的英语浸入式教学。
3. **I wish to apply for admission to your university as an undergraduate student next year.**
   我想申请明年到贵校就读大学本科课程。
4. **I am in my final year of studies at UC, where I am pursuing a B.A. in law.**
   我是加州大学的大四学生，主修法律。
5. **It has long been my dream to pursue English language studies at your college.**
   能到贵校研读英语语言课程一直是我的梦想。

必背关键单词

1. ***university*** [ˌjuːnɪˈvɜːsətɪ] n. 大学
2. ***undergraduate*** [ˌʌndəˈɡrædʒʊət] n. 大学本科生
3. ***bachelor*** [ˈbætʃələ(r)] n. 学士学位
4. ***economics*** [ˌiːkəˈnɒmɪks] n. 经济学
5. ***pursue*** [pəˈsjuː] v. 追求
6. ***institution*** [ˌɪnstɪˈtjuːʃn] n. 机构；惯例；制度
7. ***faculty*** [ˈfækltɪ] n. 全体教员；（大学的）系
8. ***confident*** [ˈkɒnfɪdənt] adj. 自信的；确信的
9. ***recommendation*** [ˌrekəmenˈdeɪʃn] n. 推荐；推荐信

02 申请假期

这样写就对了

From nancy1985@163.com
To mr.liu123@tom.com
Subject Asking for a Leave
Date April 11, 2008

Dear Boss,

I am afraid that I have to tell you the bad news.
Something ***terrible***[1] happened to me last weekend. I had a car ***accident***[2] and thus broke my legs. ***Therefore***[3], I have to ***inform***[4] you that I cannot go to work in the next few weeks because of this ***serious***[5] accident, so I want to ask for a month's leave. Thank you for your ***consideration***[6] and I'm hoping for your ***approval***[7].

Enclosed herewith is my X-ray photo for ***verification***[8].

Sincerely yours,
Nancy Li

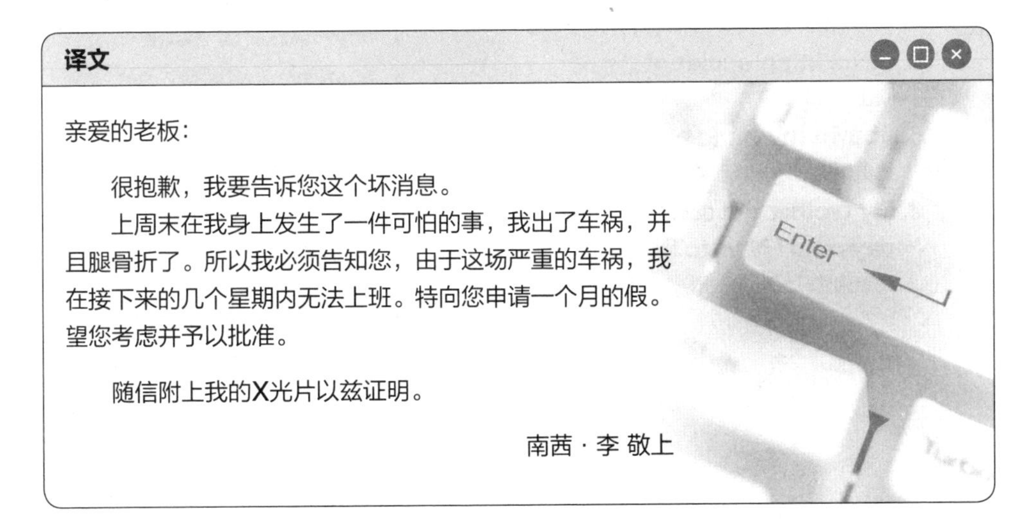

译文

亲爱的老板:

很抱歉，我要告诉您这个坏消息。

上周末在我身上发生了一件可怕的事，我出了车祸，并且腿骨折了。所以我必须告知您，由于这场严重的车祸，我在接下来的几个星期内无法上班。特向您申请一个月的假。望您考虑并予以批准。

随信附上我的X光片以兹证明。

南茜 · 李 敬上

语法重点解析

1 解析重点1 have to

have to 和 must 都是“必须”的意思，但两者有差别。have to 强调的是客观上“不得不”“只好”的意愿，主观上并非愿意。而 must 则强调主观上“必须”。请对照以下例句：

It is a pity that we have to leave now.（我们现在不得不离开了，真是遗憾。）

We must cut down the expenses.（我们必须削减开支。）

2 解析重点2 ask for a month's leave

ask for a month's leave 意为“申请一个月的假”。ask for leave 是“请假”的意思。holiday 虽然也有“假期”的意思，但是请假的时候不能说 ask for a holiday。“休假”可以说 take a holiday，请对照以下例句：

I'll ask our boss for a half-day's leave.（我要向我们的老板请半天假。）

When do you plan to take your holiday?（你打算什么时候休假？）

1. **I would like to know if I could ask for a casual leave of absence for one day.**
   我想知道我是否可以请一天的事假。
2. **I think a one-day leave this Wednesday may be the best *solution*[9].**
   我认为在这个星期三请一天假或许是最好的解决办法。
3. **I *apologize*[10] for the inconvenience my absence from work may cause.**
   我为缺勤可能对工作造成的不便表示歉意。
4. **I was in an accident.**
   我出了意外。
5. **I burnt my hands while cooking.**
   我做饭的时候烫伤了双手。
6. **My brother got seriously ill these few days and I have to look after him.**
   我弟弟这几天病得很严重，我得照顾他。

必背关键单词 

1. ***terrible*** [ˈterəbl] adj. 糟糕的；可怕的
2. ***accident*** [ˈæksɪdənt] n. 事故；意外遭遇
3. ***therefore*** [ˈðeəfɔ:(r)] adv. 因此；所以
4. ***inform*** [ɪnˈfɔ:m] v. 告诉；通知
5. ***serious*** [ˈsɪərɪəs] adj. 严重的；认真的；庄重的
6. ***consideration*** [kənˌsɪdəˈreɪʃn] n. 考虑；体贴；关心
7. ***approval*** [əˈpru:vl] n. 批准
8. ***verification*** [ˌverɪfɪˈkeɪʃn] n. 证实；证明
9. ***solution*** [səˈlu:ʃn] n. 解决方法；解答
10. ***apologize*** [əˈpɒlədʒaɪz] v. 道歉

03 申请汇款

这样写就对了

From ted001@126.com

To operation@jcb.com

Subject Application for Remittance

Date November 27, 2008

Dear Bank of China, Shanghai ***Branch***[1],

I ***hereby***[2] request you to ***effect***[3] the following ***remittances***[4] subject to the conditions ***overleaf***[5], which I have read and agreed to be bound by.

T/T　　M/T　　D/D

Date
Amount
Name of Beneficiary
Address of Beneficiary
Name of ***Remitter***[6]
Address of Remitter
Remarks
Signature[7]

In ***payment***[8] of the above remittance, please ***debit***[9] my account with you.

译文 

致中国银行上海分行：

本人已阅读并同意遵守此页背面所列条款，兹委托贵行据此办理下列汇款。

电汇　　信汇　　票汇

日期
金额
收款人姓名
收款人地址
汇款人姓名
汇款人地址
备注
签名

上述汇款请从本人在贵行开立的账户中扣除。

语法重点解析

1 解析重点1 request

request 意思是“请求；要求”，而 ask 也有“询问；要求”之意，但是在非常正式的场合下要用 request。请对照以下例句：

You must ask if you have any questions.（如果你有任何疑问一定要问。）

The bank requests us to offer guarantee.（银行要求我们提供担保。）

2 解析重点2 beneficiary

beneficiary 有“受益者；受惠者”之意，商务英语中它的意思是“受益人；收款人”。payee 也有“收款人”的意思。例如：

Is the beneficiary the same as the insured?（受益人和被保险人为同一人吗？）

Could you tell me how to spell the name of the payee?（麻烦您告诉我收款人的姓名该怎样拼写？）

1. **I have read and will *comply*[10] with all the following terms.**
   我已阅读并将遵守所有下列条款。
2. **Please write in capital letters and choose the proper method of the remittance.**
   请用大写字母填写，并选择合适的汇款方式。
3. **The bank's transfer fee outside China are to be borne by the remitter.**
   中国境外的银行转账手续费由汇款人承担。
4. **I hereby request you to effect the following remittances subject to the conditions overleaf.**
   本人在此委托贵行依据背面的条款办理下列汇款。
5. **In payment of the above remittance, please debit my account with you.**
   上述汇款请从本人在贵行的账户中扣除。
6. **I agree to be bound by all the regulations.**
   我同意遵守所有的规章制度。

必背关键单词

1. ***branch*** [brɑːntʃ] n. 分支机构；分公司
2. ***hereby*** [ˌhɪəˈbaɪ] adv. 以此方式；特此
3. ***effect*** [ɪˈfekt] v. 实现；使生效
4. ***remittance*** [rɪˈmɪtns] n. 汇款
5. ***overleaf*** [ˌəʊvəˈliːf] adv. 在背面；
6. ***remitter*** [rɪˈmɪtə] n. 汇款人
7. ***signature*** [ˈsɪɡnətʃə(r)] n. 签字
8. ***payment*** [ˈpeɪmənt] n. 付款
9. ***debit*** [ˈdebɪt] v. 记入借方
10. ***comply*** [kəmˈplaɪ] v. 遵从

04 申请商标注册

这样写就对了

From　dazhong@abc.com
To　commissioner@P&T.com
Subject　Application for Trademark Registration
Date　July 23, 2008

Dear ***Commissioner*** [1] of ***Patents*** [2] and ***Trademarks*** [3],

(Corporate Name)
(State or Country of Corporation)
(Business Address)

The above identified ***applicant*** [4] has ***adopted*** [5] and is using the trademark shown in the accompanying drawing for (common, usual or ordinary name of goods) and request that such mark be ***registered*** [6] in the United States Patent and Trademark Office on the Principal Register established by the Act of July 5, 1946.
The trademark was first used on the goods on (Date); was first used in (Type of Commerce) ***commerce*** [7] on (Date); and is now in use in such commerce.
The mark is used by applying it to (manner of application, such as the labels ***affixed*** [8] to the product). Five specimens showing the mark as actually used are presented **herewith**.

Corporate Name
By: (Signature of Corporate Officer and Official Title)

译文

尊敬的专利商标局局长：

（公司名称）
（公司所在州或国家）
（营业地址）

上述申请人已经正式采用并将附图中展示的商标用于（商品通用名称），现请求美国专利商标局根据1946年7月5日通过的法案注册该商标。
该商标于（日期）首次用于该商品，于（日期）第一次使用于（贸易类型），且现在仍在该贸易中使用。
该商标采取（使用方式，例如在产品上附标签）用在商品上。现附上5份样品，显示商标的实际使用情况。

公司名称
由：（公司要员的签名及正式职务）

语法重点解析

1 解析重点1 adopt

adopt 有好几种意思，包括“采纳；收养；正式接受；批准”等，在此语境下为“正式通过”的意思。通常通过法案、决议等都会用 adopt。请对照下列例句：

After much deliberation, the general manager decided to adopt her suggestion.（总经理再三考虑之后，决定采纳她的建议。）

Mr. Kern adopted the orphan as his own son.（克恩先生将那名孤儿收养为自己的儿子。）

The agenda was adopted after some discussion.（经过讨论，通过了议事日程。）

2 解析重点2 herewith

herewith 的意思为“与此一道；同此；随函；据此”等，一般并不常用，只在一些非常正式的信函里出现。请看以下例句：

I send you herewith two copies of the contract.（我随函附上合约书，一式两份。）

Herewith the principle is established.（原则借此确定。）

高频例句

1. **The above identified applicant has adopted and is using the trademark shown in the accompanying drawing for alcoholic beverages.**
   上述申请人已经正式采用附图中展示的商标并将之用于酒精饮料。
2. **The applicant requests that such mark be registered in State Intellectual Property Office of the People's Republic of China.**
   申请人请求中华人民共和国国家知识产权局注册该商标。
3. **The trademark was first used on the goods on October 11, 2007.**
   该商标于2007年10月11日首次用于该商品。
4. **The trademark was first used in textile goods on August 6, 2008.**
   该商标于2008年8月6日首次用于纺织品。

必背关键单词 

1. ***commissioner*** [kə'mɪʃənə(r)] n. 长官；委员
2. ***patent*** ['pætnt] n. 专利；专利权；专利品
3. ***trademark*** ['treɪdmɑːk] n. 商标
4. ***applicant*** ['æplɪkənt] n. 申请人
5. ***adopt*** [ə'dɒpt] v. 采用；收养；接受
6. ***register*** ['redʒɪstə(r)] v. 注册；挂号
7. ***commerce*** ['kɒmɜːs] n. 商业；贸易
8. ***affix*** [ə'fɪks] v. 附于；粘贴

05 申请信用证

这样写就对了

From abc2008@sina.com
To davidchen@tom.com
Subject Application for Opening L/C
Date August 1, 2008

Dear Sirs,

Thank you for your letter of June 18 enclosing ***details***[1] of your ***terms***[2].

According to your ***request***[3] for opening an irrevocable L/C, we have ***instructed***[4] Mega International ***Commercial***[5] Bank to open a ***credit***[6] for US$50,000 in your favor, ***valid***[7] until Sep. 20. Please inform us by fax when the order has been ***executed***[8].

Thank you for your cooperation!

Sincerely yours,
ABC Company

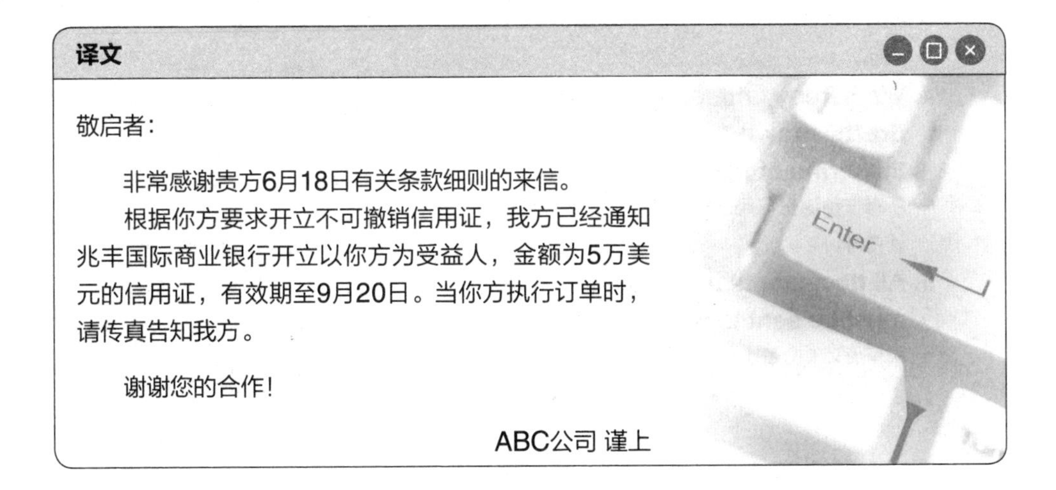

译文

敬启者：

非常感谢贵方6月18日有关条款细则的来信。

根据你方要求开立不可撤销信用证，我方已经通知兆丰国际商业银行开立以你方为受益人，金额为5万美元的信用证，有效期至9月20日。当你方执行订单时，请传真告知我方。

谢谢您的合作！

ABC公司 谨上

语法重点解析

1 解析重点1 irrevocable L/C

L/C 是信用证 letter of credit 的缩写。信用证为国际贸易中最主要、最常见的付款方式。irrevocable L/C，即“不可撤销信用证”，是指开证银行一经开出，在有效期内未经受益人或议付行等有关当事人同意，不得随意修改或撤销的信用证。它的特征是有开证银行确定的付款承诺和不可撤销性。

2 解析重点2 valid until Sep. 20

valid until Sep. 20意思是“有效期至9月20日”。这里的 valid 表示“有效的”，尤其是指“具有法律效力的”。effective 也有“有效的”的意思，但是它一般强调“有效果的；实际的”。请对照以下例句：

The letter of credit is valid until August 31st.（本信用证有效期至8月31日。）

Advertising is often the most effective method of promotion.（广告往往是最有效的促销方法。）

1. **Thank you for your letter of March 5 enclosing details of your terms.**
   非常感谢你方3月5日有关条款详细情况的来信。
2. **This credit shall remain in force until August 15, 2009.**
   这份信用证到2009年8月15日为止有效。
3. **We have instructed the Industrial & Commercial Bank to open a credit for US$20,000.**
   我方已经通知工商银行开立金额为2万美元的信用证。
4. **We hereby undertake to *honor*[9] all drafts drawn in accordance with the terms of this credit.**
   所有按照这份信用证条款开具的汇票，我们保证兑付。
5. **All documents made out in English must be sent to our bank in one lot.**
   所有英文单据必须一次性寄到我行。

必背关键单词

1. ***detail*** [ˈdiːteɪl] n. 详情
2. ***term*** [tɜːm] n. 条件；条款
3. ***request*** [rɪˈkwest] n. & v. 要求；请求
4. ***instruct*** [ɪnˈstrʌkt] v. 通知；命令；指示
5. ***commercial*** [kəˈmɜːʃl] adj. 商业的
6. ***credit*** [ˈkredɪt] n. 信用；赊购
7. ***valid*** [ˈvælɪd] adj. 有效的
8. ***execute*** [ˈeksɪkjuːt] v. 执行
9. ***honor*** [ˈɒnə(r)] v. 兑现；支持

06 申请许可证

这样写就对了

From limin_123@163.com
To ic-ceca@ceca.com
Subject Application for License
Date May 21, 2008

Dear ***President***[1] of CECA,

I, Min Li, do hereby ***apply***[2] for a ***license***[3] to ***display***[4] the trademark of CECA, “COOL” at my place of business located at Felicity Street 520, in the Tang city.
This application is in ***accordance***[5] with the regulations of the CECA. I am ***cognizant***[6] of the ***regulations***[7] of CECA that govern the display of trademark and the manner of conducting business, and I agree to abide by such regulations at all times.

Sincerely yours,
Min Li

译文

尊敬的中国电子商务协会会长:

本人，李民，在此向中国电子商务协会郑重申请位于堂城幸福大街520号公司所在地展示的“COOL”商标的使用许可证。

本申请是依据中国电子商务协会条例提出。本人清楚协会对上述商标展示和业务经营模式的规范条例，并同意永久遵守这些条例。

李民 谨上

语法重点解析

1 解析重点1 **CECA**

CECA 是 China Electronic Commerce Association（中国电子商务协会）的首字母缩写。对于一些官方机构或者大型国际组织在信件中一般不会使用全称，而直接使用缩写形式，例如：

IAEA 即 International Atomic Energy Agency 国际原子能机构
WTO 即 World Trade Organization 世界贸易组织
ICC 即 The International Chamber of Commerce 国际商会
OPEC 即 Organization of Petroleum Exporting Countries 石油输出国组织

2 解析重点2 **abide**

abide 有“遵守；居留；忍受”的意思，在此语境下意为“遵守”。表达“遵守”还可以用 observe, comply, follow 等。请对照以下例句：

Everyone must abide by the law.（所有的人都应遵守法律。）
They faithfully observed the rules.（他们忠实地遵守规则。）
It's not complying with our policy.
（这不符合我们的政策。）
These orders must be followed at once.
（这些命令必须立即执行。）

1. I do hereby apply for a license to display the trademark.
本人在此郑重申请商标使用许可证。

2. This application is in accordance with the regulations of the CIMA.
本申请依据中国仪器仪表行业协会条例提出。

3. I hereby apply for a food hygiene license.
本人在此申请食品卫生许可证。

4. I need to apply for a parking *permit*[8].
我需要申请停车许可证。

5. You can't take pictures here without a special permit.
未经特许，你不能在这里拍照。

必背关键单词 

1. ***president*** [ˈprezɪdənt] n. 总统；总裁；校长
2. ***apply*** [əˈplaɪ] v. 申请；应用
3. ***license*** [ˈlaɪsns] n. 许可；许可证
4. ***display*** [dɪˈspleɪ] v. & n. 陈列；展览
5. ***accordance*** [əˈkɔːdns] n. 一致；符合
6. ***cognizant*** [ˈkɒgnɪzənt] adj. 认知的；知晓的
7. ***regulation*** [ˌregjʊˈleɪʃn] n. 管理；规章
8. ***permit*** [pəˈmɪt] n. 许可证；许可

07 申请贷款

这样写就对了

From green@yahoo.com
To bank_branch22@icbc.com
Subject Application for Bank Loan
Date February 25, 2009

Dear Loan Section Head,

I'm writing this letter in applying US$150,000 from your bank for opening a Japanese ***cuisine*** [1] restaurant.

I've carried out a ***survey*** [2] and found that there is only a small restaurant selling Japanese foods. The ***potential*** [3] Japanese food market is large and we have ***sufficient*** [4] customers. What's more, we have employed ***excellent*** [5] cooks that can ensure the quality of the meal. The loan money will be used in the interior ***decoration*** [6]. My partner, Peter and I will provide real ***estate*** [7] of our families that is worth US$200,000 as ***guarantee*** [8].

Approving this loan will prove a wise choice. Please consider our application seriously and we are looking forward to your response!

Sincerely yours,
Ted Green

译文  

尊敬的贷款部门经理：

我写这封信是为了向贵行申请十五万美元的贷款来开一家日本料理店。

我已经做了调查，并且发现当地只有一家小型的餐馆卖日式食品。日式餐饮的潜在市场非常大，并且我们有充足的客源。此外，我们聘请了优秀的厨师来确保餐饮的品质。申请的贷款将用于室内装潢。我的合伙人彼得和我将会提供我们价值二十万美元的房产作为担保。

批准此贷款是一个明智的选择，请慎重考虑我们的申请，我们期待着您的答复!

泰德 · 格林 谨上

语法重点解析

1 解析重点1 **carry out a survey**

短语 carry out a survey 是“进行了调查”的意思。carry out 为“执行；贯彻；进行”的意思。survey 做动词时有“勘察；检查；眺望”的意思；做名词时有“调查”的意思，在此语境中则表示“调查”。表达“做调查”还可以用短语 make a survey, conduct a survey 等。请看以下例句：

Let's make a survey about fast food and home cooking.（我们来做一份关于快餐和家庭料理的调查报告吧。）

2 解析重点2 **what's more**

what's more 的意思是“此外；更有甚者；更重要的是”。一般在说明情况或原因的时候会用到，用来加强语气，增加说服力。类似的表达还有 besides, in addition（除此之外）等。请看以下例句：

It's a useful book, and what's more, not an expensive one.（这是一本实用的书，而且还不贵。）

The play had terrible actors, besides being far too long.（这出戏除了太冗长以外，演员也很差。）

She has two cars, in addition, a motorboat.（她有两辆小汽车外加一艘汽艇。）

高频例句

1. **I'm writing this letter to apply for a US$100,000 loan from your bank.**
   我写信是为了向贵行申请10万美元的贷款。
2. **I'm applying for the *loan*[9] for opening a bookstore.**
   我申请贷款是为了开一家书店。
3. **We have sufficient customers.**
   我们拥有充足的客源。
4. **I made an investigation and found that it was a great business opportunity.**
   我做了一项调查，发现这是一个很好的商机。
5. **I will provide real estate that is worth US$500,000 as guarantee.**
   我将提供价值50万美元的不动产作为担保。

必背关键单词 

1. ***cuisine*** [kwɪˈziːn] n. 烹饪
2. ***survey*** [ˈsɜːveɪ] n. 勘测；调查
3. ***potential*** [pəˈtenʃl] adj. 潜在的；可能的
4. ***sufficient*** [səˈfɪʃnt] adj. 足够的；充足的
5. ***excellent*** [ˈeksələnt] adj. 优秀的；极好的
6. ***decoration*** [ˌdekəˈreɪʃn] n. 装饰
7. ***estate*** [ɪˈsteɪt] n. 房地产
8. ***guarantee*** [ˌɡærənˈtiː] n. 担保
9. ***loan*** [ləʊn] n. 贷款

08 申请出国进修

这样写就对了

From bettyliu@tom.com
To global_biz@icbc.com
Subject Application for Further Studies Abroad
Date April 7, 2008

Dear Mr. Miller,

I am writing to you about a big plan.
I am the ***supervisor*** [1] of the ***Research*** [2] and Development Department. In order to ***improve*** [3] my ***professional*** [4] skill and offer better ***service*** [5] for our company, I think I need to learn more about current international cutting-edge technology. I hereby advance an application for further studies abroad.
I ***promise*** [6] to study hard and come back to ***contribute*** [7] more to our company's prosperous future. Your decision may affect our tomorrow. Please consider carefully.

Hoping for your support!

Sincerely yours,
Betty

译文

亲爱的米勒先生:

我写信给您是为了一项大计划。

我是研发部的主管，为了提高职业技能以便更好地为公司服务，我认为我需要学习当下国际上的尖端技术，特此向您提出出国进修的申请。

我保证会努力学习，学成归来后为公司未来的繁荣做出更大的贡献。您的决定可能会影响公司的未来，请慎重考虑。

期望您的支持!

贝蒂 谨上

语法重点解析

1 解析重点1 **advance**

advance 有动词、名词和形容词三种词性，意思分别是“前进；提出”“增长”“预先的”。在此语境中 advance 作为动词，表示“提出”。“提出”还可以用 bring forward, raise 等表示。例如：

He advanced many reasonable proposals.（他提出了许多合理的建议。）

Please bring the matter forward at the next meeting.（这个问题请在下次会议中提出。）

Why don't we raise this question?（我们何不提出这个问题呢？）

2 解析重点2 **Your decision may affect our tomorrow.**

这句话的意思是“您的决定可能影响公司的未来”。事实上，申请出国进修并不是一件容易的事情，它一定会耗费公司的财力、物力、人力等，所以在写进修申请时更需要讲究技巧。这句话就非常有说服力，暗指进修并不是为了员工自己，如果不批准进修，似乎就会影响公司的发展和未来，主管出于对公司前景的期许，应该会慎重考虑。员工很难把握主管的心思，不知道哪句话能打动对方，所以写信的时候真的需要字斟句酌，讲究技巧。

高频例句

1. I need to study *further*[8] to enhance myself.
我需要去进修来提升自己。

2. I want to know more about the outside world.
我想了解更多外面的世界。

3. I hereby advance an application for further *education*[9].
我特此提出进修申请。

4. Hoping that you could think over my suggestion.
希望您能仔细考虑我的建议。

5. I promise that I will come back to give more contribution.
我保证会回来做出更多的贡献。

6. Your decision might determine our company's future.
您的决定可能决定我们公司的未来。

7. I will absolutely study hard there.
我在那里一定会努力学习。

必背关键单词

1. ***supervisor*** [ˈsu:pəvaɪzə(r)] n. 主管
2. ***research*** [rɪˈsɜ:tʃ] n. 研究
3. ***improve*** [ɪmˈpru:v] v. 提高；改善
4. ***professional*** [prəˈfeʃənl] adj. 专业的；职业的
5. ***service*** [ˈsɜ:vɪs] n. 服务
6. ***promise*** [ˈprɒmɪs] v. 保证
7. ***contribute*** [kənˈtrɪbju:t] v. 贡献
8. ***further*** [ˈfɜ:ðə(r)] adj. 更进一步的；深层的
9. ***education*** [ˌedʒʊˈkeɪʃn] n. 教育

09 申请调换部门

From zhangxue_snow@sina.com　Date November 7, 2012
To ceca@limitco.com
Subject Application for Transfer

Dear Sir or Madam,

This is Chrissie Snow from the ***Administrative***[1] Department. I have worked here for seven months since May 2, 2012. I am always working very hard in my ***post***[2] and doing my best to pursue ***perfection***[3].
I love our company, and I really hope I could have ***long-term***[4] development here. Therefore, I want to know the company ***overall***[5]. As the Market Department is the leading department, which is in ***charge***[6] of the main business, I want to enter this department to learn more. I promise I will work as hard in the new department as in my ***current***[7] department.

I desperately hope for your permission!

Sincerely yours,
Chrissie Snow

译文

敬启者：

我是行政部门的克丽茜·斯诺，于2012年5月2日开始在这个部门工作，至今已有7个月的时间了。在这一岗位，我始终坚持努力工作，每项工作都力求完美。

我热爱我们的公司，并希望能够在这里得到长远的发展，所以我想更全面地了解公司。营销部是负责公司主要业务的重要部门，因此我希望能够进入这个部门学习更多知识。我保证一定会像在现在这个部门一样努力工作。

热切盼望着您的批准！

克丽茜·斯诺 敬上

语法重点解析

1 解析重点1 promise

promise 既有名词词性，又有动词词性，做名词是“诺言；前途”的意思；做动词则是“允诺；答应”的意思。在此语境中是做动词“允诺”，可以引导宾语从句。I promise that...意为“我答应……；我承诺……”。请看以下例句：

Life is a promise, fulfill it.（生活是承诺，去实现它。）

I promise that it won't happen again.（我保证这种事情不会再发生。）

2 解析重点2 I desperately hope for your permission!

这句话的意思是“我殷切盼望着您的批准！”desperately 这个单词虽然是 desperate（绝望的）的副词形式，但在此并不作“绝望地”理解，而是“极度地；非常地”的意思。一句普通的请求加上这个单词，语气就会变得更强烈、更有诚意。类似的程度副词还有 badly，它也是“极度地”的意思，用来加强语气。请看以下例句：

We desperately need that money.（我们实在非常需要那笔钱。）

He was wounded badly.（他伤得非常重。）

高频例句

1. **I am Bill from the Personnel Department.**
   我是人事部的比尔。
2. **I am looking forward to *entering*[8] another department.**
   我希望进入另一个部门。
3. **I have worked in the department for two years.**
   我已经在这个部门工作两年了。
4. **I am a *creative*[9] and diligent employee.**
   我是个有创造力并且勤奋的员工。
5. **Could you give me an *opportunity*[10] to know more about our company?**
   你能给我一个机会让我更加了解公司吗？
6. **The Planning Department is the department I dream of.**
   企划部正是我向往的部门。

必背关键单词 

1. ***administrative*** [ədˈmɪnɪstrətɪv] adj. 行政的
2. ***post*** [pəʊst] n. 职位
3. ***perfection*** [pəˈfekʃn] n. 完美
4. ***long-term*** [lɒŋ tɜːm] adj. 长远的
5. ***overall*** [ˌəʊvərˈɔːl] adv. 全部地；总体上
6. ***charge*** [tʃɑːdʒ] v. 负责
7. ***current*** [ˈkʌrənt] adj. 现在的
8. ***enter*** [ˈentə(r)] v. 进入
9. ***creative*** [krɪˈeɪtɪv] adj. 有创造力的
10. ***opportunity*** [ˌɒpəˈtjuːnətɪ] n. 机会

10 申请员工宿舍

这样写就对了

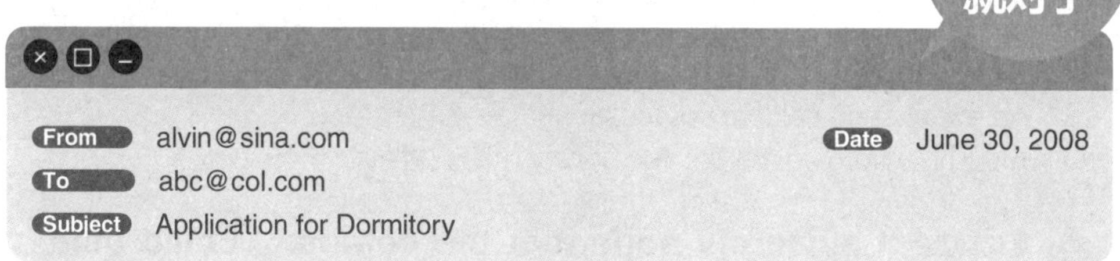

From alvin@sina.com
Date June 30, 2008
To abc@col.com
Subject Application for Dormitory

Dear Boss,

I am writing to you for solving a problem.
My name is Alvin Hsu, a new ***staff***[1] member of the company. I come from Xianghe ***County***[2], and came to the city alone, which is ***completely***[3] ***strange***[4] to me. I have no ***relatives***[5] here and I have no money to ***rent***[6] a house. It is definitely difficult for me to get ***accommodations***[7]. **I sincerely hope that the company could help solve my big problem.** I hereby apply for a ***dormitory***[8].

Hoping for your approval!

Truly yours,
Alvin Hsu

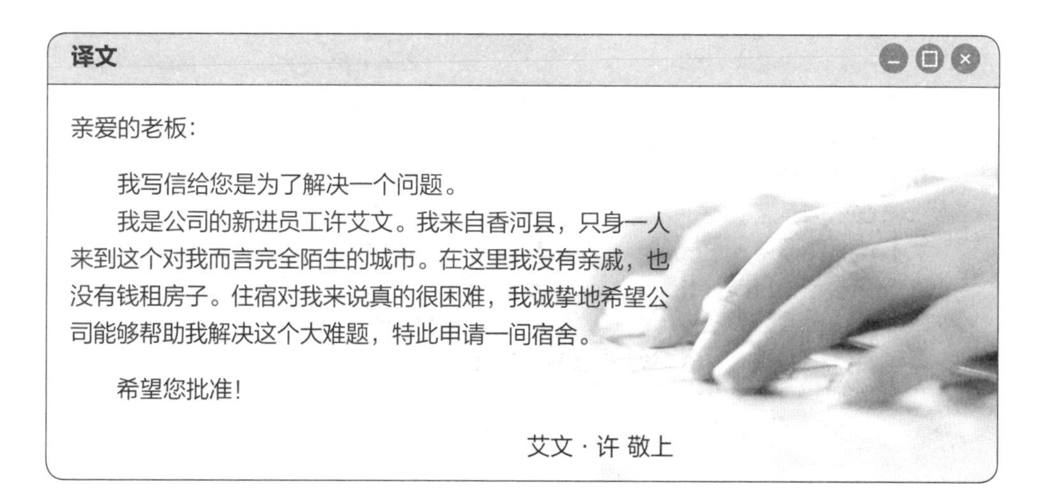

译文

亲爱的老板：

我写信给您是为了解决一个问题。

我是公司的新进员工许艾文。我来自香河县，只身一人来到这个对我而言完全陌生的城市。在这里我没有亲戚，也没有钱租房子。住宿对我来说真的很困难，我诚挚地希望公司能够帮助我解决这个大难题，特此申请一间宿舍。

希望您批准！

艾文·许 敬上

语法重点解析

1 解析重点1 alone

alone 为副词词性，意思是“独自地；单独地”，还有一个跟它很相近的形容词 lonely，意为“孤独的；凄凉的”。要注意的是这两个词有区别，不能混用。alone 指的是一种客观上“单独”的状态，而 lonely 则强调的是主观上“孤独”的感受。请对照以下例句：

An evil chance seldom comes alone.（祸不单行。）

After his wife died, he felt very lonely.（妻子去世后，他很孤独。）

2 解析重点2 I sincerely hope that the company could help solve my big problem.

这句话的意思是“我诚挚地希望公司能够帮助我解决这个大难题”。注意句子中的几个单词 sincerely，could，big，都无形中使句子充满了感情色彩。sincerely 表现了真诚；could 表达了委婉；big 强调了问题的严重性。如果去掉了这些单词，句子就会显得平淡无奇，也就缺乏了说服力。

1. **I am writing to you for a *pure-hearted*[9] proposal.**
   我写信给您是为了提出一个真诚的建议。
2. **I come from a remote village.**
   我来自一个偏远的乡村。
3. **I am here for the first time.**
   我第一次来到这里。
4. **I am not familiar with the city at all.**
   我一点也不熟悉这个城市。
5. **Renting a house is too expensive for me.** 租房子对我来说太贵了。
6. **I can't afford to rent a flat alone.**
   我自己一个人租不起一间公寓。
7. **I really hope our company could give more consideration to employees like us.**
   我真的希望公司能够给予像我们这样的员工更多关心。

必背关键单词 

1. ***staff*** [stɑ:f] n. 员工（集体名词）
2. ***county*** [ˈkaʊntɪ] n. 县
3. ***completely*** [kəmˈpli:tlɪ] adv. 完全地
4. ***strange*** [streɪndʒ] adj. 奇怪的；陌生的
5. ***relative*** [ˈrelətɪv] n. 亲戚
6. ***rent*** [rent] v. 租用
7. ***accommodation*** [əˌkɒməˈdeɪʃn] n. 住宿；适应
8. ***dormitory*** [ˈdɔ:mətrɪ] n. 宿舍
9. ***pure-hearted*** [ˈpjʊəˈhɑ:tɪd] adj. 真诚的

Unit 3 求职 Applying for a job

01 应聘行政助理

这样写就对了

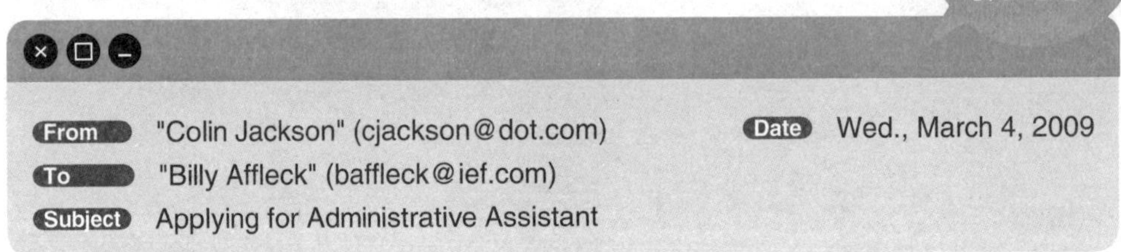

From "Colin Jackson" (cjackson@dot.com)
Date Wed., March 4, 2009
To "Billy Affleck" (baffleck@ief.com)
Subject Applying for Administrative Assistant

Dear Mr. Affleck,

I am replying to your advertisement in the *New York Times* for an ***administrative***[1] ***assistant***[2].
I worked for a big multinational company for one year as an administrative assistant and such experience has ***prepared***[3] me for the work you are calling for. **I believe I am the best man for this *position***[4].
Enclosed is my ***resume***[5]. I hope you will consider my ***application***[6]. Thank you for your time.

Looking forward to hearing from you soon!

Yours sincerely,
Colin Jackson
(Enclosure)

译文

亲爱的阿弗莱克先生：

我写信是为了回应贵公司在《纽约时报》上的一则招聘行政助理的广告。

我曾在一家大型跨国公司做过一年行政助理的工作。这一经历使我对你们所需的工作要求有所准备。我相信我是这个职位的最佳人选。

附件中有我的个人简历。敬请考虑。谢谢您宝贵的时间！

期待您能尽快回复！

科林·杰克森 谨上
（附件）

语法重点解析

1 解析重点1 **I am replying to your advertisement in the *New York Times* for an administrative assistant.**

一般我们在应聘的时候，是先看到对方的招聘信息才发出应聘请求的。我们在信中也应告知对方是在什么地方看到的招聘信息，以避免唐突。句子中的 your advertisement in the *New York Times* for an administrative assistant（贵公司在《纽约时报》上所刊登的行政助理的招聘广告）就详细说明了自己是在《纽约时报》上看到了招聘广告，而且是招聘一名行政助理。这样的说法很具体（Concreteness），也有利于引出接下来要说的求职意向。

2 解析重点2 **I believe I am the best man for this position.**

虽然应聘的时候，最后的决定权掌握在别人手中，但是也不用太低声下气，或是表现出乞求的样子。求职者应该充满自信，这样你的印象分数才会比别人高。例如在应聘某一职位的时候，我们就可以说：

I believe I am the best man for this position.（我相信我是这个职位的最佳人选。）

I believe I am the right man for the job.（我相信我是这份工作最好的人选。）

1. **I am writing to apply for the position of computer *engineer*[7].**
   我想申请贵公司的计算机工程师一职。
2. **I am confident that my experience and references will show that I can *fulfill*[8] the particular requirements of this position.**
   我相信我的经验和推荐信会向您证明，我能够符合这一职位的特定需要。
3. **Thank you very much for your time and *consideration*[9].**
   谢谢您抽出时间对我予以考虑。
4. **Please refer to my resume for more information.**
   详情请见我的个人简历。

必背关键单词

1. ***administrative*** [ədˈmɪnɪstrətɪv] adj. 行政上的；管理上的
2. ***assistant*** [əˈsɪstənt] n. 助手；助理
3. ***prepare*** [prɪˈpeə(r)] v. 预备；准备
4. ***position*** [pəˈzɪʃn] n. 位置；职位；形势
5. ***resume*** [rɪˈzjuːm] n. 简历；履历
6. ***application*** [ˌæplɪˈkeɪʃn] n. 应用；申请
7. ***engineer*** [ˌendʒɪˈnɪə(r)] n. 工程师
8. ***fulfill*** [fʊlˈfɪl] v. 实践；实现；履行
9. ***consideration*** [kənˌsɪdəˈreɪʃn] n. 考虑

02 咨询职位空缺

From "Andy Carter" (acarter@cao.com)　Date Thurs., March 5, 2009
To "Simon Jackman" (sjackman@plk.com)
Subject Inquiry for Open Positions

Dear Mr. Jackman,

I am writing to inquire if your company has any *opening*[1] in the area of food engineering. I have long been interested in working in your company. Although I am a ***recent***[2] ***graduate***[3] with some ***intern***[4] experience, **I still want to *pursue*[5] a job which I find *fascinating*[6].**
If possible, please reply and I will send you a resume ***via***[7] e-mail. Thank you for your time.

Looking forward to hearing from you soon!

Yours sincerely,
Andy Carter

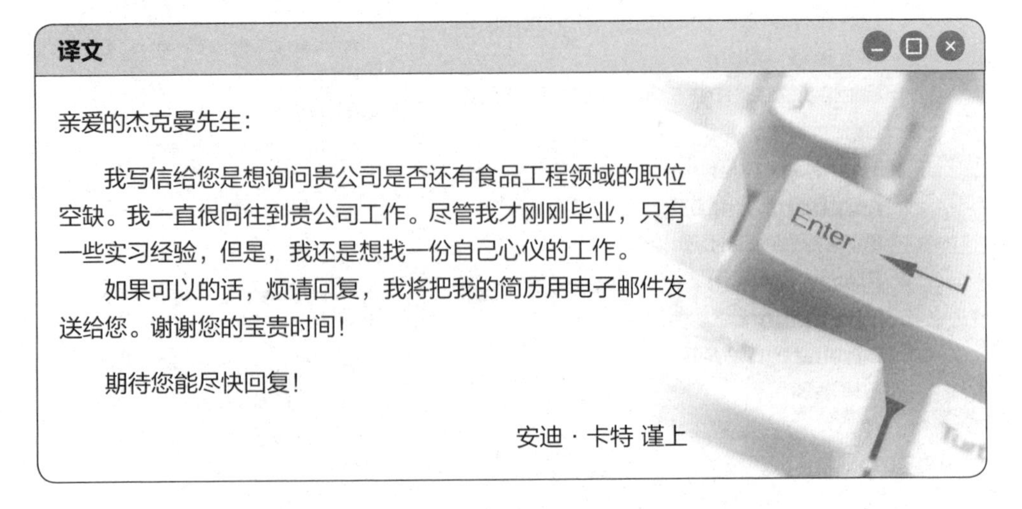

译文

亲爱的杰克曼先生：

　　我写信给您是想询问贵公司是否还有食品工程领域的职位空缺。我一直很向往到贵公司工作。尽管我才刚刚毕业，只有一些实习经验，但是，我还是想找一份自己心仪的工作。

　　如果可以的话，烦请回复，我将把我的简历用电子邮件发送给您。谢谢您的宝贵时间！

　　期待您能尽快回复！

安迪 · 卡特 谨上

语法重点解析

1 解析重点1 **I am writing to inquire if your company has any opening in the area of food engineering.**

询问职位空缺时，开头有几点是需要特别留意的。首先，我们要注意询问招聘公司是否有某一方面的职位空缺，然后再继续下文。I am writing to inquire if your company has any opening（我写信给您是想询问贵公司是否还有职位空缺）这句话是在询问是否有职位空缺，而 in the area of food engineering（食品工程领域）则是指明是哪方面的职位。这样一来大方向就先设定好了，意思清楚明白（Clearness），对方也不至于一头雾水了。

2 解析重点2 **I still want to pursue a job which I find fascinating.**

找工作的时候，也许我们才刚踏出校门不久，也许我们是初出茅庐的新手，有时候不得不先行就业再选择适当的职业。但是，坚持才会胜利，兴趣才是最好的老师。正如上面的句子所说的，I still want to pursue a job which I find fascinating（我还是想找一份自己心仪的工作）中 pursue 和 fascinating 这两个单词不禁让人觉得写信人很有毅力，也很有自己的想法。同时，写信人也间接夸赞了该公司的职位很吸引人。

高频例句

1. **If you accept my application, please reply.**
   如果您接受我的申请，烦请回复。
2. **I have *skills*[8] which could be of use to your company.**
   我的技能会对贵公司有所帮助。
3. **I would like to find out more about the opening.**
   我想知道空缺职位的详细信息。
4. **I am a recent graduate of Washington University.**
   我是一名刚从华盛顿大学毕业的学生。
5. **I would be very grateful for your consideration.**
   我将十分感谢您的考虑！
6. **I would appreciate it if you can *grant*[9] me an interview.**
   如果您能给我一个面试的机会，我将不胜感激。

必背关键单词 

1. ***opening*** [ˈəʊpnɪŋ] n. （职位的）空缺
2. ***recent*** [ˈriːsnt] adj. 最近的；近代的
3. ***graduate*** [ˈɡrædʒʊət] n. 毕业生
4. ***intern*** [ɪnˈtɜːn] n. 实习医师；实习生
5. ***pursue*** [pəˈsjuː] v. 追捕；追求
6. ***fascinating*** [ˈfæsɪneɪtɪŋ] adj. 迷人的；有极大吸引力的
7. ***via*** [ˈvaɪə] prep. 经由
8. ***skill*** [skɪl] n. 技能
9. ***grant*** [ɡrɑːnt] v. 答应；允许；授予

03 推荐信

From "Adam Bennett" (ab1245@cpr.com)
Date Thurs., March 5, 2009
To "Sean Reynolds" (sreynolds@nnk.com)
Subject Recommendation Letter

Dear Mr. Reynolds,

I am ***honored***[1] to provide this letter of ***recommendation***[2] for Ronan Cruise. I have known him for several years as his team ***leader***[3]. **He is an *excellent*[4] assistant with high *responsibility*[5]**, **team work spirit and *positive*[6] working attitude**, and **I believe he would be a valuable *asset*[7] to any company**.

If you need any further information, please don't hesitate to contact me at 9876-5431.

Yours sincerely,
Adam Bennett

译文

亲爱的雷诺兹先生：

我很荣幸能为罗南·克鲁斯写这封推荐信。作为他的团队领导人，我们相识多年。他是一位很出色的助手，有着高度的责任感、团队精神和积极的工作态度。我相信，他对任何一个公司而言都将是一位宝贵的人才。

如果你们想了解更多的信息，请尽管与我联系，我的电话是9876-5431。

亚当·贝内特 谨上

1 解析重点1 He is an excellent assistant with high responsibility, team work spirit and positive working attitude.

帮别人写推荐信的时候，要尽量具体（Concreteness）地展现出被推荐人的优点和特点，这样才有说服力。上面的邮件内容中就列举了被推荐人的不少的优点：an excellent assistant, with high responsibility, team work spirit and positive working attitude（一名很出色的助手，有着强烈的责任感、团队精神和积极的工作态度）。在短短的信件中，利用几个简短（Conciseness）的词组，勾画出了被推荐人积极的态度和稳重的性格。

2 解析重点2 I believe he would be a valuable asset to any company.

对一个员工而言，没有谁的评价能比他上司的评价更具说服力。如果连他的上司都对他赞赏有加的话，那么这个人必定是个很出色的员工。身为推荐人，对于优秀的员工不要吝啬于赞扬的语句。I believe he would be a valuable asset to any company（我相信，他在任何一个公司都将会是一个宝贵的人才），从这句话中，我们可以感受到上司的那份骄傲，以及他对员工的赞赏。

1. It's my pleasure to provide this letter of recommendation for Famke Janssen.
我很荣幸能为法姆克·詹森写这封推荐信。

2. Please feel free to contact me at 2123-4567.
请随时联系我，我的电话是2123-4567。

3. He possesses exceptional skills in this field.
他在这一领域拥有出色的技能。

4. He is gifted at leading groups.
他很有领导团队的天赋。

5. I am *confident*[8] that he will be an asset to your company.
我相信他会是贵公司的一笔财富。

必背关键单词 

1. ***honored*** [ˈhɒnɔːd] adj. 荣幸的
2. ***recommendation*** [ˌrekəmenˈdeɪʃn] n. 推荐
3. ***leader*** [ˈliːdə(r)] n. 领袖；领导者
4. ***excellent*** [ˈeksələnt] adj. 优秀的
5. ***responsibility*** [rɪˌspɒnsəˈbɪlətɪ] n. 责任
6. ***positive*** [ˈpɒzətɪv] adj. 肯定的；积极的；正面的
7. ***asset*** [ˈæset] n. 有价值的人；资产
8. ***confident*** [ˈkɒnfɪdənt] adj. 确信的；有信心的；自信的

04 自荐信

这样写就对了

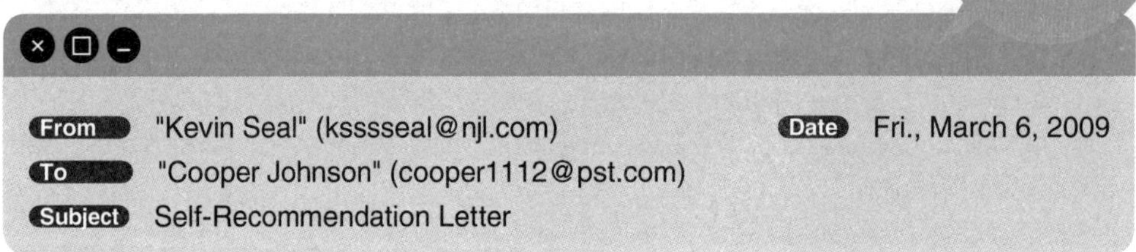

From "Kevin Seal" (ksssseal@njl.com)
Date Fri., March 6, 2009
To "Cooper Johnson" (cooper1112@pst.com)
Subject Self-Recommendation Letter

Dear Mr. Johnson,

I worked as an English ***editor***[1] for two years, but I have held a ***particular***[2] ***interest***[3] in teaching English since I was an English ***Department***[4] student in University.
With my knowledge in English, work experience, ***patience***[5] and ***passion***[6] for teaching, I believe I am ***capable***[7] of doing this job. Please consider me for the teaching position. If possible, you can contact me at 02-2987-6543.

Looking forward to hearing from you!

Yours sincerely,
Kevin Seal

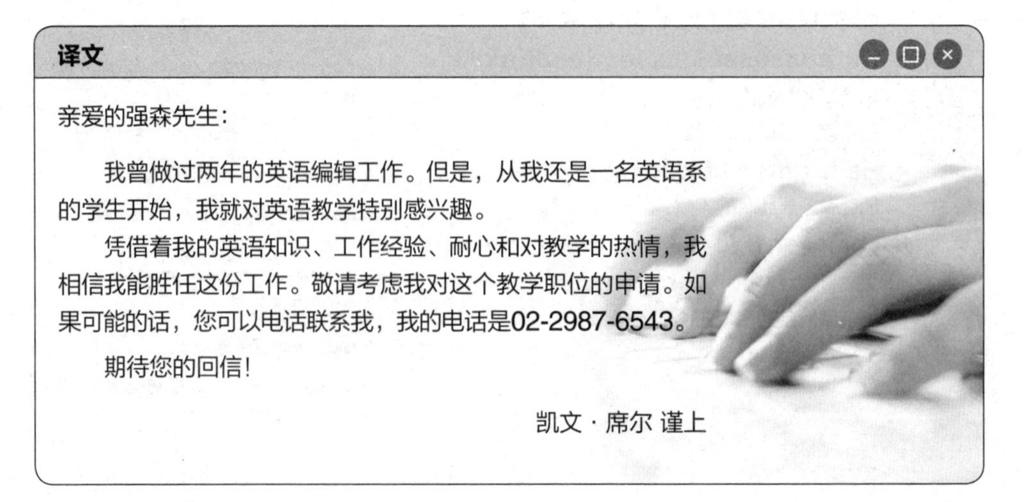

译文

亲爱的强森先生：

我曾做过两年的英语编辑工作。但是，从我还是一名英语系的学生开始，我就对英语教学特别感兴趣。

凭借着我的英语知识、工作经验、耐心和对教学的热情，我相信我能胜任这份工作。敬请考虑我对这个教学职位的申请。如果可能的话，您可以电话联系我，我的电话是02-2987-6543。

期待您的回信！

凯文 · 席尔 谨上

语法重点解析

1 解析重点1 I have held a particular interest in teaching English since I was an English Department student.

自荐的时候，我们尤其需要突出自己对工作的强烈兴趣，说明自己非常渴望拥有这份工作。要是连对工作的兴趣都没有或是不高，别人怎么可能会考虑录用你呢？I have held a particular interest in teaching English（我对英语教学特别感兴趣）中 particular 这个词道出了自荐者对这个职业的特殊情感，而后面的 since I was an English Department student（从我还是一个英文系的学生开始……）更是说明自荐者对这份工作的热爱与执着。

2 解析重点2 With my knowledge in English, work experience, patience and passion for teaching, I believe...

自荐的时候，重点是要突出自己的能力，说明自己有哪些优势，这样才能说服别人。my knowledge in English, work experience, patience and passion for teaching（我的英语知识、工作经验、耐心和对教学的热情），这些都是一名教育者必须具备的能力和品质。把自己与这份工作所要求的能力列举出来，才能有的放矢，并让人充分相信你能胜任这份工作。

高频例句

1. **I am interested in working in your company.**
   对于在贵公司工作，我很感兴趣。
2. **I believe I am capable of doing such a job.**
   我相信我能胜任这样一份工作。
3. **I think I am well qualified for this position.**
   我认为我能胜任这一职位。
4. **I do hope you will consider me for this position.**
   我真的希望您能考虑让我担任这个职位。
5. **This would be a great opportunity for me.**
   对我来说，这将会是一个绝佳的机会。
6. **I have a wealth of *professional*[8] knowledge.**
   我有丰富的专业知识。

必背关键单词

1. ***editor*** [ˈedɪtə(r)] n. 编辑
2. ***particular*** [pəˈtɪkjələ(r)] adj. 特别的
3. ***interest*** [ˈɪntrəst] n. 兴趣
4. ***department*** [dɪˈpɑːtmənt] n. 系；部门
5. ***patience*** [ˈpeɪʃns] n. 耐心
6. ***passion*** [ˈpæʃn] n. 热情
7. ***capable*** [ˈkeɪpəbl] adj. 有能力的
8. ***professional*** [prəˈfeʃənl] adj. 专业的

05 发送电子简历

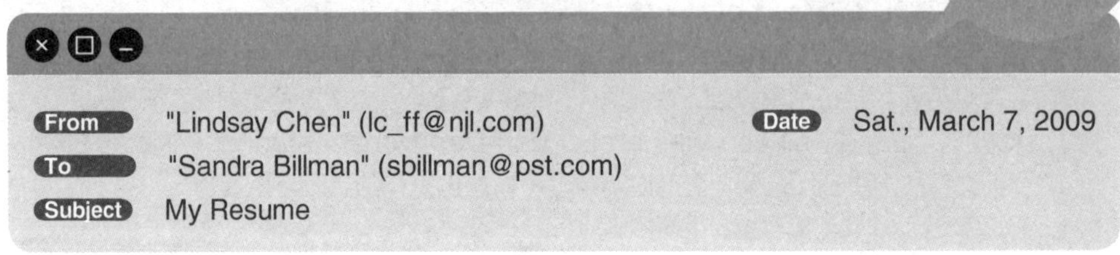

From "Lindsay Chen" (lc_ff@njl.com)   Date Sat., March 7, 2009
To "Sandra Billman" (sbillman@pst.com)
Subject My Resume

Dear Ms. Billman,

I am responding to your job offer ***announcement***[1] in job.com and applying for the position of computer ***clerk***[2]. I ***majored***[3] in computer ***science***[4] with a ***minor***[5] in English. I believe all these will ***qualify***[6] me for this position described on this website.
Enclosed is my resume, which will detail my ***qualifications***[7] for this position. Thank you for your time.

Looking forward to hearing from you!

Yours sincerely,
Lindsay Chen

译文

亲爱的比尔曼女士：

我看到您在求职网上发布的招聘启示，特此回复，想应聘贵公司的电脑职员。我大学期间主修计算机科学，辅修英语。我相信这些将使我能够胜任网站上所描述的职位。
附件中是我的个人履历，里面有我各项资历的详细描述。谢谢您的宝贵时间！

期待您的回信！

琳赛 · 陈 谨上

语法重点解析

1 解析重点1 **I am responding to your job offer announcement in talent recruitment website.**

我们在网络上投递简历时，最好要说明得知招聘信息的来源。由于现在网络发展迅猛，很多公司会选择在一些专门负责企业招聘的网站上发布招聘信息。因此，我们利用电子邮件发送简历的时候，最好提到是从哪一个网站上看到了招聘信息。同时，从一个可靠的网站获取到的信息，对求职者而言也比较安全可信。

2 解析重点2 **I majored in computer science with a minor in English.**

不管我们是刚出校门没多久的学生，还是已经工作一段时间、正在寻找新的发展机会的求职者，我们的专业或多或少都会在我们今后的工作履历中留下印记。因此，针对应聘的职位介绍自己的专业是必不可少的。我们通常可以说：I majored in...（我主修……），My major is...（我的主修科目是……），I have a degree in...（我有……学位）。而 a minor in...则是“辅修……”的意思。

1. **I have a *degree*[8] in engineering science in New York University.**
   我有纽约大学的工程科学学位。
2. **As requested, I have enclosed the following materials.**
   按照要求，我附上了以下资料。
3. **I studied history in Harvard University, and got a *bachelor's*[9] degree.**
   我在哈佛大学学习历史，获得了学士学位。
4. **If you would like more information, please e-mail me at mike666@yahoo.com.**
   如果您想知道更多信息，请发邮件至mike666@yahoo.com 联系我。
5. **If you would like to set up an interview, please contact me at 02-2789-0123.**
   若您想安排面试，请电话联系我，我的电话是02-2789-0123。

必背关键单词

1. ***announcement*** [əˈnaʊnsmənt] n. 通知；布告
2. ***clerk*** [klɑːk] n. 职员
3. ***major*** [ˈmeɪdʒə(r)] v. 主修
4. ***science*** [ˈsaɪəns] n. 科学
5. ***minor*** [ˈmaɪnə(r)] n. 辅修科目
6. ***qualify*** [ˈkwɒlɪfaɪ] v. 使合格
7. ***qualification*** [ˌkwɒlɪfɪˈkeɪʃn] n. 资格
8. ***degree*** [dɪˈɡriː] n. 学位；程度
9. ***bachelor*** [ˈbætʃələ(r)] n. 单身汉；学士

06 推荐人发函确认

这样写就对了

From "Orlando Knight" (ok_1981@xma.com)
Date Sun., March 8, 2009
To "Hugh Jackson" (hjackson@kal.com)
Subject Confirmation Letter

Dear Mr. Jackson,

Stanley Donen has been in our ***employ***[1] for the past three years. **He is a man with a *pleasant*[2] *personality*[3] and really did a great job those years**.
We are ***indeed***[4] sorry to lose Stanley's services, but he leaves us to find a position with great opportunities for ***advancement***[5]. **We wish him a *brilliant*[6] future ahead of him**.

Yours sincerely,
Orlando Knight

译文

亲爱的杰克森先生:

斯坦利·多南曾在我们公司工作三年之久。他性格开朗，这些年工作也相当出色。

我们对他的离开深感遗憾，但是我们也深知，他是为了谋求更好的发展空间而寻找新的工作。在此，我们预祝他前程似锦。

奥兰多·奈特 谨上

语法重点解析

1 解析重点1 **He is a man with a pleasant personality and really did a great job those years.**

为挑选到合适的申请者我们可以从个人性格和工作表现两方面来考察。现在的团队合作越来越紧密，良好的性格有利于你融入一个团队，从而在团队中发挥更大的作用。a pleasant personality（性格开朗）和 really did a great job those years（工作出色）都大大地说明了申请者在前一份工作中的表现。

2 解析重点2 **We wish him a brilliant future ahead of him.**

优秀的员工，有时候为了争取更大的发展空间，也会需要更换工作。这就应验了一句俗语：人往高处走，水往低处流。上司或是主管也要在一定程度上理解员工的选择。同时，祝福他们有一个好的前程。We wish him a brilliant future ahead of him.（我们祝他前程似锦。）

高频例句

1. **Her *duties*[7] were confined to answering the telephone.**
   她的工作仅仅是接听电话。
2. **He has been a great asset to us in dealing with customers.**
   他在处理客户关系方面表现出色。
3. **I am really sorry that he has to leave us.**
   对于他要离开，我深感遗憾。
4. **We hope he will *succeed*[8] in the future.**
   我们希望他将来能成功。
5. **We believe he will do well at any task he *undertakes*[9].**
   我们相信他能做好每一项工作。
6. **She is a very intelligent young woman with a bright personality.**
   她是一位性格开朗、年轻聪明的女士。
7. **She is proficient in accounting.**
   她精通做账。
8. **He acquired rich experience in market operation.**
   他拥有丰富的市场运作经验。

必背关键单词

1. ***employ*** [ɪmˈplɔɪ] v. 雇用
2. ***pleasant*** [ˈpleznt] adj. 愉快的
3. ***personality*** [ˌpɜːsəˈnæləti] n. 个性；人格
4. ***indeed*** [ɪnˈdiːd] adv. 确实；的确
5. ***advancement*** [ədˈvɑːnsmənt] n. 进步；提升
6. ***brilliant*** [ˈbrɪlɪənt] adj. 光辉的；辉煌的
7. ***duty*** [ˈdjuːtɪ] n. 责任；义务
8. ***succeed*** [səkˈsiːd] v. 成功
9. ***undertake*** [ˌʌndəˈteɪk] v. 承担；担保

07 请求安排面试

这样写就对了

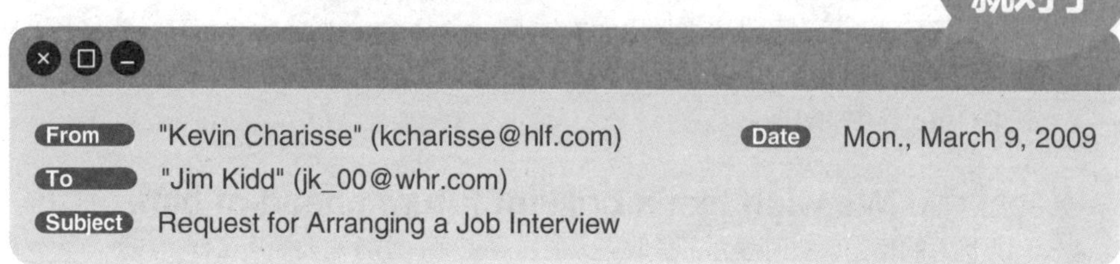

From "Kevin Charisse" (kcharisse@hlf.com) Date Mon., March 9, 2009
To "Jim Kidd" (jk_00@whr.com)
Subject Request for Arranging a Job Interview

Dear Mr. Kidd,

I am writing to apply for the position of a ***secretary***[1], as ***advertised***[2] in the *New York Times.*
I have more than five years of experience in ***government***[3] ***agencies***[4]. **I am sure my employment experience is *perfect*[5] for the secretary position. I hope I may be granted an interview, when I can fully *explain*[6] my qualifications**.

Looking forward to hearing from you soon!

Yours sincerely,
Kevin Charisse

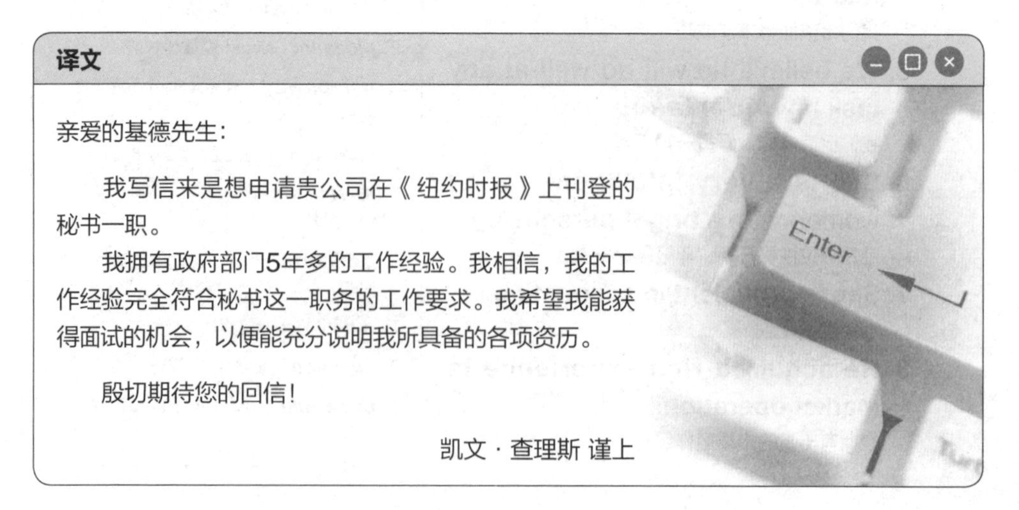

译文

亲爱的基德先生:

　　我写信来是想申请贵公司在《纽约时报》上刊登的秘书一职。

　　我拥有政府部门5年多的工作经验。我相信，我的工作经验完全符合秘书这一职务的工作要求。我希望我能获得面试的机会，以便能充分说明我所具备的各项资历。

　　殷切期待您的回信!

凯文·查理斯 谨上

语法重点解析

1 解析重点1 **I am sure my employment experience is perfect for the secretary position.**

我们在寻找工作时，要充满自信，相信自己可以胜任所申请的职位，这样才能让他人有理由去相信你的才干，给你面试的机会。上面的邮件内文中是这样说的：I am sure my employment experience is perfect for the secretary position（我相信，我的工作经验完全符合秘书这一职务的工作要求）。在这里，be perfect for...（对……完美的），充分表达了申请者自身的信心，有力地帮助申请者提高了求职成功的几率。

2 解析重点2 **I hope I may be granted an interview, when I can fully explain my qualifications.**

无论是自我介绍，还是发送简历，我们都希望对方能给予自己一个面试的机会。表现出高姿态固然不可取，但是我们也不能表现得太低声下气。而是应该在赞美招聘公司的同时，仍旧保持不卑不亢的态度。I hope I may be granted an interview（我谨希望获得面试机会）中，grant 这个单词使用得很恰当。后面的 I can fully explain my qualifications（充分地说明我所具备的各项资历）更有力地道出了面试的重要性，表明求职者也有自信能成功被录用。

1. **The job, as described, sounds very much like what I am looking for.**
   这份工作所描述的职务内容，听起来正是我一直在寻找的工作。
2. **I am available to come for an interview at your convenience.**
   我愿意在您方便的时候来面试。
3. **I am ready for new *challenges*[7].**
   我已经准备好迎接新的挑战。
4. **I have been an accountant for the past four years.**
   我已经从事会计师工作4年了。
5. **I have more than ten years of experience in major *corporations*[8].**
   我有10多年的大型企业工作经验。

必背关键单词

1. ***secretary*** [ˈsekrətrɪ] n. 秘书
2. ***advertise*** [ˈædvətaɪz] v. 为……做广告；为……宣传
3. ***government*** [ˈɡʌvənmənt] n. 政府
4. ***agency*** [ˈeɪdʒənsɪ] n. 部；处；机构
5. ***perfect*** [ˈpɜːfɪkt] adj. 完美的
6. ***explain*** [ɪkˈspleɪn] v. 解释；说明
7. ***challenge*** [ˈtʃæləndʒ] n. 挑战
8. ***corporation*** [ˌkɔːpəˈreɪʃn] n. 公司；企业；社团

08 询问面试结果

这样写就对了

From "David Reynolds" (dreynolds@adr.com)　Date Tues., March 10, 2009
To "Mike Mitchell" (mmitchell@app.com)
Subject Request for the Result of the Interview

Dear Mr. Mitchell,

You ***mentioned***[1] that you would be getting back to me concerning the ***outcome***[2] of our meeting by March 5, 2009. Since that ***date***[3] has ***passed***[4], I think I should make sure that you have no ***additional***[5] questions.
Your ***prompt***[6] consideration would be greatly appreciated and you can reach me at 02-2645-3280.

Looking forward to hearing from you soon!

Yours sincerely,
David Reynolds

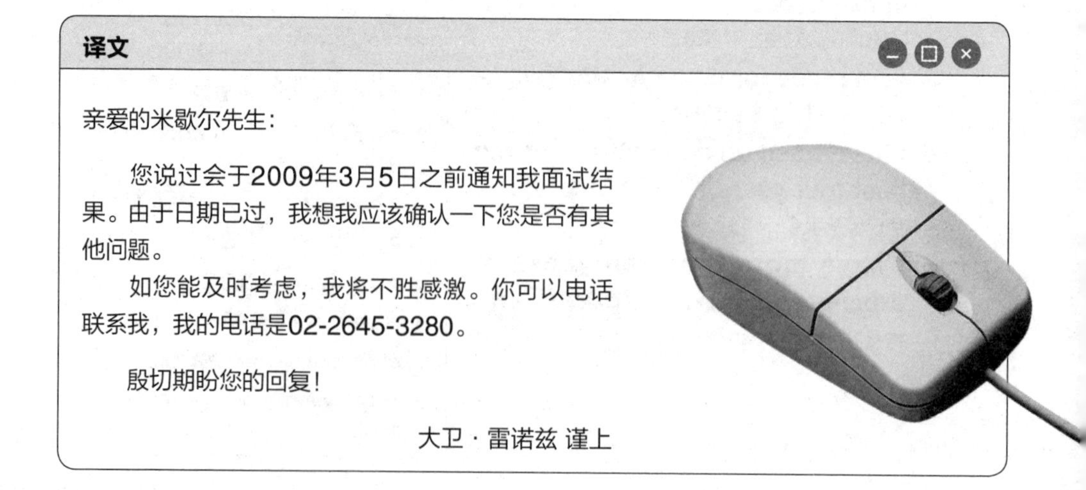

译文

亲爱的米歇尔先生：

您说过会于2009年3月5日之前通知我面试结果。由于日期已过，我想我应该确认一下您是否有其他问题。

如您能及时考虑，我将不胜感激。你可以电话联系我，我的电话是02-2645-3280。

殷切期盼您的回复！

大卫·雷诺兹 谨上

语法重点解析

1 解析重点1 **You mentioned that you would be getting back to me concerning the outcome of our meeting by March 5, 2009.**

招聘公司答应给予回复，却迟迟没有回音，这时候我们可以写封邮件进行询问，作为一种间接的提醒。这个时候，我们务必要说清楚（Clearness）写信的目的和当初约定的面试结果回复时间。上面的邮件内文中就涉及到这两个方面，写信人是为了 the outcome of our meeting（面试结果）而来的，而当初约定的回复时间是 by March 5, 2009（2009年3月5日之前）。

2 解析重点2 **Your prompt consideration would be greatly appreciated and you can reach me at 02-2645-3280.**

虽然对方没有及时回复我们，我们还是要很客气地请求对方再考虑一下并回电给我们：Your prompt consideration would be greatly appreciated（如您能及时考虑，我将不胜感激）。并且，后面还告知了自己的电话号码：You can reach me at 02-2645-3280（你可以电话联系我，我的电话是02-2645-3280），从而方便招聘公司联系。

1. **I am *anxious*[7] to know about the outcome.**
   我很想知道面试的结果。
2. **Does the general manager consider me acceptable?**
   总经理是否考虑录用我了？
3. **Please contact me at 02-2345-6789 at your earliest convenience.**
   请您尽早在方便的时候拨打02-2345-6789联系我。
4. **Your company is right at the top of my list for employment consideration.**
   贵公司是我就业的首选。
5. **I would like to *conclude*[8] the job search process soon.**
   我想尽快结束求职过程。
6. **I enjoyed the meeting with you last time.**
   上次参加您的面试，我感到很开心。

必背关键单词 

1. ***mention*** [ˈmenʃn] v. 提起
2. ***outcome*** [ˈaʊtkʌm] n. 结果；成果
3. ***date*** [deɪt] n. 日期；约会
4. ***pass*** [pɑːs] v. 经过；消逝；通过
5. ***additional*** [əˈdɪʃənl] adj. 额外的；附加的；另外的
6. ***prompt*** [prɒmpt] adj. 立刻的；迅速的
7. ***anxious*** [ˈæŋkʃəs] adj. 忧心的
8. ***conclude*** [kənˈkluːd] v. 结束；终止

09 感谢信

这样写就对了

From "Nicole Jackson" (njackson@art.com)

Date Wed., March 11, 2009

To "Kevin Urban" (kurban@htt.com)

Subject Letter of Thanks

Dear Mr. Urban,

Thank you for your offer of employing me as your secretary.
I am very *excited*[1] about being *part*[2] of your *team*[3], and *eager*[4] to start work on March 15, 2009. I will work harder than before and get along with the team ***members***[5].

Yours sincerely,
Nicole Jackson

译文

亲爱的厄本先生：

谢谢您录用我作为您的秘书。

能成为你们团队中的一员我感到非常兴奋，并且渴望在2009年3月15日开始到公司上班。我会比以前更加努力工作，并且与团队成员融洽相处。

妮可·杰克森 谨上

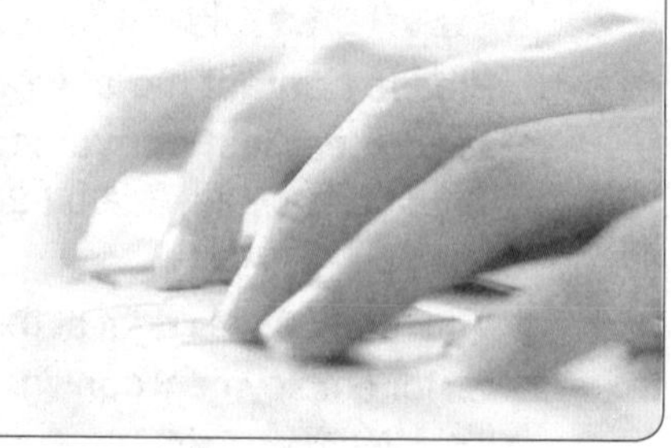

语法重点解析

1 解析重点1 **Thank you for your offer of employing me as your secretary.**

得知自己已被录用，邮件开头肯定是感谢的话语。可以说 Thank you for your offer（谢谢您的录用），也可以说成 I really appreciate your offer（非常感谢贵公司的录用）。你也可以再详细一些，提一提自己的职位，再次确认录用相关信息。employing me as your secretary（录用我作为您的秘书），这样就万无一失了。

2 解析重点2 **I am very excited about being part of your team, and eager to start work on March 15, 2009.**

表达了谢意，我们当然也要不失时机地表达一下自己的激动心情。这样一来，对方也会觉得他的选择没有错。I am very excited about being part of your team（我很高兴能够成为你们团队中的一份子），并且，还需要再次确认一下具体（Concreteness）的上班时间：I am eager to start work on March 15, 2009（我很期待在2009年3月15日那天到公司开始上班）。

1. **Thank you for your offer of employing me as your electrical engineer.**
   谢谢贵公司录用我为你们的电子工程师。
2. **I am *grateful*[6] to you for employing me as your assistant.** 我很感激您录用我为您的助理。
3. **I am so glad to hear such good news from you.**
   从您那听到这个好消息，我十分高兴。
4. **I'd like to express my *gratitude*[7] to you.**
   我想在此向您表达由衷的感谢。
5. **I look forward to the start of work on Tuesday this week.**
   我很期待这个星期二开始到公司上班。
6. **I really think the position suits my education background best.**
   我确实觉得这个职位最适合我的教育背景。
7. **I will try my best to finish any task *assigned*[8] to me.**
   我会竭尽所能完成分配给我的任务。

必背关键单词 

1. ***excited*** [ɪk'saɪtɪd] adj. 兴奋的；激动的
2. ***part*** [pɑ:t] n. 部分
3. ***team*** [ti:m] n. 团队；队
4. ***eager*** ['i:gə(r)] adj. 渴望的
5. ***member*** ['membə(r)] n. 成员
6. ***grateful*** [greɪtfl] adj. 感激的；感谢的
7. ***gratitude*** ['grætɪtju:d] n. 感激
8. ***assign*** [ə'saɪn] v. 分派；指定

10 拒绝信

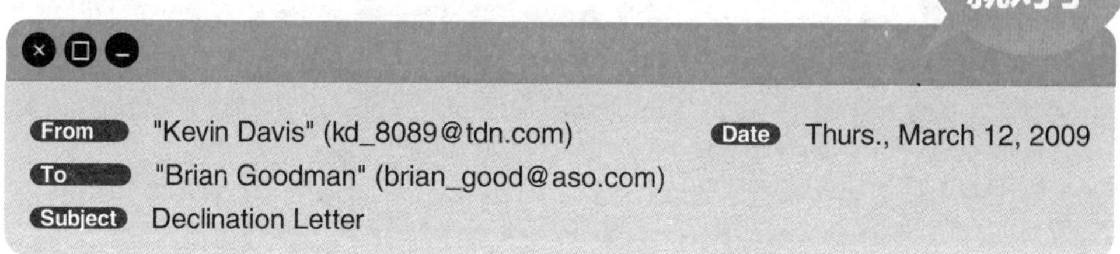

From "Kevin Davis" (kd_8089@tdn.com)　Date Thurs., March 12, 2009
To "Brian Goodman" (brian_good@aso.com)
Subject Declination Letter

Dear Mr. Goodman,

Thank you for your letter of March 12, in which you offered me the position of ***auditor***[1].
I am extremely sorry but I just ***accepted***[2] another offer that I feel is more ***interesting***[3] to me. That position ***suits***[4] my education background better.
I hope you will find the ***right***[5] man soon and thank you for your offer again.

Yours sincerely,
Kevin Davis

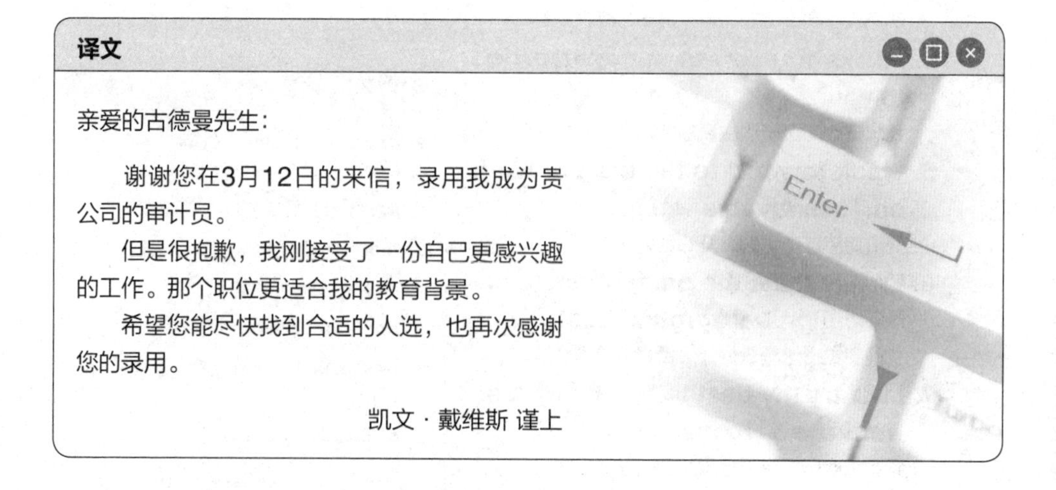

译文

亲爱的古德曼先生：

谢谢您在3月12日的来信，录用我成为贵公司的审计员。
但是很抱歉，我刚接受了一份自己更感兴趣的工作。那个职位更适合我的教育背景。
希望您能尽快找到合适的人选，也再次感谢您的录用。

凯文 · 戴维斯 谨上

语法重点解析

1 解析重点1 **I am extremely sorry but I just accepted another offer that I feel is more interesting to me.**

当对方决定录用我们，而我们后来觉得自己不适合，或是有了更好的选择时，我们就需要回函拒绝对方的录用。一般我们可能这样说：I am sorry that I can't accept your offer, because I just accepted another offer which is better（抱歉，我不能接受你的录用，因为我刚接受了一份更好的工作）。首先，这种说法很繁琐；其次，会使人觉得很不礼貌，感觉好像在说对方的工作不够好。我们可以跟上面的邮件内文一样，用 I am extremely sorry but...（很抱歉，但是……）来表达同样的意思，这样就简洁多了（Conciseness）。而且，我们说另外的那份工作 more interesting（更感兴趣）也会显得比较委婉、礼貌（Courtesy）。

2 解析重点2 **I hope you will find the right man soon and thank you for your offer again.**

拒绝对方并说明拒绝缘由之后，我们还要再次感谢对方：thank you for your offer again（也再次感谢你们的录用）。并且祝福对方尽快找到合适的人选，不要耽误工作进度或是打乱他们的计划：I hope you will find the right man soon（希望你们能尽快找到合适的人选）。

1. **I really appreciate your offer, but I must *decline*[6] it.**
   非常感谢贵公司的录用，但我不得不拒绝这份工作。
2. **Thank you for your offer. But I am sorry to tell you I have accepted another position.**
   感谢贵公司的录用，不过非常遗憾地告诉您，我已经接受了另一个职位。
3. **I don't want to accept a position that offers me too little *salary*[7].**
   我不想接受一份薪水太少的工作。
4. **I am inclined to find a job which could supply me with an *apartment*[8].**
   我倾向于找一份能为我提供住宿的工作。

必背关键单词 

1. ***auditor*** [ˈɔːdɪtə(r)] n. 审计员；查账员
2. ***accept*** [əkˈsept] v. 接受
3. ***interesting*** [ˈɪntrəstɪŋ] adj. 有趣的
4. ***suit*** [suːt] v. 适合
5. ***right*** [raɪt] adj. 合适的
6. ***decline*** [dɪˈklaɪn] v. 下降；衰败；婉拒
7. ***salary*** [ˈsæləri] n. 薪水
8. ***apartment*** [əˈpɑːtmənt] n. 公寓

Unit 4 感谢 Gratitude

01 感谢咨询

From hmcorporation@yahoo.com
To tedsmith_11@sina.com
Subject Thanks for the Inquiry
Date March 4, 2008

Dear Mr. Smith,

Thank you for your ***inquiry***[1] ***regarding***[2] the ***humidifiers***[3] UL9982001.
The ***information***[4] you want has been ***sent***[5] to you today. Please kindly ***check***[6].
If you have ***further***[7] questions about this product, please contact Miss Lee at our Service Department. Her ***direct***[8] number is 666-1234.

Thanks again for your attention and support.

Sincerely yours,
HM Corporation

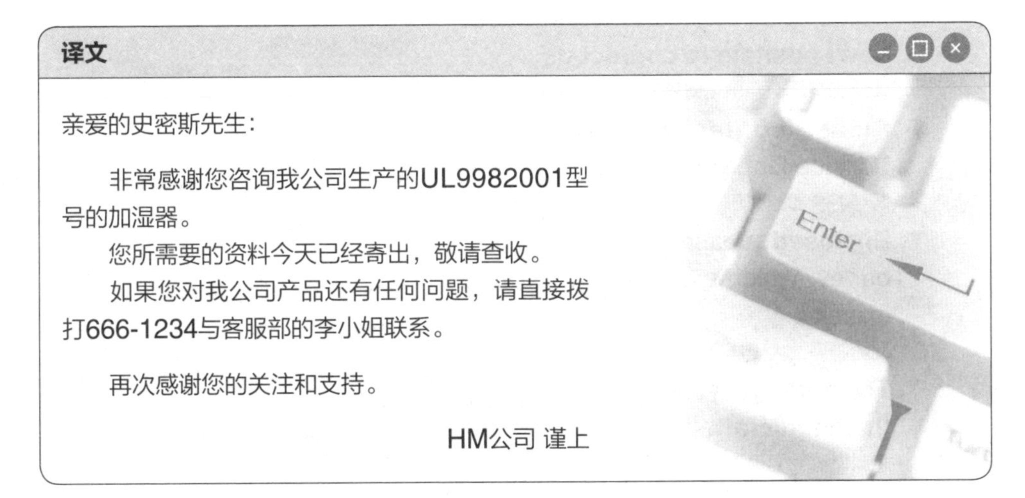
译文

亲爱的史密斯先生：

非常感谢您咨询我公司生产的UL9982001型号的加湿器。

您所需要的资料今天已经寄出，敬请查收。

如果您对我公司产品还有任何问题，请直接拨打666-1234与客服部的李小姐联系。

再次感谢您的关注和支持。

HM公司 谨上

语法重点解析

1 解析重点1 Thank you for your inquiry regarding...

一般表达“感谢您……的咨询”可以说 Thank you for asking about...书面一般用 Thank you for your inquiry regarding...或 Thanks for your interest in...表达，会显得更地道。

2 解析重点2 Please kindly check.

这句话的意思是“敬请查收”。无论是发送电子邮件还是寄送包裹，一般都会提醒对方查看是否收到，这时可以说 please check（请查收）或 please find enclosed（请查收附件）。类似的表达还有：

Please find the attachment that includes the new drawings.（附件包含新图稿，敬请查收。）

Please find enclosed my CV and a letter of recommendation from my school principal.（附件包含我的简历和校长的推荐信，敬请查收。）

1. **Thank you very much for your inquiry regarding our product.**
   非常感谢您对我们产品的咨询。
2. **Thank you for asking the detail of the therapeutic apparatus.** 感谢您询问治疗仪器的详细信息。
3. **We have sent to you the information you need.**
   我们已经把您需要的资料寄过去了。
4. **If you have any questions, please don't hesitate to contact us.**
   如果您有任何问题，请随时与我们联系。
5. **Please contact Mr. Cole at the Service Department.**
   请与客服部的科尔先生联系。
6. **Enclosed please find a sample of barley, which we would like you to examine.**
   现随函寄上大麦样品，敬请查收。
7. **Thank you for your inquiry regarding the handicraft.**
   感谢您对我们手工艺品的咨询。

必背关键单词

1. ***inquiry*** [ɪnˈkwaɪərɪ] n. 咨询
2. ***regarding*** [rɪˈɡɑːdɪŋ] prep. 关于；至于
3. ***humidifier*** [hjuːˈmɪdɪfaɪə(r)] n. 加湿器
4. ***information*** [ˌɪnfəˈmeɪʃn] n. 信息；资料
5. ***send*** [send] v. 发送；寄送
6. ***check*** [tʃek] v. 检查；核对
7. ***further*** [ˈfɜːðə(r)] adj. 更多的；进一步的
8. ***direct*** [dəˈrekt] adj. 直接的

02 感谢来信

From orsan_snow@yahoo.com Date November 12, 2008
To alan_ice@sina.com
Subject Thanks for the Mail

Dear Alan,

I feel so ***delighted*** [1] to ***receive*** [2] your mail. It has been a ***couple*** [3] of months since we met in ***Bali*** [4].
I've been concentrating on my work like a ***workaholic*** [5] since I came back from Bali. What about you? Have you ***finished*** [6] your ***novel*** [7]?
Please tell me what's happening in London. I am ***planning*** [8] to visit you ***sometime*** [9] soon.

Let's stay in touch. Bye for now!

Best wishes,
Orsan

译文

亲爱的艾伦:

收到你的来信，我感到非常开心。自从上次在巴厘岛一别已经有好几个月的时间了。

我一从巴厘岛回来就像个工作狂一样专心投入工作。不知道你的近况如何呢？写完你的小说了吗？

请告诉我你在伦敦的近况，我打算不久去拜访你。

保持联络哦，再见！

献上最诚挚的祝福，

奥森

语法重点解析

1 解析重点1 **I feel so delighted to receive your mail.**

这句话的意思是“收到您的邮件，我感到非常开心”。这是英文书信中常见的开头句。其中 delighted（愉快的）可以换成 happy, glad, pleasant。表达“我很高兴收到您的来信”还可以说：

I am happy to receive your letter.

I am so glad to hear from you.

It's so pleasant to obtain your mail.

2 解析重点2 **Let's stay in touch. Bye for now!**

这是英文书信中的结束语，意思是“让我们保持联络，再见”。表达“保持联络”还可以说 keep in touch with each other。Bye for now 事实上就相当于中文书信中的“不多说了，就此止笔”。当然这通常只在熟悉的朋友之间才会使用，一般不会用在非常正式的书信当中。

1. **I am so glad to receive your letter.**
   很高兴收到你的来信。
2. **I feel joyous to have the mail from you.**
   收到你的来信我非常开心。
3. **How are you doing recently?**
   你最近过得如何?
4. **Have you found a new job?**
   你找到新的工作了吗?
5. **It's been three weeks since we met last time.**
   上次一别已经三个星期了。
6. **Please tell me what happened to you these days.**
   请告诉我你的近况。
7. **I am going to see you some day.**
   我改天要去看你。
8. **Let's stay in touch with each other.**
   让我们彼此保持联络。

必背关键单词  

1. ***delighted*** [dɪˈlaɪtɪd] adj. 高兴的；愉快的；欣喜的
2. ***receive*** [rɪˈsiːv] v. 收到
3. ***couple*** [ˈkʌpl] n. 几个
4. ***Bali*** [ˈbɑːlɪ] n. 巴厘岛
5. ***workaholic*** [ˌwɜːkəˈhɒlɪk] n. 工作狂；工作优先的人
6. ***finish*** [ˈfɪnɪʃ] v. 完成
7. ***novel*** [ˈnɒvl] n. 小说
8. ***plan*** [plæn] v. 计划；打算
9. ***sometime*** [ˈsʌmtaɪm] adv. 改天；未来某时

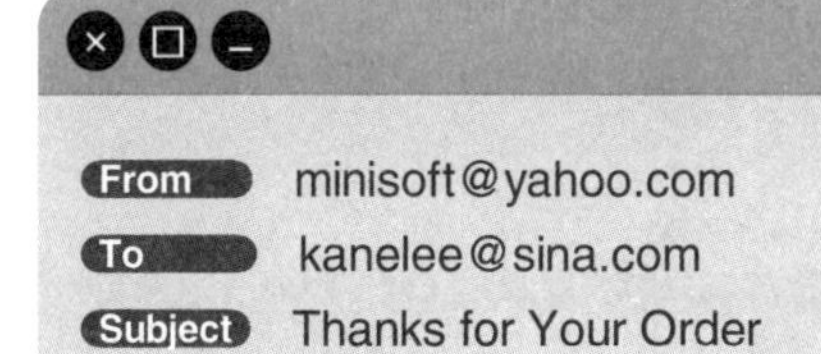

03 感谢订购

From minisoft@yahoo.com　Date September 14, 2008
To kanelee@sina.com
Subject Thanks for Your Order

Dear Mr. Kane,

Thank you very much for ***ordering***[1] our ***software***[2] "Mini Diary".
Your ***purchase***[3] information is attached. We will send the goods to you upon ***receipt***[4] of your ***payment***[5].
If there are any other ***commodities***[6] you are interested in, **please feel *free*[7] to contact us for further information**.

Sincerely yours,
Minisoft Co.

译文

亲爱的凯恩先生:

非常感谢您订购我公司的“迷你日记”软件。

您的购买资料已经随函附上，在您确认付款后，商品将会寄出。

如果您对我公司的其他产品也有兴趣，请随时与我们联系，获取更多信息。

迷你软件公司 谨上

语法重点解析

1 解析重点1 receipt

receipt 是 receive 的名词形式，它的一般含义是“收到；接收”。receipt 通常指收到别人给的或邮寄来的款项、货物、信件，这时其含义为“收到”或“接收”；也可以指“收据”或“发票”。on receipt of 这个短语的意思是“收到……后”，例如：

On receipt of your check, we shall ship the goods immediately.（收到贵方支票后，我方会立即出货。）

2 解析重点2 Please feel free to contact us for further information.

这句话的意思是“如果需要进一步信息，请随时联系我们。”这是英文书信中经常会用到的句型。feel free 意思是“感到自由；随自己意愿（做某事）”。类似的表达还有：

Please don't hesitate to contact us if you need any information.（如果需要任何信息，请随时联系我们。）

Please feel free to call me if you want my service.（如果需要我服务，请随时打电话给我。）

1. **Thank you very much for ordering our textiles.**
   感谢您订购我们的纺织品。
2. **The purchase confirmation has been attached.**
   订购确认书已随函附上。
3. **Thank you for your interest in our software.**
   感谢您对我们的软件感兴趣。
4. **We will send the *merchandise*[8] to you upon receipt of your payment.**
   收到您的款项后，我们会将商品寄出。
5. **If you have any interest in other products, please contact us directly.**
   如果您对其他商品感兴趣，请直接与我们联系。

必背关键单词 

1. ***order*** [ˈɔːdə(r)] v. 订购
2. ***software*** [ˈsɒftweə(r)] n. 软件
3. ***purchase*** [ˈpɜːtʃəs] n. & v. 购买
4. ***receipt*** [rɪˈsiːt] n. 收据；收到
5. ***payment*** [ˈpeɪmənt] n. 付款
6. ***commodity*** [kəˈmɒdətɪ] n. 商品
7. ***free*** [friː] adj. 自由的；免费的
8. ***merchandise*** [ˈmɜːtʃəndaɪs] n. 商品

04 感谢提供样品

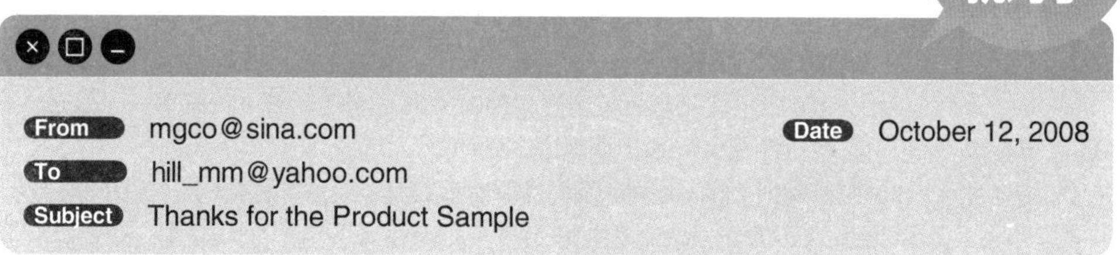

From mgco@sina.com
To hill_mm@yahoo.com
Subject Thanks for the Product Sample
Date October 12, 2008

Dear Mr. Hill,

Thank you ever so much for your ***kindness***[1] and ***assistance***[2] in sending the ***electronic***[3] ***component***[4] sample to us from your company so soon.
We are in the ***process***[5] of trying it out at ***present***[6], and we will ***connect***[7] with you any time if there are any ***findings***[8] and questions we may have about it.

Thanks for your cooperation.

Your sincerely,
MG Co.

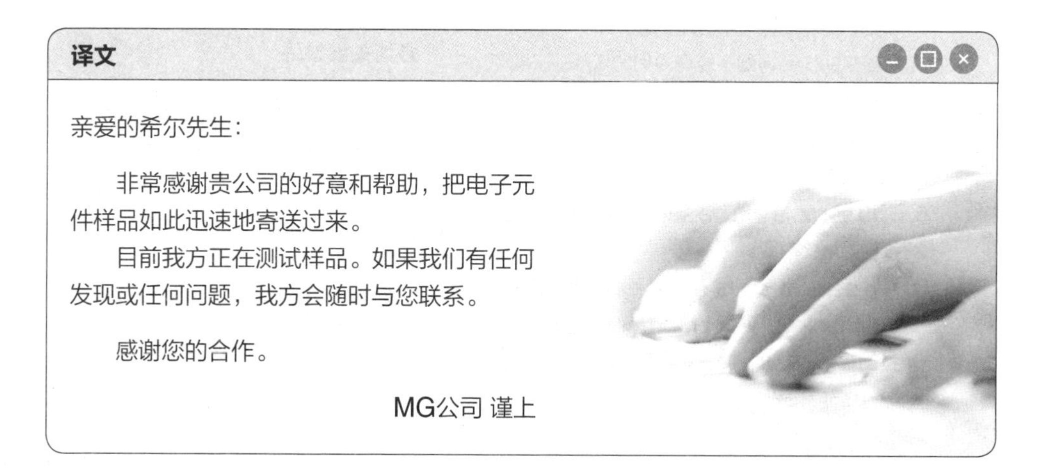

译文

亲爱的希尔先生：

非常感谢贵公司的好意和帮助，把电子元件样品如此迅速地寄送过来。

目前我方正在测试样品。如果我们有任何发现或任何问题，我方会随时与您联系。

感谢您的合作。

MG公司 谨上

语法重点解析

1 解析重点1 try it out

try it out 表示“尝试、试用”。例如：

Try it out and let us know what you think.（请试试这项服务，并告知我们您的想法。）

2 解析重点2 connect with

connect with 意思是“与……连接；使有关系；用电话与……联系”，表示电话联络还可以用 call, contact。请看以下例句：

Please connect with him soon.（请立即与他联系。）

She promised to call at noon.（她承诺中午打电话。）

For further information, contact the local agent.（想进一步了解信息，请与当地代理商联系。）

高频例句

1. **Thank you very much for sending us the sample we want.**
   非常感谢您寄送我们要的样品。
2. **The sample will be presented to you immediately upon request.**
   只要您需要，样品会立即寄给您。
3. **The distributors looked with favor on your sample shipment.**
   您寄来的样品得到了经销商们的青睐。
4. **The goods delivered were very different from the sample.**
   运送的货物与样品大不相同。
5. **The sample will be sent free of charge.**
   该样品免费赠送。
6. **We sent you four samples of textile and will thank you for an order for them.**
   我们今天寄出四种纺织品样本，贵公司若能订购，则非常感谢。

必背关键单词

1. ***kindness*** [ˈkaɪndnəs] n. 仁慈；善良
2. ***assistance*** [əˈsɪstəns] n. 协助
3. ***electronic*** [ɪˌlekˈtrɒnɪk] adj. 电子的
4. ***component*** [kəmˈpəʊnənt] n. 零件；成分
5. ***process*** [ˈprəʊses] n. 过程
6. ***present*** [ˈpreznt] adj. 现在的
7. ***connect*** [kəˈnekt] v. 连接
8. ***finding*** [ˈfaɪndɪŋ] n. 研究结果；发现物

05 感谢馈赠

From lucychen@163.com　Date December 24, 2008
To cecily_gg@yahoo.com
Subject Thanks for the Christmas Gift

Dear Cecily,

I was so ***excited***[1] yesterday evening that I am sure I did not thank you ***adequately***[2] for the ***beautiful***[3] Christmas gift. The ***woolen***[4] ***shawl***[5] is lovely and exactly the color I would have ***selected***[6] myself.
Now that winter has come, it is indeed the right time to wear the warm shawl. I will think of you with ***gratitude***[7] and ***affection***[8] every time I wear it.

Many thanks to you again and my best wishes to you for the New Year!

Affectionately[9] yours,
Lucy

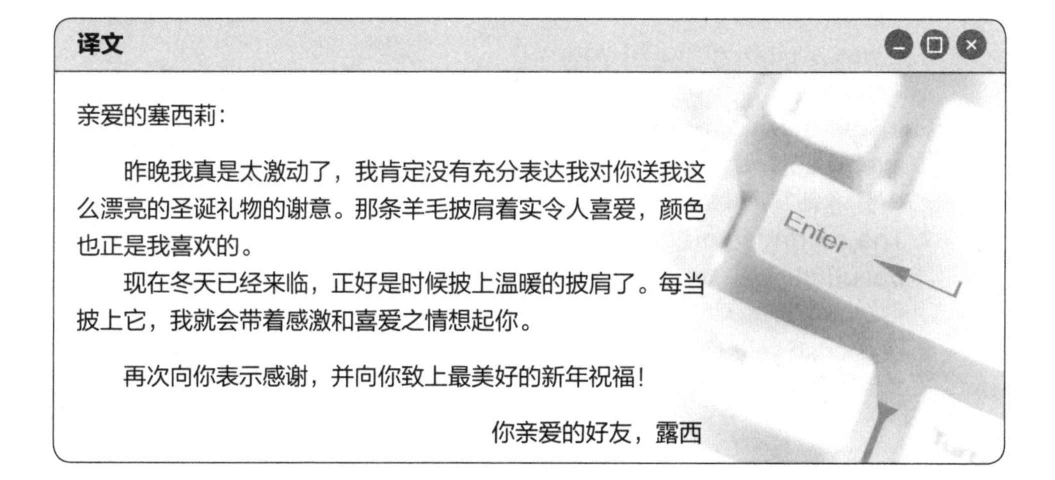

译文

亲爱的塞西莉：

昨晚我真是太激动了，我肯定没有充分表达我对你送我这么漂亮的圣诞礼物的谢意。那条羊毛披肩着实令人喜爱，颜色也正是我喜欢的。
现在冬天已经来临，正好是时候披上温暖的披肩了。每当披上它，我就会带着感激和喜爱之情想起你。

再次向你表示感谢，并向你致上最美好的新年祝福！

你亲爱的好友，露西

语法重点解析

1 解析重点1 **would have**

would have done sth 是虚拟语气，意思是“本来会做某事”，表示对过去事情的假设，一般表示与过去事实相反。在此语境中是指“即使我自己去买，也会挑选这个颜色”，而事实上不是自己买的。例如：

I would have told you all about the boy's story, but you didn't ask me.（我本来会告诉你这个小男孩的故事，但是你没有问我。）

Without your help, I wouldn't have achieved so much.（没有你的帮助，我不会取得如此大的成就。）

2 解析重点2 **Affectionately yours,**

Affectionately yours, 是英文书信中的结尾敬语（Complimentary Close），意思是“你亲爱的”或“挚爱你的”，一般对关系比较亲近的人才会使用。对一般人，可用 Sincerely yours；对较友好的人，可用 Truly yours；对上级或长辈，可用 Respectfully yours。

1. I am so happy to receive your present.
我很高兴收到你的礼物。

2. Thank you for the beautiful card.
感谢你送的美丽贺卡。

3. Thank you for the blue sweater you sent me.
感谢你送我的蓝色毛衣。

4. Thanks again for your kindness.
再次感谢你的好意。

5. It was so thoughtful of you.
你人真体贴。

6. It was the most precious gift I've ever received.
这是我收到过的最珍贵的礼物。

7. The earrings you gave me are so beautiful.
你送我的耳环真是太漂亮了。

8. I'll cherish your gift for years to come.
我会永远珍惜你送我的礼物。

必背关键单词

1. ***excite*** [ɪk'saɪt] v. 使兴奋；使激动
2. ***adequately*** ['ædɪkwətlɪ] adv. 足够地
3. ***beautiful*** ['bju:tɪfl] adj. 美丽的；漂亮的
4. ***woolen*** ['wʊlɪn] adj. 羊毛的
5. ***shawl*** [ʃɔ:l] n. 围巾；披肩
6. ***select*** [sɪ'lekt] v. 选择；挑选
7. ***gratitude*** ['grætɪtju:d] n. 感激；感谢
8. ***affection*** [ə'fekʃn] n. 慈爱；喜爱；感情
9. ***affectionately*** [ə'fekʃənətlɪ] adv. 热情地；体贴地

06 感谢关照

From catherine@163.com
Date December 11, 2008
To marylady@yahoo.com
Subject Thanks for Your Kindness and Hospitality

Dear Mary,

This is to express my ***appreciation***[1].
I would like to thank you for your ***enthusiasm***[2] and ***hospitality***[3] during my visit to your ***country***[4]. It was my first ***experience***[5] staying overseas for two months, so I was a little nervous in the beginning. However, you treated me like a member of your family and **made me feel at home**.
Please give my pure-hearted regards and *special*[6] thanks to your family.

I look forward to hearing from you again as soon as possible.

Kindest regards,
Catherine

译文

亲爱的玛丽：

写这封信是为了表达感谢之情。
在我到访贵国期间，多谢您的热情款待。因为这是我第一次在海外待两个月之久，所以刚开始还有些紧张。但是您待我如您的家人一般，让我无拘无束。
请代我向您的家人致上诚挚的问候及特别的感谢。

期待尽快收到您的来信。

献上最诚挚的问候，

凯瑟琳

语法重点解析

1 解析重点1 make me feel at home

make sb feel at home 意思是“使某人（像在家里一样）自在、不拘束”。主人为了尽量让客人放松，时常会用到这句话。与它相近的用语还有：take it easy（放松；别紧张）。

例如：

They go out of their way to make me feel at home.（他们尽心尽力地让我有宾至如归的感觉。）

Tell him to take it easy.（告诉他放松些，别紧张。）

2 解析重点2 Please give my pure-hearted regards and special thanks to your family.

这句话的意思是“请代我向您的家人致上诚挚的问候及特别的感谢”。Please give my regards to... 是“向……转达我的问候”的意思。其偏口语化的表达方式是 Please give my best to...如果去掉前面的 please 就会显得更生活化。如果是非常亲密的关系，还可以说 Say hello to...for me。

高频例句

1. **Thank you so much for your warm hospitality during my stay.** 非常感谢您在我逗留期间对我的热情款待。
2. **Thank you for your kindness.** 感谢您的体贴。
3. **Please give my sincere regards to you and your staff.** 请允许我向您和您的员工致上诚挚的问候。
4. **Thanks again for your kindness and *friendliness*[7].** 再次感谢你的善良和友好。
5. **I hereby give my *infinite*[8] thanks for you.** 我在此向您致以无尽的感谢。
6. **I am extremely *grateful*[9] for your friendship.** 我非常感谢您的友善。
7. **I would like to express my appreciation for your kind treatment.** 感谢您对我的悉心照顾。
8. **Thank you for making so much for me.** 感谢您为我做这么多。

必背关键单词 

1. ***appreciation*** [əˌpriːʃɪˈeɪʃn] n. 感激
2. ***enthusiasm*** [ɪnˈθjuːzɪæzəm] n. 热情；热忱
3. ***hospitality*** [ˌhɒspɪˈtælətɪ] n. 好客；殷勤
4. ***country*** [ˈkʌntrɪ] n. 国家
5. ***experience*** [ɪkˈspɪərɪəns] n. 经验；经历
6. ***special*** [ˈspeʃl] adj. 特别的；特殊的
7. ***friendliness*** [ˈfrendlɪnəs] n. 友好；亲切
8. ***infinite*** [ˈɪnfɪnət] adj. 无限的；无穷的
9. ***grateful*** [ˈgreɪtfl] adj. 感激的；感谢的

07 感谢款待

From Jessica_cool@yahoo.com
To gengwg@126.com
Subject Thanks for Entertaining Us!
Date April 26, 2008

Dear Mr. Geng,

Thank you very much for your ***wonderful***[1] hospitality **during our stay** in Beijing. I would like to express my appreciation to you for making our ***trip***[2] such an ***enjoyable***[3] and ***successful***[4] one.
We sincerely appreciate your **taking time out of your busy** ***schedule***[5] to show us around your ***office***[6] and ***factory***[7]. I also hope that you will have a ***chance***[8] to visit us in London in the near future.

Yours truly,
Jessica

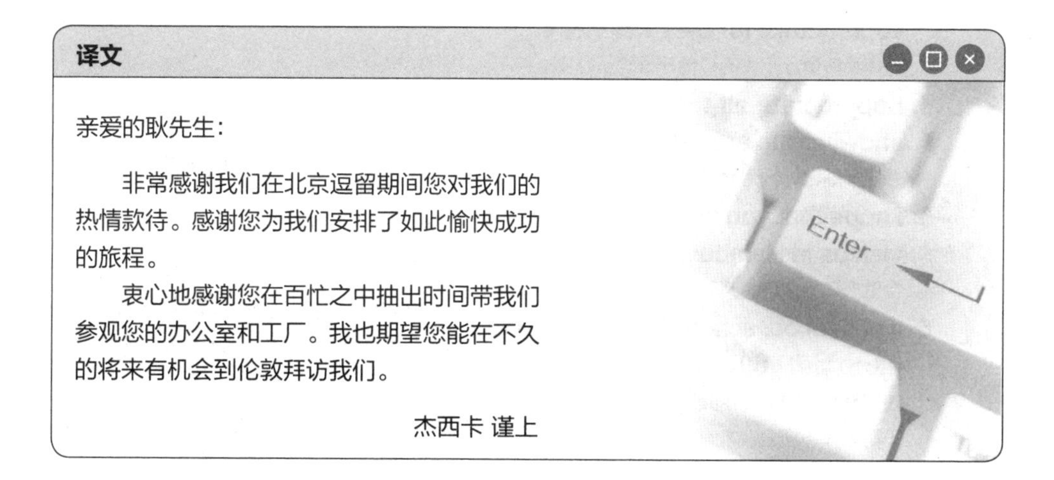

译文

亲爱的耿先生：

非常感谢我们在北京逗留期间您对我们的热情款待。感谢您为我们安排了如此愉快成功的旅程。

衷心地感谢您在百忙之中抽出时间带我们参观您的办公室和工厂。我也期望您能在不久的将来有机会到伦敦拜访我们。

杰西卡 谨上

语法重点解析

1 解析重点1 **during our stay**

during our stay 意思是“在我们逗留期间”。stay 既有动词词性又有名词词性，意思是“停留”，在此语境中 stay 做名词“逗留”使用。请对照以下 stay 作为动词和名词的用法：

I will stay here until tomorrow.（我要在这儿待到明天。）

We have accomplished a great deal during our brief stay in your country.（我们在贵国短暂停留期间取得了丰硕的成果。）

2 解析重点2 **taking time out of your busy schedule**

take...out of... 的意思是“把……从……取出、拿出”。taking time out of your busy schedule 可以直译为“把时间从繁忙的时间表中拿了出来”，也就是中文常说的“从百忙之中抽出空来”。这是一种非常礼貌客气的说法，因而非常适合用于感谢信中。

1. **Thank you for all your kindness and support during my trip.**
   感谢您在我旅行期间对我的照顾和支持。
2. **It was a pleasure to meet you and your colleagues in New York.**
   很荣幸能在纽约与您和您的同事会面。
3. **Please also convey my thanks to all the workers at your factory for their kindness.**
   请代我向贵厂工人的热情友好表示感谢。
4. **I appreciate all the time you spent showing me around your factory.**
   感谢您抽出时间带我参观您的工厂。
5. **I hope that you will have a chance to visit us in Hamburg sometime soon.**
   希望您在不久的将来有机会来汉堡找我们。
6. **If both of our schedules permit, I would like to visit you again this summer.**
   如果我们双方的时间都允许的话，我想今年夏天再去拜访您。

必背关键单词

1. ***wonderful*** ['wʌndəfl] adj. 极好的；精彩的
2. ***trip*** [trɪp] n. 旅行
3. ***enjoyable*** [ɪn'dʒɔɪəbl] adj. 愉快的
4. ***successful*** [sək'sesfl] adj. 成功的
5. ***schedule*** ['ʃedju:l] n. 时间表；进度表
6. ***office*** ['ɒfɪs] n. 办公室
7. ***factory*** ['fæktrɪ] n. 工厂
8. ***chance*** [tʃɑ:ns] n. 机会

08 感谢慰问

From elizabeth22@21cn.com Date December 14, 2008
To anneangel@yahoo.com
Subject Thanks for Your Consolation

Dear Anne,

I ***shall***[1] always remember with gratitude the e-mail you sent me when you learned of Jane's ***death***[2]. No one but you knew my sister so well and loved her as her own family did. Only you could have written that letter. It ***brought***[3] me ***comfort***[4], Anne, when I needed it ***badly***[5].
Thank you from the bottom of my heart for your e-mail and for your kindness to Jane during her long ***illness***[6].

Affectionately,
Elizabeth

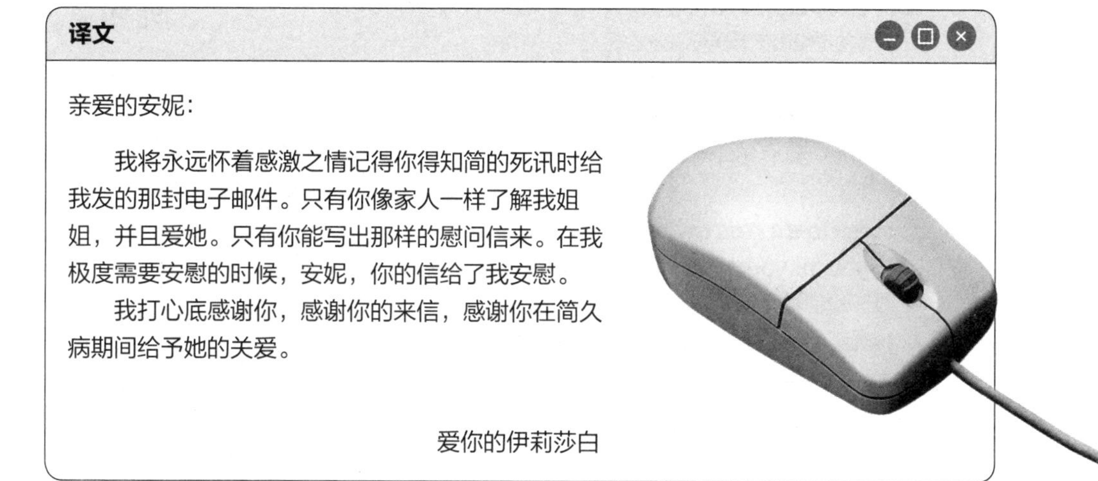

译文

亲爱的安妮：

我将永远怀着感激之情记得你得知简的死讯时给我发的那封电子邮件。只有你像家人一样了解我姐姐，并且爱她。只有你能写出那样的慰问信来。在我极度需要安慰的时候，安妮，你的信给了我安慰。

我打心底感谢你，感谢你的来信，感谢你在简久病期间给予她的关爱。

爱你的伊莉莎白

语法重点解析

1 解析重点1 learn of

learn of 是“知道，获悉”的意思，相当于 know, be aware of。因为表达“知道”的时候总是用 know 显得太过普通，可以偶尔换一些同义词来表达同样的意思。请看以下例句：

I'm sorry to learn of his illness.（听说他病了，我很难过。）

He knew where she was hiding.（他知道她藏在哪里。）

John is aware of having done something wrong.（约翰已经知道自己做错事情了。）

2 解析重点2 bring me comfort

bring sb sth 意思是“带给某人某物”。bring 是“带来”的意思。除此之外，take, carry 也可以表达“带”的意思，但是也略有差别：bring 侧重于“带来”；take 侧重于“带走”；carry 是“搬运”的意思。请对照以下例句：

Can I bring my friend with me?（我可以带朋友一起来吗？）

She went out of the room, taking the flowers with her.（她带着花走出了房间。）

The box is too heavy for me carry.（这个箱子太重了，我搬不动。）

1. A thousand thanks for your kind comfort.
非常感谢您的亲切安慰。

2. I can't sufficiently express my thanks for your *thoughtful*[7] kindness.
对于您无微不至的关怀，我的感激之情溢于言表。

3. I really don't know how to thank you enough.
我真的不知道该怎样感谢你才好。

4. If I can in any way return the favor, it will give me great pleasure to do so.
如果我能做点什么来报答您的话，我将非常乐意。

5. I wish to express my deep indebtedness to you for your kindness.
对于您给我的帮助，我表示深深的感激。

6. I shall always remember with gratitude the favor you did me.
我将永远感激您对我的帮助。

必背关键单词 

1. ***shall*** [ʃəl] conj. 将要；死亡
2. ***death*** [deθ] n. 死亡
3. ***bring*** [brɪŋ] v. 带来
4. ***comfort*** [ˈkʌmfət] n. 舒适；安慰
5. ***badly*** [ˈbædlɪ] adv. 严重地；极度地
6. ***illness*** [ˈɪlnəs] n. 疾病
7. ***thoughtful*** [ˈθɔːtfl] adj. 深思熟虑的；体贴的

09 感谢介绍客户

这样写就对了

From eastjustice_law@lawf.com　Date April 26, 2008
To burns_bs@yahoo.com
Subject Thanks for the Introduction

Dear Mr. Burns,

Thank you for your ***introduction*** [1], which helped us to establish a new business relationship with E-Trade USA.
In a ***cozy*** [2] ***atmosphere*** [3], we have met with Mr. Tom Billy, ***Executive*** [4] Director of E-Trade in Shanghai this week regarding their ***legal*** [5] needs in China. **Mr. Billy holds you in high regard** and is ***particularly*** [6] interested in several of our ***attorneys*** [7] educated and trained in the U.S. We look forward to providing E-Trade with the finest and most effective services.

We *owe* [8] you the greatest *debt* [9] of gratitude.

Sincerely yours,
East Justice Law Firm

译文

亲爱的伯恩斯先生:

非常感谢您把美国电子贸易公司介绍给我们，这帮助我们建立了新的业务关系。

这个星期，在轻松惬意的氛围下，我们在上海与美国电子贸易公司的执行董事汤姆·比利先生会面，谈了一些关于他们在中国所需的法律服务。比利先生非常尊敬您，并且对我们曾在美国接受过教育和培训的几位律师尤其感兴趣。我们期望能为美国电子贸易公司提供最优质与最高效的服务。

我们在此向您表示由衷的感谢。

东方正义律师事务所 谨上

语法重点解析

1 解析重点1 Mr. Billy holds you in high regard.

hold sb in high regard 的意思是“对某人极为重视；对某人十分尊重”。那么相反的 hold sb in low regard 就是“蔑视某人”的意思。请对照以下例句：

The doctor is held in high regard by his patients.（那位医生备受他的病人们尊敬。）

We hold her in low regard because she is not honest.（我们看不起她，因为她不诚实。）

2 解析重点2 We owe you the greatest debt of gratitude.

这句话的意思是“我们向您表示由衷的感谢”，这是一种表示感谢的常用语。类似的表达方式还有：

I will be forever indebted to you for your help.（对于您的帮助，我会一辈子铭记在心。）

I will always remember your kindness.（您的友好我会一直铭记于心。）

1. **Thank you so much for introducing us to ACB Company.**
   多谢您把我们介绍给ACB公司。
2. **I am pleased to say it looks like we will build a new relationship with them.**
   我很开心地说我们将和他们建立新关系。
3. **Again, we appreciate your continued support.**
   再次感谢您一直以来的支持。
4. **Thank you for the confidence you have shown in us.**
   感谢您对我们表现出的信心。
5. **I look forward to returning the favor at the earliest opportunity.**
   我希望能有机会尽快报答您的恩情。
6. **We owe you the greatest debt of thankfulness.**
   我们向您表示衷心的感谢。

必背关键单词 

1. ***introduction*** [ˌɪntrəˈdʌkʃn] n. 介绍
2. ***cozy*** [ˈkəʊzɪ] adj. 舒适的
3. ***atmosphere*** [ˈætməsfɪə(r)] n. 气氛；氛围
4. ***executive*** [ɪɡˈzekjətɪv] adj. 执行的
5. ***legal*** [ˈliːɡl] adj. 法律的；法定的；合法的
6. ***particularly*** [pəˈtɪkjələlɪ] adv. 尤其；特别；异乎寻常地
7. ***attorney*** [əˈtɜːnɪ] n. 律师
8. ***owe*** [əʊ] v. 欠债；归功于；应感谢
9. ***debt*** [det] n. 债务

10 感谢协助

From qamagazine@vogue.com Date May 11, 2008
To dearcustomers@yahoo.com
Subject Thanks for Sending the Questionnaire

Dear ***Subscriber***[1],

I am writing this to express my gratitude for your cooperation and kind assistance.
The ***questionnaire***[2] you ***filled***[3] out is a great help to us in our ***effort***[4] to improve the ***efficiency***[5] of our customer service departments. In addition, you will be ***rewarded***[6] by our improved services.
With your help, I ***believe***[7] that we can **make it**! Thanks again from the ***bottom***[8] of our heart.

Sincerely yours,
Q&A Magazine

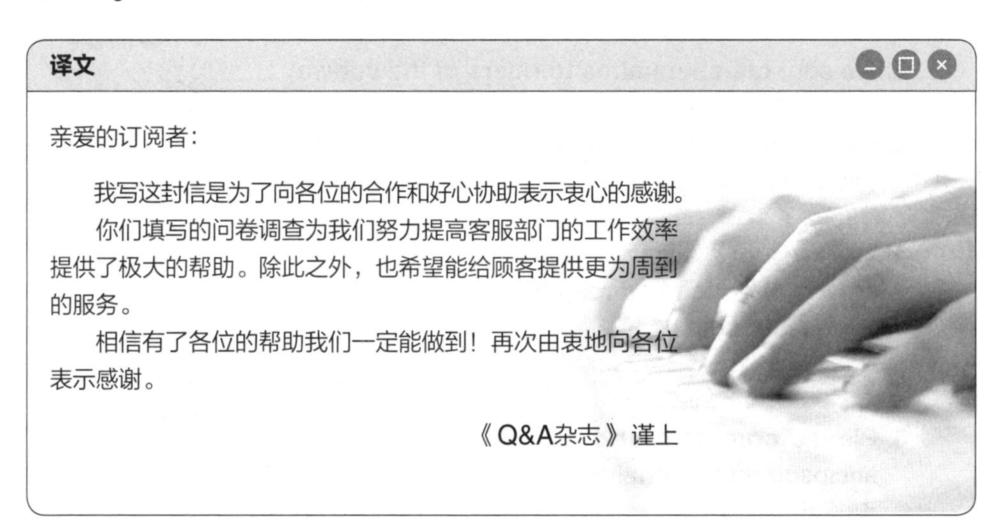

译文

亲爱的订阅者:

我写这封信是为了向各位的合作和好心协助表示衷心的感谢。
你们填写的问卷调查为我们努力提高客服部门的工作效率提供了极大的帮助。除此之外，也希望能给顾客提供更为周到的服务。
相信有了各位的帮助我们一定能做到！再次由衷地向各位表示感谢。

《Q&A杂志》谨上

语法重点解析

1 解析重点1 questionnaire

“调查问卷”可以用 questionnaire 表示，若要表示“调查”，还可以用 survey, investigation，但是这两个单词也有区别，survey 一般偏向于指“测量；勘测；民意调查”，而 investigation 一般侧重于比较正式的官方调查研究，所以英文书信中简单的调查问卷一般用 questionnaire 比较合适。请对照以下例句：

It took me quite a while to fill out the questionnaire.（填写那份问卷花了我好长一段时间。）

A recent survey of public opinion showed that most people were worried about the increasing crime rate.（一份最近的民意调查显示，大多数人对不断增长的犯罪率表示担忧。）

The investigation into the accident was carried out by two policemen.（两名警察对这一事故展开了调查。）

2 解析重点2 make it

make it 可以表达很多种意思：“达到预定目标”“完成某事”“及时抵达”“走完路程”“（病痛等）好转”“成功”等。在此语境下是“提高客服部门的效率，为顾客提供更周到的服务”的意思。例如：

You'll make it if you hurry.（如果你快一点便能及时赶到。）

I feel sure that I can make it.（我坚信我一定能成功。）

高频例句

1. **Thank you for all your kindness and support.**
   感谢各位的友好和支持。
2. **We sent questionnaires to riders of the subway.**
   我们向地铁乘客发了调查问卷表。
3. **I am writing this to express my thanks to all of you.**
   此信是为了表达我对各位的感谢之情。
4. **We are very sorry if this letter arrived after you had responded to our questionnaire.**
   如果此信在你们已经答复了我们的问卷调查表之后才收到，敬请原谅。
5. **Please complete and return the enclosed questionnaire.**
   随附问卷请填妥交回。

必背关键单词

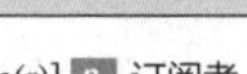

1. ***subscriber*** [səbˈskraɪbə(r)] n. 订阅者
2. ***questionnaire*** [ˌkwestʃəˈneə(r)] n. 问卷；调查表
3. ***fill*** [fɪl] v. 填充；填写
4. ***effort*** [ˈefət] n. 努力
5. ***efficiency*** [ɪˈfɪʃnsɪ] n. 效率
6. ***reward*** [rɪˈwɔːd] v. 奖赏；酬谢
7. ***believe*** [bɪˈliːv] v. 相信
8. ***bottom*** [ˈbɒtəm] n. 底部

11 感谢陪同

这样写就对了

From luke_wang@163.com **Date** August 23, 2008
To brown_witty@yahoo.com
Subject Thanks for Accompanying Me

Dear Brown,

This is to tell you that I've arrived home now. I wish to express my thanks to you for the wonderful ***vacation***[1] I ***spent***[2] with you and your family. ***During***[3] the vacation, you ***taught***[4] me how to swim, boat and ***fish***[5]. I really appreciate your **taking time off work** to show me around.
Your wife is such a *terrific*[6] **cook**! I think I must have ***gained***[7] 10 pounds just in the week I spent with you. I've had a happy and ***memorable***[8] vacation. Thanks again.

I hope you will be able to visit us sometime. Let's keep in touch.

Truly yours,
Luke Wang

译文

亲爱的布朗先生:

我已经回到家了！非常感谢您和您的家人陪我度过了一个如此美好的假期。假期中，您教了我如何游泳、划船和钓鱼，还抽空陪我四处参观，我真的很感激。
您的妻子厨艺真棒！我想和您一起度过的这一周，我一定长了10磅肉。我度过了一个愉快而难忘的假期。再次表示感谢。

希望您也能有机会来我们这里玩。保持联络哦！

卢克·王 谨上

语法重点解析

1 解析重点1 **take time off work**

take time off work 的意思是“抽出时间”，类似于“从工作中抽出时间”或者更深程度的“从百忙之中抽出时间”。这是一种非常礼貌客气的说法，一方面表现出了主人的热情款待，另一方面表现出了客人的感激之情，因而非常适合用于感谢信之中。类似的表达还有：Thank you for taking time out of your busy schedule to accompany me.（感谢您从百忙之中抽出时间陪我。）

2 解析重点2 **Your wife is such a terrific cook!**

这句话可以直接译为“您的妻子是个很棒的厨师！”但是考虑到情境，发现写信者是去对方家做客而感受到了女主人高超的厨艺，所以事实上，女主人的职业并非一定是厨师，而是拥有好的厨艺。所以正确的理解应该是“您妻子的厨艺真棒！”

1. **Thank you for all your attendance during my stay.**
   感谢您在我逗留期间对我的照顾。
2. **You treat me like a member of your family.**
   您待我如家人一般。
3. **It's my first time staying overseas for so long.**
   这是我第一次在海外待这么久。
4. **I appreciate all the time you spent showing me around.**
   非常感谢您抽出时间带我四处游览。
5. **I hope that you will have a chance to visit us sometime.**
   希望某天您也有机会来拜访我们。
6. **I was totally overwhelmed by your hospitality.**
   非常感激你的盛情款待。
7. **I appreciate that you introduced us so many places of interest there.**
   感谢您向我们介绍了众多的名胜古迹。
8. **Thank you for making our trip so wonderful.**
   感谢您使我们的旅程如此美妙。

必背关键单词

1. ***vacation*** [vəˈkeɪʃn] n. 假期；休假
2. ***spend*** [spend] v. 度过；花费
3. ***during*** [ˈdjʊərɪŋ] prep. 在……期间
4. ***taught*** [tɔ:t] v. 教（teach 的过去式和过去分词）
5. ***fish*** [fɪʃ] v. 钓鱼
6. ***terrific*** [təˈrɪfɪk] adj. 极好的；非常的
7. ***gain*** [geɪn] v. 得到
8. ***memorable*** [ˈmemərəbl] adj. 值得纪念的；难忘的

12 感谢参访

Dear Mr. Zhang,

I am writing this letter to thank you for your warm hospitality to us in your beautiful country.
During the ***entire*** [1] visit, we were ***overwhelmed*** [2] by the ***enthusiasm*** [3] expressed by your business ***representatives*** [4]. I sincerely hope we could have more ***exchanges*** [5] like this one so that we would be able to continue our ***discussion*** [6], ***expand*** [7] our ***bilateral*** [8] ***economic*** [9] and trade relations and bring benefits to our people.

I am looking forward to your early visit here.

With kind regards,
UV delegation

译文  

亲爱的张先生：

此信是为了感谢在贵国时您的盛情款待。
整个参访过程中，我们都被贵国业务代表的热情所感染。真诚地希望我们能有更多像此次一样的交流，使得我们能继续深入探讨，扩展双边经贸关系，以造福两国人民。

期待您尽早访问我国。

献上最诚挚的问候，

UV代表团

语法重点解析

1 解析重点1 overwhelm

overwhelm 有好几种意思，例如“战胜”“压倒”“覆盖”“使不知所措”等，一般用于被动语态，在此语境下则是“（在心理和情感上）深深受到影响”的意思。由于写信者被接待者的热情深深打动和感染，所以选择了这个有感染力的词，用来表达深深的感激之情。请看以下 overwhelm 的相关用法：

The defense was overwhelmed by superior numbers.（防守被具有优势的兵力摧垮了。）

The village was overwhelmed by ash from the volcano.（村子被火山灰覆盖了。）

I was overwhelmed by his generosity.（他的慷慨令我感激难言。）

2 解析重点2 expand bilateral economic and trade relations

该短语的意思是“扩展双边经济贸易关系”，是涉外贸易关系中的常用语。bilateral 意思是“双边的”，指的是涉及两国之间的；而单词 multilateral 的意思是“多边的”，指的是涉及多国之间的。例如：

Such programs exist on a bilateral and multilateral basis all over the world.（这些项目在全球各地有双边和多边的合作模式。）

1. **I would like to express my thanks for your warm hospitality.**
   我要感谢您的盛情接待。
2. **I really appreciate your taking time off work to show me around.**
   我真的非常感谢您抽出时间陪我游览。
3. **I would also like to thank you for your interesting discussion with me.**
   我还要感谢您跟我进行了有趣的讨论。
4. **Thank you for taking time out of your busy schedule to visit our company.**
   对于您在百忙之中参观本公司，我们在此表示诚挚的谢意。
5. **We were pleased to be able to show you our facilities.**
   我们非常高兴您能参观我们的各项设施。

必背关键单词

1. ***entire*** [ɪnˈtaɪə(r)] adj. 整个的
2. ***overwhelm*** [ˌəʊvəˈwelm] v. 压倒
3. ***enthusiasm*** [ɪnˈθjuːzɪæzəm] n. 热情
4. ***representative*** [ˌreprɪˈzentətɪv] n. 代表
5. ***exchange*** [ɪksˈtʃeɪndʒ] v. 交换；交流
6. ***discussion*** [dɪˈskʌʃn] n. 讨论
7. ***expand*** [ɪkˈspænd] v. 扩大
8. ***bilateral*** [ˌbaɪˈlætərəl] adj. 双边的
9. ***economic*** [ˌiːkəˈnɒmɪk] adj. 经济的

13 感谢建议

这样写就对了

From peterjunior@yahoo.com
To steven_101@21n.com
Subject Thanks for Your Suggestion!
Date September 12, 2008

Dear Mr. Steven,

Thank you very much for meeting with Ted and giving him your ***helpful***[1] ***advice***[2] about the ***legal***[3] ***profession***[4]. As I am sure you could tell, he is very ***enthusiastic***[5] about law and ***eager***[6] to begin in this field.
He got a lot out of your talk and **I can't think of a better *example*[7] for him to follow.**

I appreciate your kind assistance and suggestions **on *behalf*[8] of** my son.

Yours sincerely,
Peter

译文

亲爱的史蒂芬先生：

非常感谢您能够与泰德见面，并且给他一些法律专业上有益的建议。我相信您能看的出来，我的儿子对法律充满热情并且渴望从事法律行业。

他从和您的谈话中获益良多。对他来说，我再也找不到比你更好的榜样了。

我代表我的儿子向您给予的协助和支持表示感谢。

彼得 谨上

语法重点解析

1 解析重点1 **I can't think of a better example for him to follow.**

这句话的意思是“对他来说，我再也找不到比你更好的榜样了”。在感谢信中向提供了有价值的建议的人表达感谢。

2 解析重点2 **on behalf of**

这个短语的意思是“代表”。可以表达“代表”的短语和单词还有：stand for，represent。但是它们之间有些差别：on behalf of 一般是代表某个或某些人；stand for 一般是代表缩写；represent 一般是代表某个个人或团体，是更为正式的用语。请对照以下例句：

I wrote this letter on behalf of my father.（我代表我父亲写这封信。）

What does UE stand for?（UE 是什么的缩写？）

We chose a committee to represent us.（我们选出一个委员会来代表我们。）

1. **I would like to say how grateful I am for your information.**
   我谨对您提供的消息表达我深深的谢意。
2. **I'd like to express my gratitude along with my best wishes.**
   我想在此向您表示感谢，并献上我最美好的祝福。
3. **I owe you a thousand thanks for your friendly advice.**
   对你友好的建议，我万分感激。
4. **I appreciate it that you gave me such helpful advice.**
   非常感谢您给我这么有帮助的建议。
5. **Your suggestions will benefit me all my life.** 您的建议将使我受益终生。
6. **I've gained a lot from your suggestions.**
   我从您的建议中获益良多。
7. **Can you give me any suggestions on this matter?**
   关于这件事，你能给我一些建议吗?
8. **Good advice is beyond price.**
   好的忠告是无价的。

必背关键单词

1. ***helpful*** [ˈhelpfl] adj. 有益的；有帮助的
2. ***advice*** [ədˈvaɪs] n. 建议；忠告
3. ***legal*** [ˈliːgl] adj. 法律上的
4. ***profession*** [prəˈfeʃn] n. 职业
5. ***enthusiastic*** [ɪnˌθjuːzɪˈæstɪk] adj. 热情的；热心的
6. ***eager*** [ˈiːgə(r)] adj. 渴望的
7. ***example*** [ɪgˈzɑːmpl] n. 榜样
8. ***behalf*** [bɪˈhɑːf] n. 利益；维护

14 感谢邀请

这样写就对了

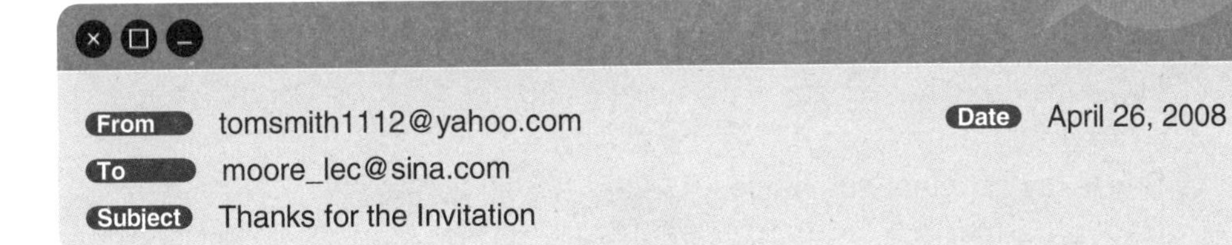

From tomsmith1112@yahoo.com
Date April 26, 2008
To moore_lec@sina.com
Subject Thanks for the Invitation

Dear Mr. Moore,

My wife ***joins***[1] me in thanking you and your kind wife for a ***delightful***[2] night at your house last Sunday.
The ***delicious***[3] food, ***pleasant***[4] **company** and ***intelligent***[5] conversation made the ***occasion***[6] an ***unforgettable***[7] one for both of us.
Once again we'd like to express our thankfulness to you for your ***zealous***[8] invitation and warm friendship. **I wish you and your family good health and happiness forever**!

Sincerely yours,
Tom Smith

译文

亲爱的摩尔先生:

我和我的夫人在此向您和您的夫人表示感谢，感谢上个星期日在您家中度过的愉快夜晚。

晚宴上可口的食物、令人愉快的客人以及充满智慧的交谈都给我们留下了难忘的印象。

再次感谢您的热情邀请和友好款待，并祝福您和您的家人永远健康、幸福!

汤姆 · 史密斯 谨上

语法重点解析

1 解析重点1 company

company 不仅有“公司”的意思，而且还有“同伴；伙伴”的意思。此外，companion, partner, fellow 也可以表示“同伴；伙伴”的意思。companion 意思是“同伴；伴侣”；partner 的意思是“合作者；搭档”；fellow 通常指“同伴，伙伴”。在使用这些词的时候，一定要注意场合和正式的程度。请对照以下例句：

I take my daughter for company while going out.（我出门的时候带上女儿作个伴。）

She is my constant companion.（她是我始终如一的伴侣。）

He is a partner in a law firm.（他是律师事务所的合伙人。）

2 解析重点2 I wish you and your family good health and happiness forever!

这句话的意思是“祝福您和您的家人永远健康幸福！”动词 wish“希望”后面可以接双宾语，即 wish sb sth。虽然 hope 也是“希望”的意思，但两者的用法有一定区别：wish 能接双宾语，表示祝福，hope 则不能。请对照以下例句：

I wish you success.（我祝你成功。）

I hope you will get well soon.（我希望你早日康复。）

高频例句

1. It was indeed a pleasure to have dinner with you.
与您共进晚餐确实是我的荣幸。

2. Thank you for your invitation and I look forward to our next interaction.
感谢您的邀请并期待下一次的交流。

3. Please convey my thanks to all your family members.
请代我向您的家人们表示感谢。

4. I hope you will be able to attend.
我希望您能出席。

5. I hope that you will have a chance to visit our home soon.
期望您在不久之后有机会来我们家做客。

6. Many thanks for your warm hospitality.
非常感谢您的盛情款待。

必背关键单词

1. ***join*** [dʒɔɪn] v. 加入；参加
2. ***delightful*** [dɪˈlaɪtfl] adj. 令人愉快的；讨人喜欢的
3. ***delicious*** [dɪˈlɪʃəs] adj. 美味的
4. ***pleasant*** [ˈpleznt] adj. 令人愉快的
5. ***intelligent*** [ɪnˈtelɪdʒənt] adj. 聪明的；智慧的
6. ***occasion*** [əˈkeɪʒn] n. 场合；机会
7. ***unforgettable*** [ˌʌnfəˈgetəbl] adj. 难忘的
8. ***zealous*** [ˈzeləs] adj. 热情的；热心的

15 感谢合作

From chemical_t@211c.com　Date December 30, 2008
To bobwells@yahoo.com
Subject Thanks for Your Kind Cooperation

Dear Mr. Wells,

This is to express great gratitude for your ***close***[1] ***collaboration***[2] with us.
I would like to thank you for your cooperation for our ***business***[3]. We have had a ***profitable***[4] year. Therefore, we are ***keenly***[5] ***desirous***[6] of ***enlarging***[7] our trade in ***various***[8] kinds of ***chemical***[9] textiles. We do hope we could have further cooperation with each other and achieve co-prosperity in the future. Thank you so much again.

With thanks and regards.

Yours truly,
CT Corporation

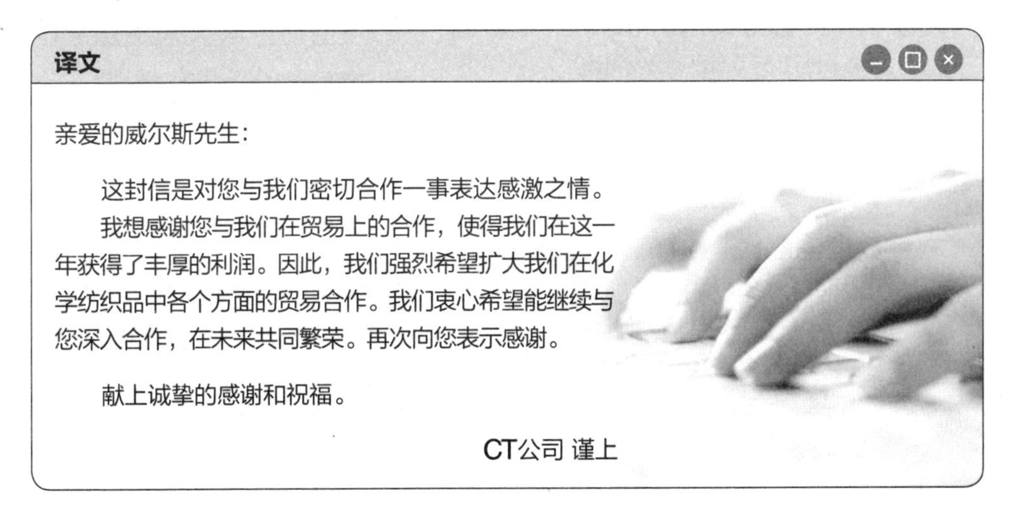

译文

亲爱的威尔斯先生:

　　这封信是对您与我们密切合作一事表达感激之情。

　　我想感谢您与我们在贸易上的合作，使得我们在这一年获得了丰厚的利润。因此，我们强烈希望扩大我们在化学纺织品中各个方面的贸易合作。我们衷心希望能继续与您深入合作，在未来共同繁荣。再次向您表示感谢。

　　献上诚挚的感谢和祝福。

CT公司 谨上

语法重点解析

1 解析重点1 **We are keenly desirous of enlarging our trade in various kinds of chemical textiles.**

这句话的意思是“我们强烈希望扩大我们在化学纺织品各个方面的贸易合作”。短语 be desirous of 意思是“渴望；想要”，相当于 be eager to（渴望……）。keenly 则表示“敏锐地；强烈地”，这个词表达了写信者想要合作的强烈愿望，如果少了这个词，那么句子就会归于平淡。一起来看以下例句：

She has always been desirous of fame.（她一直想成名。）

He keenly felt that he should do something to help.（他深感他应当助其一臂之力。）

2 解析重点2 **co-prosperity**

co-prosperity 的意思是“共同繁荣”，可以理解为合作中的 win-win（双赢），即在合作中追求双方共同发展、共同进步。请看以下例句：

We pursue a result of co-prosperity.（我们追求共同繁荣。）

We want a win-win cooperation.（我们想要共赢的合作模式。）

1. **Thank you for your kind cooperation during this year.**
   感谢您在这一年中友好的合作。
2. **It was a pleasure to collaborate with you.**
   能够跟您合作是我的荣幸。
3. **Please also convey my thanks to all the staff of your company.**
   请代我向贵公司所有员工转达谢意。
4. **We have made a lot of money after cooperating with you.**
   跟您合作了之后，我们公司赚了很多钱。
5. **I hope that we will have a chance for further cooperation.**
   希望我们有机会展开更深入的合作。
6. **We pursue co-prosperity for both of us.**
   我们追求共同繁荣。

必背关键单词

1. ***close*** [kləʊz] adj. 亲密的，密切的
2. ***collaboration*** [kəˌlæbəˈreɪʃn] n. 合作
3. ***business*** [ˈbɪznəs] n. 商业；生意
4. ***profitable*** [ˈprɒfɪtəbl] adj. 有利可图的；有益的
5. ***keenly*** [kiːnlɪ] adv. 强烈地；敏锐地
6. ***desirous*** [dɪˈzaɪərəs] adj. 渴望的
7. ***enlarge*** [ɪnˈlɑːdʒ] v. 扩大
8. ***various*** [ˈveərɪəs] adj. 各种各样的
9. ***chemical*** [ˈkemɪkl] adj. 化学的

Unit 5 邀请 Invitation

01 邀请参加聚会

From fiona012@yahoo.com　　Date June 25, 2008

To bob_wit@21n.com

Subject Party Invitation

Dear Bob,

Tom and I have recently moved to ***Purple***[1] ***Vine***[2] Town and would like to ***invite***[3] all of our friends over for a ***housewarming***[4] party.
Please join us **at 4:00 p.m. on Sunday, June 29, 2008**. ***Directions***[5] are enclosed.

We hope you and your wife will be able to ***attend***[6] **on time**.

Yours truly,
Fiona

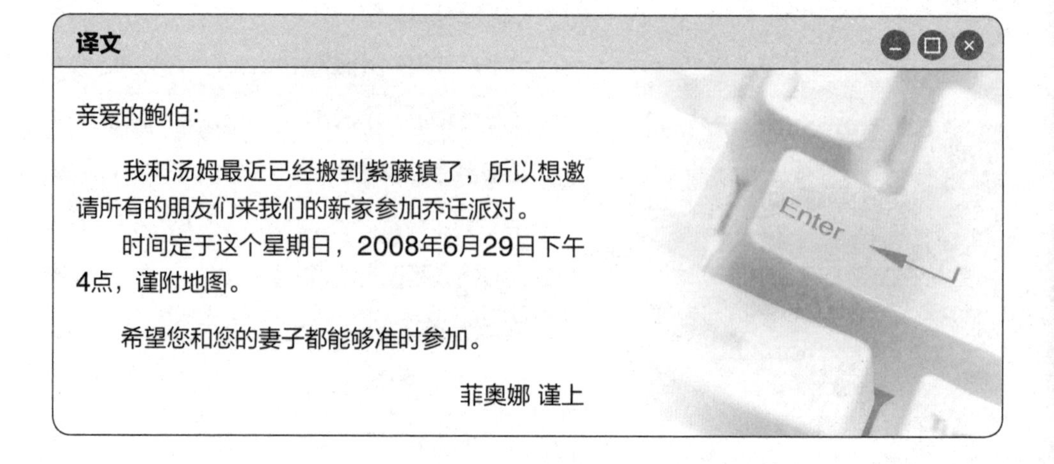

译文

亲爱的鲍伯：

我和汤姆最近已经搬到紫藤镇了，所以想邀请所有的朋友们来我们的新家参加乔迁派对。
时间定于这个星期日，2008年6月29日下午4点，谨附地图。

希望您和您的妻子都能够准时参加。

菲奥娜 谨上

语法重点解析

1 解析重点1 at 4:00 p.m. on Sunday, June 29, 2008

英文书信写作中有一个原则为 Correctness（准确）原则。它的意思是在英文书信写作过程中如果提到具体日期、地点等内容时，要准确地表达，以免发生歧义。所以在此，应该把聚会时间的年月日、星期几、具体时间点等内容都说清楚，如此就不会产生歧义。

2 解析重点2 on time

on time 的意思是“准时”，还有一个容易和它混淆的短语为 in time，它的意思是“按时；及时”。请对照以下例句：

Will the train arrive on time?（火车会准时到达吗？）

They were just in time for the bus.（他们及时赶上了公交车。）

1. **I would like you to attend our party.**
   我想邀请你来参加我们的聚会。
2. **You are invited to a special showing of our new line of *cosmetics*[7].**
   邀请您出席此次新上市的化妆品系列特别发布会。
3. **Can you join us this Friday evening for dinner?**
   这个星期五晚上和我们一起吃饭好吗?
4. **Let me know if you would like to attend.**
   如果您愿意参加，请联系我。
5. **I hope you could take part in our anniversary party.**
   我希望您能参加我们的周年纪念派对。
6. **Information on transportation is enclosed.**
   已附上交通信息。
7. **Kindly *respond*[8] on or before December 10, 2013.**
   请于2013年12月10日之前回复。
8. **Please respond to this invitation by July 8 so that we can prepare well.**
   为了能准备妥善，请于7月8日前回复这次邀请，以便我们能充分准备。

必背关键单词 

1. ***purple*** ['pɜ:pl] adj. 紫色的
2. ***vine*** [vaɪn] n. 藤
3. ***invite*** [ɪn'vaɪt] v. 邀请
4. ***housewarming*** ['haʊzwɔ:mɪŋ] n. 乔迁庆宴
5. ***direction*** [də'rekʃn] n. 方向；用法说明
6. ***attend*** [ə'tend] v. 参加
7. ***cosmetic*** [kɒz'metɪk] n. 化妆品
8. ***respond*** [rɪ'spɒnd] v. 回应；回复

02 邀请参加发布会

From pressconttact@prc.com Date November 15, 2014
To sirsmadam@renmin.com
Subject Invitation for Press Conference

Dear Sir or Madam,

Co-organized[1] by CFTC and United Nations ***Industrial***[2] Development ***Organization***[3] (UNIDO), the ***World***[4] Industrial and ***Commercial***[5] Organizations (WICO) ***Summit***[6] will be ***held***[7] on December 19-20, 2014 in Beijing.

To facilitate your better understanding of the Summit, a press conference will be held at ABC Building, at 3:00-4:30 p.m. on November 20th, 2014. Please refer to http://www. abcbuilding.com for further information of the WICO Summit.

We sincerely hope you will be with us.

Sincerely yours,
Contact person: Jason Wu

译文 

亲爱的女士们、先生们：

由中国对外贸易中心和联合国工业开发组织共同主办的世界工商协会峰会将于2014年12月19至20日于北京召开。

为了便于您更了解此次的峰会，将于2014年11月20日下午3:00-4:30，在ABC 大楼举行新闻发布会。想要进一步了解世界工商协会峰会详情，请查询网站：http://www.abcbuilding.com。

我们真诚地希望各位能够共襄盛举。

会议联系人：杰森 · 吴 谨上

语法重点解析

1 解析重点1 **hold**

hold 做动词时有多种释义，如：“握住”“持有”“抱”“保持”“举行”等，在此语境中是指“举行”。表达“举行”还可以说 take place, stage。请对照以下例句：

The meeting will be held on Thursday.（这次会议将于星期四举行。）

The fete will take place on Sunday.（游园会定于星期日举行。）

The union decided to stage a one-day strike.（工会决定举行一天的罢工。）

2 解析重点2 **facilitate**

facilitate 的意思是“使容易；帮助；促进”，在此语境下是“使容易”的意思。事实上， To facilitate your better understanding of the Summit, ... 还可以表达为 In order to help you understand the Summit well, ...

1. **The Press Conference is approved by the State Council.**
   此次记者招待会经过了国务院的批准。
2. **I hope you will be able to *attend*[8] on time.**
   我希望您能准时出席。
3. **We look forward to hearing soon that you can be with us.**
   我们期待尽快得到您参加的消息。
4. **I sincerely hope you will be with us.**
   我诚挚地希望您能与我们一起。
5. **You are invited to join us for the session as our guest.**
   我们邀请您以嘉宾的身份参加此次会议。
6. **I do hope you could attend the conference.**
   我真心希望您能出席这场会议。
7. **The press conference will be held at the Hilton Hotel.**
   此次记者招待会将于希尔顿酒店举行。
8. **Please refer to www.wico-summit.org for further information.**
   有关详情请登录www.wico-summit.org网站查看。

必背关键单词

1. ***co-organize*** [kəʊˈɔ:gənaɪz] v. 联合组织
2. ***industrial*** [ɪnˈdʌstrɪəl] adj. 工业的
3. ***organization*** [ˌɔ:gənaɪˈzeɪʃn] n. 组织
4. ***world*** [wɜ:ld] n. 世界
5. ***commercial*** [kəˈmɜ:ʃl] adj. 商业的
6. ***summit*** [ˈsʌmɪt] n. 峰会；顶点；高层会议
7. ***hold*** [həʊld] v. 举行
8. ***attend*** [əˈtend] v. 出席；参加

03 邀请担任发言人

这样写就对了

From nada@conf.com
Date October 12, 2014
To janeeyre@yahoo.com
Subject Invitation to Serve as Our Speaker

Dear Ms. Jane,

Would you like to ***serve***[1] as our speaker on ***National***[2] Advertising Directors Association (NADA)? **I can think of no one more *qualified*[3] to fill this role than you.** NADA is ***prepared***[4] to pay all your ***expenses***[5]. The ***media***[6] panel is scheduled to begin at 3 p.m. on Thursday, October 16, 2014 and end no later than 5 p.m.

I do hope it will be ***possible***[7] for you to ***undertake***[8] this ***assignment***[9]. Let me know as soon as you can, please. If your response is **favorable**, I'll send other information to you.

Sincerely yours,
NADA Organizer

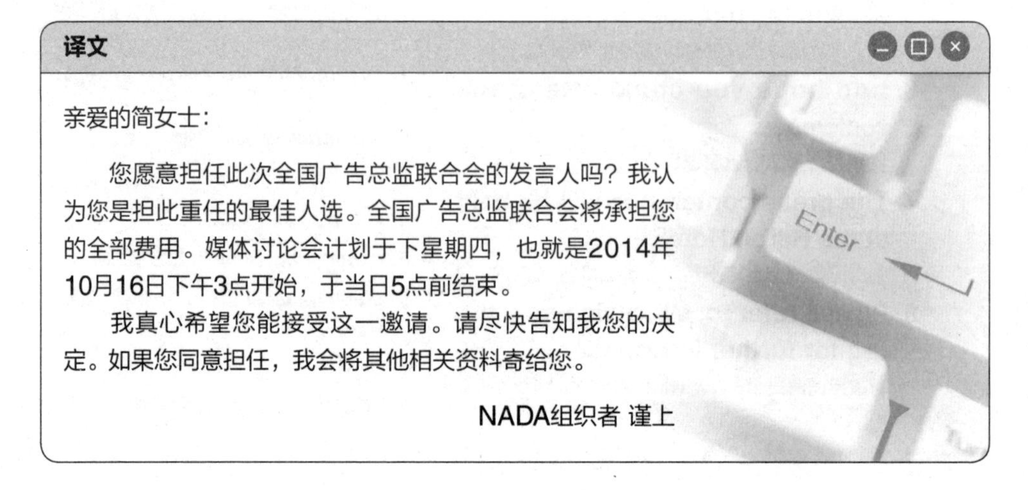

译文

亲爱的简女士：

您愿意担任此次全国广告总监联合会的发言人吗？我认为您是担此重任的最佳人选。全国广告总监联合会将承担您的全部费用。媒体讨论会计划于下星期四，也就是2014年10月16日下午3点开始，于当日5点前结束。

我真心希望您能接受这一邀请。请尽快告知我您的决定。如果您同意担任，我会将其他相关资料寄给您。

NADA组织者 谨上

语法重点解析

1 解析重点1 I can think of no one more qualified to fill this role than you.

这句话的意思是“我认为您是担此重任的最佳人选”或者“除了您之外，我想不出更适合担此重任的人选了”。它属于英文中形式上肯定而意义上否定的用法。我们在表达否定的时候，可以灵活地运用一些句型，使句子更新颖。例如：

He is the last man I want to see.（他是我最不想见的人。）

2 解析重点2 favorable

这个单词有“赞成的；有利的；良好的；讨人喜欢的；起促进作用的”等意思。在此语境下则是指“赞成的，同意的”。所以除了我们通常使用的 agreeable, consenting 外，还可以用这个词来增加表达的多样性。请看以下例句：

They give us a favorable answer.（他们给了我们一个令人满意的答复。）

He was quite agreeable to accepting the plan.（他相当乐意接受这计划。）

My father will not be consenting to our marriage.（我父亲不会同意我们的婚事。）

1. **Would you serve as our speaker on *direct*[10] mail?**
   您愿意担任我们“直递邮件”的发言人吗?
2. **I would like to invite you as our speaker of the conference.**
   我想邀请您担任我们会议的发言人。
3. **I cannot think of a better person than you.**
   我想不出比您更适合的人选了。

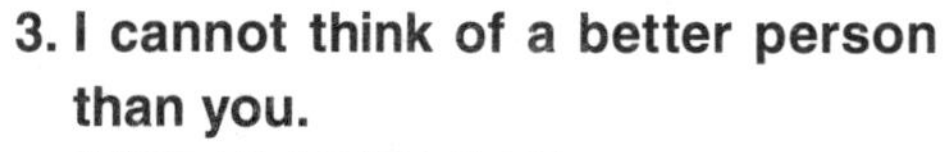

4. **We are prepared to pay all your expenses.**
   我们会支付您的全部开销。
5. **I can think of nobody except you.**
   除了您我想不出别人了。
6. **I do hope you could give me your consent.**
   我真的希望您能同意。
7. **I am sure you can undertake this important assignment.**
   我相信您能担此重任。

必背关键单词

1. ***serve*** [sɜːv] v. 作……用；服务；对待
2. ***national*** [ˈnæʃnəl] adj. 全国的
3. ***qualify*** [ˈkwɒlɪfaɪ] v. 使有资格；胜任
4. ***prepare*** [prɪˈpeə(r)] v. 准备
5. ***expense*** [ɪkˈspens] n. 花费；支出
6. ***media*** [ˈmiːdɪə] n. 媒体
7. ***possible*** [ˈpɒsəbl] adj. 可能的
8. ***undertake*** [ˌʌndəˈteɪk] v. 承担
9. ***assignment*** [əˈsaɪnmənt] n. 任务
10. ***direct*** [dəˈrekt] adj. 直接的

04 邀请参加研讨会

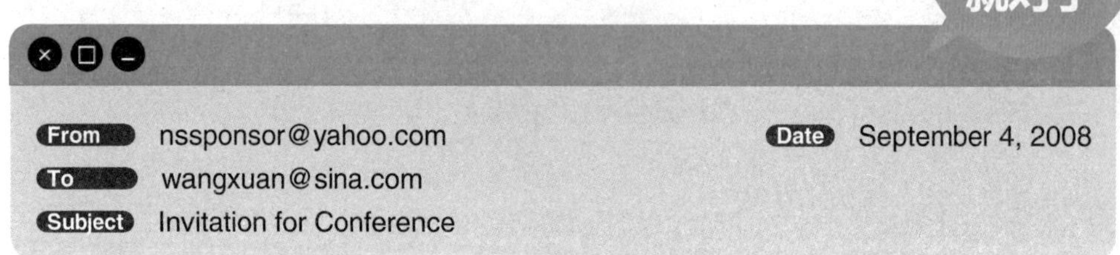

From nssponsor@yahoo.com Date September 4, 2008
To wangxuan@sina.com
Subject Invitation for Conference

Dear Professor Wang,

I am pleased to inform you that you are ***cordially***[1] invited to ***participate***[2] in the conference of North America as our ***guest***[3]. **Your round-trip air ticket, accommodations and meal expenses will be *subsidized***[4]. Should you be interested, **please let us know at your earliest *convenience***[5].
I am looking forward to seeing you in this conference, and I am sure you will play an important role in the ***event***[6]. If your response is ***consenting***[7], I'll send the ***relevant***[8] information to you.

Sincerely yours,
NS Sponsor

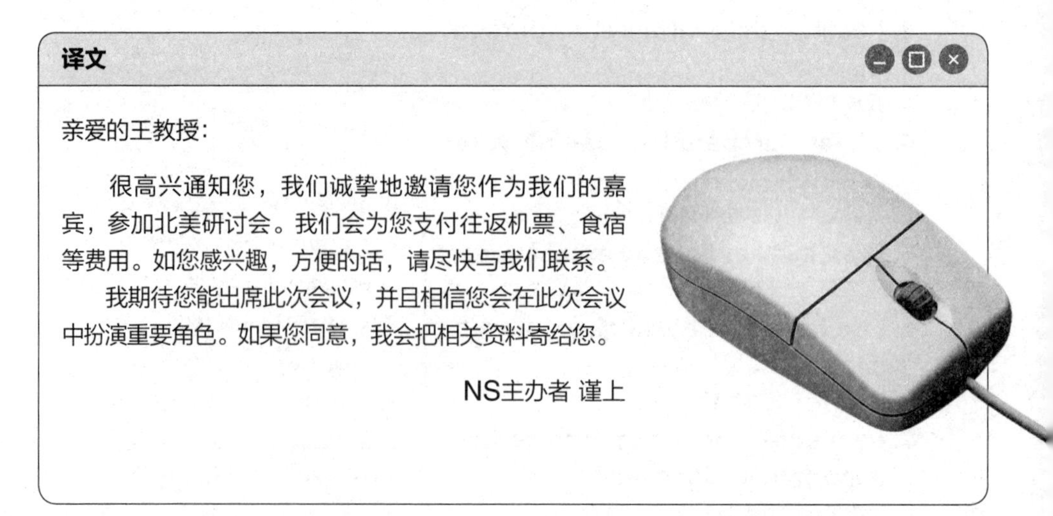

译文

亲爱的王教授：

很高兴通知您，我们诚挚地邀请您作为我们的嘉宾，参加北美研讨会。我们会为您支付往返机票、食宿等费用。如您感兴趣，方便的话，请尽快与我们联系。

我期待您能出席此次会议，并且相信您会在此次会议中扮演重要角色。如果您同意，我会把相关资料寄给您。

NS主办者 谨上

语法重点解析

1 解析重点1 **Your round-trip air ticket, accommodations and meal expenses will be subsidized.**

这句话是被动语态，意思是“我们将为您报销往返机票及食宿费用”。round-trip air ticket 的意思是“往返机票”。subsidize 意思是“给……津贴或补贴；资助；补助”。这句话还可以换成主动语态：

We will pay the round-trip air ticket, accommodations and meal expenses for you.（我们将为您支付往返机票及食宿费用。）

2 解析重点2 **Please let us know at your earliest convenience.**

这是英文书信中常会用到的结束语。意思是“方便的话，请尽快通知我”。类似的表达还有：

I expect your earliest reply.（期待您尽快回复。）

Expecting your immediate response.（期待您的立即回应。）

We look forward to your reply at your earliest convenience.（方便的话，我们期待您尽快的回复。）

1. **Would you like to take part in our meeting?**
   你愿意参加我们的会议吗？
2. **I would like to invite you to participate in the conference.**
   我想邀请您参加此次会议。
3. **You are cordially invited to attend the conference.**
   我们诚挚地邀请您参加这次会议。
4. **We will pay all your expenses.**
   我们会支付您的全部开销。
5. **I look forward to seeing you in the conference.** 期望您能出席会议。
6. **I do hope you could *agree*[9] to come.**
   我真心希望您能同意前来。
7. **I am sure you will play an important role in the meeting.**
   我确信您会在此次会议中扮演重要的角色。

必背关键单词

1. ***cordially*** [ˈkɔːdɪəlɪ] adv. 诚恳地；诚挚地；友善地
2. ***participate*** [pɑːˈtɪsɪpeɪt] v. 参加
3. ***guest*** [gest] n. 客人；贵宾
4. ***subsidize*** [ˈsʌbsɪdaɪz] v. 给……补助、津贴；资助
5. ***convenience*** [kənˈviːnɪəns] n. 方便；便利
6. ***event*** [ɪˈvent] n. 大事；事件；活动
7. ***consenting*** [kənˈsentɪŋ] adj. 同意的
8. ***relevant*** [ˈreləvənt] adj. 相关的
9. ***agree*** [əˈgriː] v. 同意

05 邀请参加访问

这样写就对了

From unm@unimelb.edu.au **Date** June 28, 2008
To chenzhuo@sina.com
Subject Invitation for Visit

Dear Mr. Chen,

I am delighted to invite you to be a visiting ***scholar***[1] in Australia. You will be ***based***[2] at the Melbourne ***campus***[3] but may visit other campuses ***whilst***[4] you are in Australia.
As a visiting scholar, you will be able to ***pursue***[5] your ***specific***[6] research ***agenda***[7] and to collaborate and communicate with the University of Sydney ***faculty***[8] during your stay.
Please let me know if you have any further inquiries and we look forward to your visit.

Yours sincerely,
The University of Melbourne

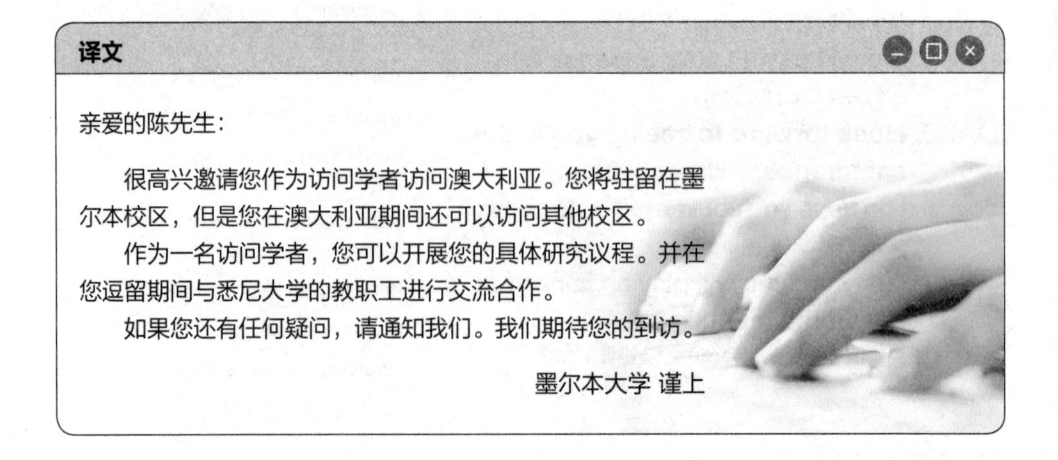

译文

亲爱的陈先生：

很高兴邀请您作为访问学者访问澳大利亚。您将驻留在墨尔本校区，但是您在澳大利亚期间还可以访问其他校区。

作为一名访问学者，您可以开展您的具体研究议程。并在您逗留期间与悉尼大学的教职工进行交流合作。

如果您还有任何疑问，请通知我们。我们期待您的到访。

墨尔本大学 谨上

语法重点解析

1

解析重点1 **You will be based at the Melbourne campus.**

这句话的意思是“您将要驻留在墨尔本大学”。be based 的原意是“以……为基础；基于……”。但在此语境下是指“驻留；住在”，相当于 live 或者 stay。请对照以下例句：

You will mostly be based in Beijing.（你大部分时间将驻留在北京。）

Jack decided to live in college during his freshman year.（杰克决定大一时住校。）

You could stay in Hilton Hotel.（你可以住在希尔顿酒店。）

2

解析重点2 **collaborate**

collaborate 既可以是褒义词，意为“合作”；也可以是贬义词，意为“勾结”。在此语境中指“合作”，相当于 cooperate。而当作“勾结”解释时，相当于 collude。请看以下例句：

Will you collaborate with me to finish the project?（你愿意和我合作完成这个项目吗？）

He was suspected of collaborating with the enemy.（他被怀疑与敌人勾结。）

1. **I am pleased to *invite*[9] you to our college.**
   我很高兴邀请您来我们学院。
2. **I would like to invite you as a visiting scholar.**
   我想邀请您做访问学者。
3. **You are cordially invited to our university as a visiting scholar.**
   我们真诚邀请您作为访问学者来我校参访。
4. **All your expenses will be subsidized.**
   我们会支付您的全部开销。
5. **I look forward to seeing you then.**
   期望届时您能来访。
6. **I do hope you could consent to come.**
   我真心希望您能同意前来。
7. **I am sure you will gain a lot.**
   我相信您会有很多的收获。

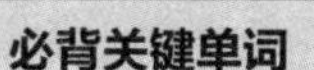

1. ***scholar*** [ˈskɒlə(r)] n. 学者
2. ***base*** [beɪs] v. 以……作基础
3. ***campus*** [ˈkæmpəs] n. 大学校园
4. ***whilst*** [waɪlst] conj. 当……时
5. ***pursue*** [pəˈsjuː] v. 进行；从事
6. ***specific*** [spəˈsɪfɪk] adj. 特殊的；明确的
7. ***agenda*** [əˈdʒendə] n. 议事日程
8. ***faculty*** [ˈfæklti] n. 才能；全体职工
9. ***invite*** [ɪnˈvaɪt] v. 邀请

06 邀请赴宴

From maysmith@yahoo.com　　Date December 7, 2008

To sarah_nice@yahoo.com

Subject Invitation for Dinner

Dear Sarah,

My husband and I should be very much ***pleased*** [1] if you and your daughter would **dine with** us at 6:30 p.m. next Sunday, on the eleventh floor of the ***Emperor*** [2] Building. I am inviting a few other people, and I hope we may sing after ***dinner*** [3]. If you would ***consent*** [4] to bring some ***video*** [5] tapes, I am sure we would have a most wonderful evening.

We do hope you can come and are expecting to see you then.

Yours cordially,
May Smith

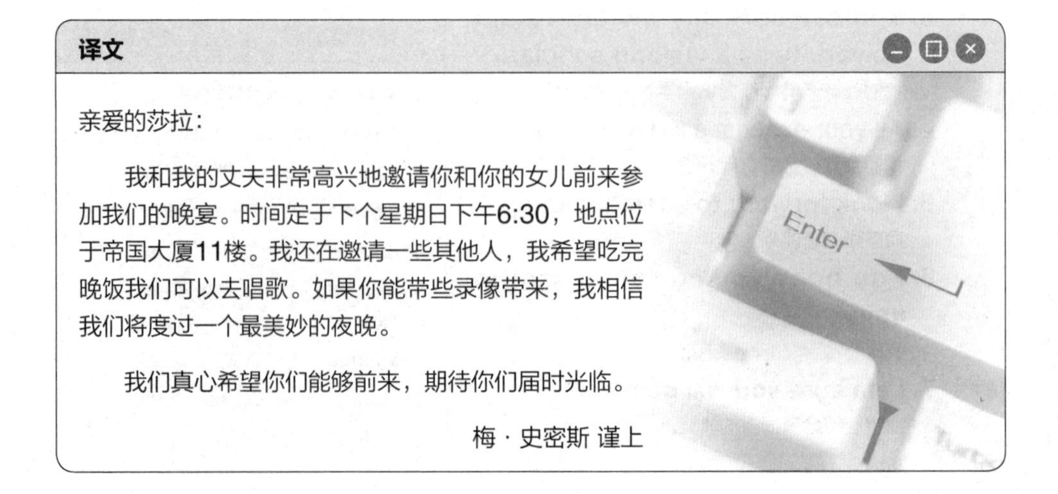

译文

亲爱的莎拉：

我和我的丈夫非常高兴地邀请你和你的女儿前来参加我们的晚宴。时间定于下个星期日下午6:30，地点位于帝国大厦11楼。我还在邀请一些其他人，我希望吃完晚饭我们可以去唱歌。如果你能带些录像带来，我相信我们将度过一个最美妙的夜晚。

我们真心希望你们能够前来，期待你们届时光临。

梅 · 史密斯 谨上

语法重点解析

1 解析重点1 dine with

dine with sb 的意思是“和某人进餐”，就相当于 have dinner with sb。为了不使句子过于平淡，在英文书信写作中要试着转变一下表达方式。请看以下例句：

She wants to dine with me tonight.（她想跟我共进晚餐。）

I'd like to have dinner with you.（我想跟你一起吃晚饭。）

2 解析重点2 consent

consent 有动词和名词两种词性，意思是“同意”，在此语境中做动词用。如果做名词用，一般会使用这个短语：give one's consent。表达“同意”还可以用 agree, approve。请看以下例句：

I definitely will give my consent to your plan.（我当然同意你的计划。）

I asked him to come with me and he agreed.（我邀请他和我一起来，他同意了。）

His parents did not approve of his companions.（他的父母不赞成他所结交的同伴。）

1. **Would you like to come to our dinner party?**
   你愿意来参加我们的晚宴吗？
2. **I would like to invite you to *dine*[6] with us.**
   我想邀请您与我们一起进餐。
3. **You are cordially invited to our dinner party on Saturday.**
   我们诚挚地邀请您参加星期六的晚宴。
4. **We are going to sing after dinner.**
   晚餐后我们要去唱歌。
5. **I look forward to seeing you then.**
   我期望届时您能出席。
6. **I do hope you could come with your *husband*[7].**
   我真心希望您和您的丈夫能够前来。
7. **I am sure it would be a wonderful night.**
   我确信我们会度过一个美好的夜晚。

必背关键单词 

1. ***pleased*** [pliːzd] adj. 高兴的
2. ***emperor*** [ˈempərə(r)] n. 皇帝；君王
3. ***dinner*** [ˈdɪnə(r)] n. 正餐；晚宴
4. ***consent*** [kənˈsent] v. 同意；许可
5. ***video*** [ˈvɪdɪəʊ] n. 录像
6. ***dine*** [daɪn] v. 进餐；吃饭
7. ***husband*** [ˈhʌzbənd] n. 丈夫

07 邀请参加婚礼

From robinrichard@yahoo.com Date August 6, 2008
To mary_lady@yahoo.com
Subject Invitation for Wedding

Dear Robin,

On Sunday, August 11, at three o'clock p.m., **Richard and I are taking the important step in life**. We are getting ***married***[1] at St. Peter's, that ***charming***[2] little ***church***[3]—you know it—at 26 Freeway Drive.
We have sent the ***invitation***[4] card to you. **I *hardly*[5] need to tell you that we would not *consider*[6] it a real *wedding*[7] if you were not present**. There will be an ***informal***[8] ***reception***[9] in the church parlor afterward and we want you there, too.

Affectionately yours,
Mary

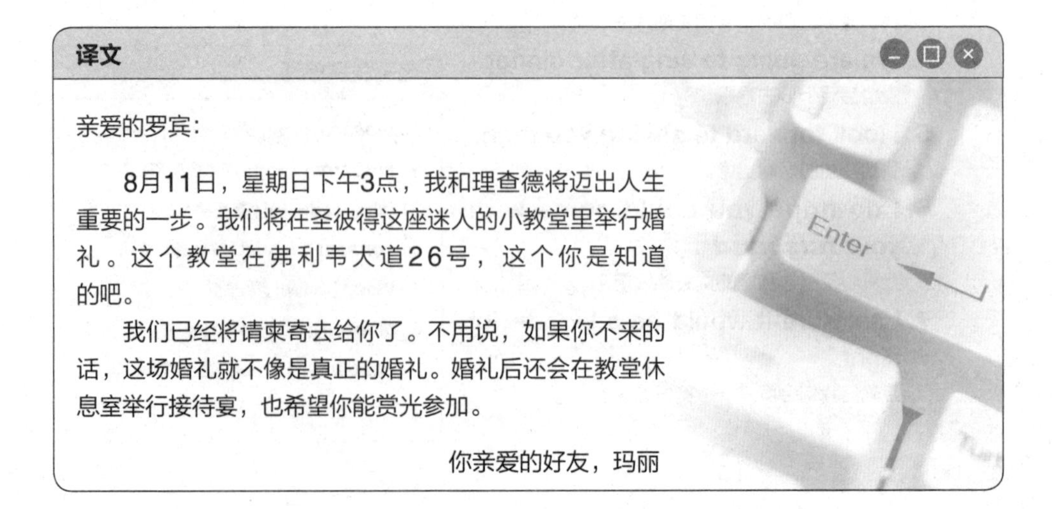

译文

亲爱的罗宾:

8月11日，星期日下午3点，我和理查德将迈出人生重要的一步。我们将在圣彼得这座迷人的小教堂里举行婚礼。这个教堂在弗利韦大道26号，这个你是知道的吧。

我们已经将请柬寄去给你了。不用说，如果你不来的话，这场婚礼就不像是真正的婚礼。婚礼后还会在教堂休息室举行接待宴，也希望你能赏光参加。

你亲爱的好友，玛丽

1

解析重点1 **Richard and I are taking the important step in life.**

这句话的意思是"我和理查德将迈出人生重要的一步"。相当于 Richard and I are getting married（我和理查德要结婚了），这是表达"结婚"的一种委婉的用法。

2

解析重点2 **I hardly need to tell you that we would not consider it a real wedding if you were not present.**

这句话的意思是"不用说，如果您不来的话，这场婚礼就不像是真正的婚礼"或者"不用说，如果您不来的话，我们会觉得这场婚礼缺了些什么"。hardly need to tell you（不用说）可以表明邀请者和被邀请者之间非常亲密友好的关系。we would not consider it a real wedding if you were not present（如果您不来的话，这场婚礼就不像是真正的婚礼）显示出这是非常诚挚的邀请，还表现了被邀请者的受重视程度。

1. **You are cordially invited to our wedding on Saturday.**
   诚挚地邀请您参加我们星期六的婚礼。
2. **I would like to invite you to take part in our wedding.**
   我想邀请您参加我们的婚礼。
3. **Aaron and I are going to take the important step in life.**
   我和艾伦将迈出人生重要的一步。
4. **I am going to get married with John.**
   我要跟约翰结婚了。
5. **We are getting married at the charming church.**
   我们要在那个迷人的教堂结婚了。
6. **We have sent you the invitation card.**
   我们已经将邀请函寄给你了。
7. **There will be an informal reception in the lobby after the wedding.**
   婚礼后将在大厅进行简单的接待。

必背关键单词

1. ***marry*** [ˈmærɪ] v. 结婚
2. ***charming*** [ˈtʃɑːmɪŋ] adj. 迷人的
3. ***church*** [tʃɜːtʃ] n. 教堂
4. ***invitation*** [ˌɪnvɪˈteɪʃn] n. 邀请
5. ***hardly*** [ˈhɑːdlɪ] adv. 几乎不
6. ***consider*** [kənˈsɪdə(r)] v. 考虑；认为
7. ***wedding*** [ˈwedɪŋ] n. 婚礼
8. ***informal*** [ɪnˈfɔːml] adj. 非正式的
9. ***reception*** [rɪˈsepʃn] n. 接待

08 邀请参加生日派对

From hannah@yahoo.com
Date August 6, 2008
To juniorlouise@yahoo.com
Subject Invitation to My Birthday Party

Dear Louise,

Please come to my ***birthday***[1] party!
It's my fifth birthday. I am going to have a ***huge***[2] birthday cake and we will play games ***together***[3]! Please call my mom to let us know if you can come. If the ***answer***[4] is yes, don't ***forget***[5] to bring me a beautiful ***present***[6]!

The date: Sunday, June 29 from 3 to 6 p.m.
P.S. Please tell your parents that there will be a ***barbecue***[7] for the grownups, too!

Sincerely yours,
Hannah

译文

亲爱的露易丝：

请来参加我的生日派对！
这是我的第5个生日。我会有一个大大的生日蛋糕，我们还要一起玩游戏哦！请打电话告诉我妈妈你会不会来。如果会来的话，别忘了带一份漂亮的礼物给我哦！

时间：星期日，6月29日，下午3点到6点。
附言：请告诉你的父母，大人可以烧烤哦！

汉娜 谨上

1 解析重点1 **It's my fifth birthday.**

看得出，这是一封邀请参加小孩子生日派对的请柬。请柬实际上是父母以五岁孩子的口吻写给孩子的朋友的。这种写法使得请柬更为活泼，语言并不必拘泥于某种形式，而且非常符合五岁孩子过生日的情境。

2 解析重点2 **P.S.**

P.S. 是 post script 的缩写，意思是“（信末的）附言；附笔”。一般缩写形式 P.S. 附于信末尾，来补充说明未说完的事情。例如：

P.S. Can you send me your photos?（你能把你的照片寄给我吗？）

She added a post script to her letter.（她在信末加上了附言。）

1. **I want to invite you to my birthday party on next Wednesday night.**
   我想邀请您参加我下星期三晚上的生日派对。
2. **I would like to invite you to my birthday party.**
   我想邀请您参加我的生日派对。
3. **I would like to have a cake sent to Catherine's room on April 11.**
   我想要订一个蛋糕在4月11日那天送到凯瑟琳的房间。
4. **I am going to have a birthday party next week.**
   我下周要举行生日派对。
5. **Remember to bring me a *gift*[8].**
   记得给我带一份礼物哦。
6. **I have sent the birthday party invitation to you.**
   我已经把生日派对的邀请函寄给你了。
7. **There will be a barbecue in the yard for us.**
   我们还可以在院子里烧烤。
8. **Please be sure to be present at my birthday party.**
   请务必出席我的生日派对。

必背关键单词

1. ***birthday*** [ˈbɜ:θdeɪ] n. 生日
2. ***huge*** [hju:dʒ] adj. 巨大的
3. ***together*** [təˈgeðə(r)] adv. 一起
4. ***answer*** [ˈɑ:nsə(r)] n. 答案
5. ***forget*** [fəˈget] v. 忘记
6. ***present*** [ˈpreznt] n. 礼物
7. ***barbecue*** [ˈbɑ:bɪkju:] n. 户外烧烤
8. ***gift*** [gɪft] n. 礼物

09 邀请参加周年庆典

From jackmiller@yahoo.com
To allresident_7@yahoo.com
Subject Invitation to the Anniversary Celebration
Date May 15, 2008

Dear ***residents*** [1] of building#7,

You are cordially invited to share in a ***celebration*** [2] of George and Emma's 50th Wedding ***Anniversary*** [3].
Fifty years ago, a beautiful woman Emma and a ***handsome*** [4] George promised to love and ***cherish*** [5] each other for the rest of their lives. Now, as their dear friends, we will celebrate that ***commitment*** [6] **once more** for them.
Please join us in **SURPRISE** reception for George and Emma in the ***hall*** [7] on May 21, from 3 to 5 p.m. Thank you for your company.

Sincerely yours,
Jack Miller

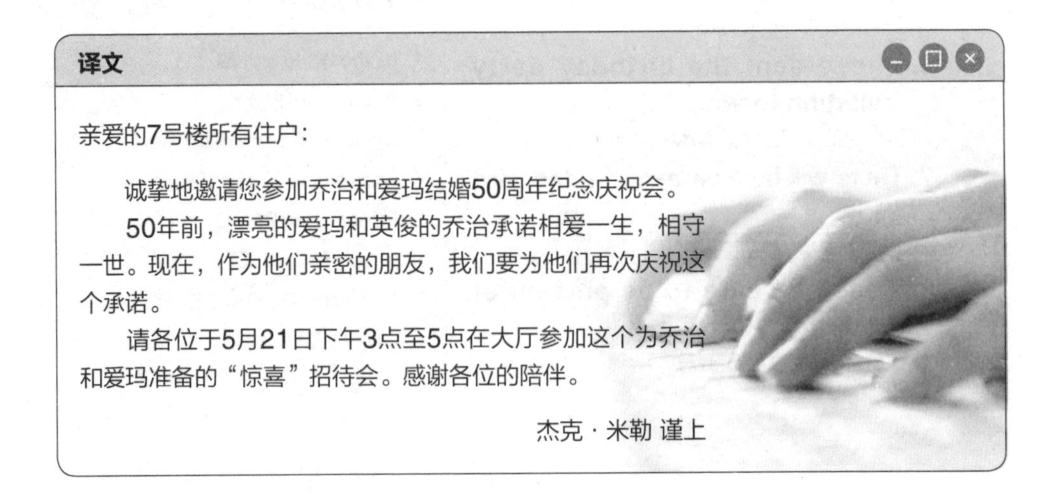

译文

亲爱的7号楼所有住户：

诚挚地邀请您参加乔治和爱玛结婚50周年纪念庆祝会。

50年前，漂亮的爱玛和英俊的乔治承诺相爱一生，相守一世。现在，作为他们亲密的朋友，我们要为他们再次庆祝这个承诺。

请各位于5月21日下午3点至5点在大厅参加这个为乔治和爱玛准备的“惊喜”招待会。感谢各位的陪伴。

杰克 · 米勒 谨上

语法重点解析

1 解析重点1 once more

once more 意思是“再一次；又一次”，相当于 a second time, once again。我们可以选择多个同义词来表达同样的意思，以使句子呈现多样性的特点。请看以下例句：

They are going to try their fortune once more.（他们想再碰一次运气。）

You'll need to type in your password a second time to confirm it.（你将需要再次输入你的密码来验证它。）

Once again we extended to them our warmest welcome.（我们再次向他们表示最热烈的欢迎。）

2 解析重点2 SURPRISE

SURPRISE 用大写，是希望引起人们的注意，强调这是一个“惊喜”，即对方事先并不知情。当需要强调某一事物时，可将词语全部大写。例如：

All applications must be submitted IN WRITING before January 31.（所有申请都必须在1月31日以前以“书面的形式”提交。）

1. **I sincerely invite you to attend our celebration.**
   诚挚邀请您参加我们的庆祝会。
2. **This is an important *event*[8].**
   这是一件非常重要的大事。
3. **I would like to invite you to take part in our 5-year Opening Anniversary.**
   我想邀请您参加我们的开业5周年庆典。
4. **You are definitely present on the anniversary day.**
   你一定要出席这次周年庆祝会。

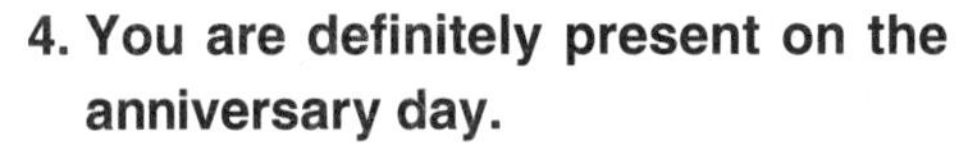

5. **Let's celebrate the 10-year anniversary.**
   我们来庆祝10周年纪念日吧。
6. **You are cordially invited to share the celebration.**
   诚挚地邀请您参加此次庆祝会。
7. **I have sent you the formal invitation card.**
   我已经将正式邀请函寄给您了。

必背关键单词

1. ***resident*** [ˈrezɪdənt] n. 居民
2. ***celebration*** [ˌselɪˈbreɪʃn] n. 庆祝会；庆典
3. ***anniversary*** [ˌænɪˈvɜːsərɪ] adj. 周年纪念的
4. ***handsome*** [ˈhænsəm] adj. 英俊的
5. ***cherish*** [ˈtʃerɪʃ] v. 珍惜
6. ***commitment*** [kəˈmɪtmənt] n. 承诺
7. ***hall*** [hɔːl] n. 大厅
8. ***event*** [ɪˈvent] n. 事件；大事

10 正式接受邀请函

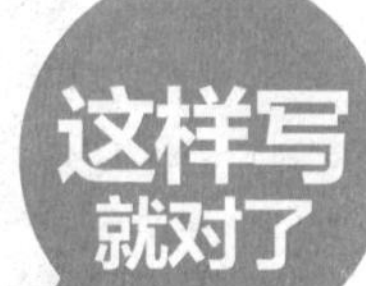

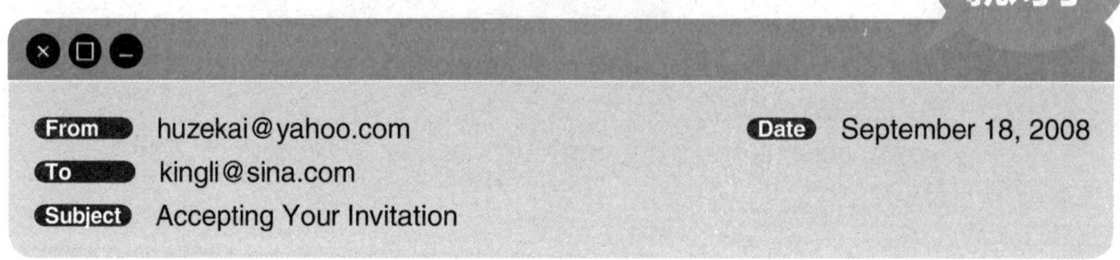

From huzekai@yahoo.com
Date September 18, 2008
To kingli@sina.com
Subject Accepting Your Invitation

Dear Mr. Li,

Thank you so much for inviting me to address the ***Chamber***[1] of ***Commerce***[2] monthly ***luncheon***[3] at 12 p.m. on September 21, at the ***Mayors'***[4] Building on the **subject** of "The ***Status***[5] of Expert Systems in Business During the Next Decade." I am pleased to **accept** your invitation.

I look forward to this ***opportunity***[6] of being with you and the members of Chamber of Commerce once again.

Sincerely yours,
Kate Hu

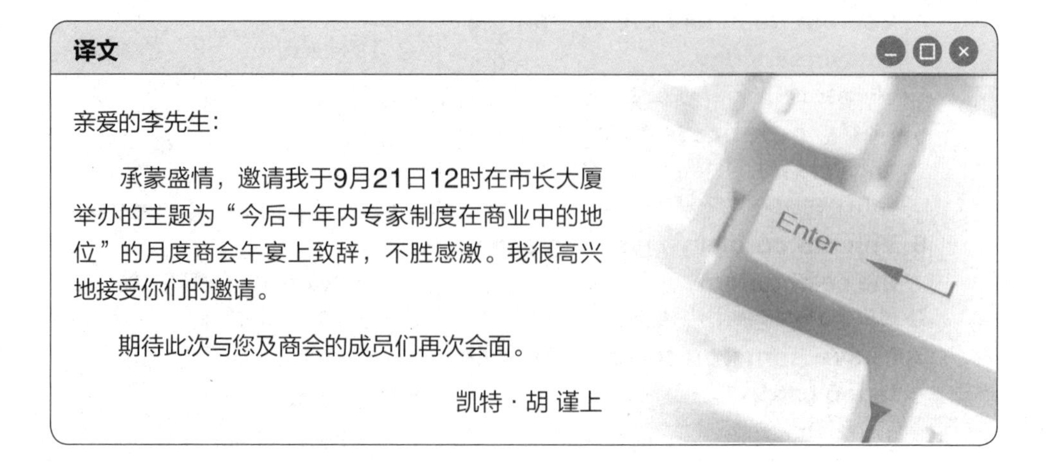

译文

亲爱的李先生：

承蒙盛情，邀请我于9月21日12时在市长大厦举办的主题为“今后十年内专家制度在商业中的地位”的月度商会午宴上致辞，不胜感激。我很高兴地接受你们的邀请。

期待此次与您及商会的成员们再次会面。

凯特 · 胡 谨上

语法重点解析

1 解析重点1 subject

subject 有三种词性：名词、动词、形容词。当作名词是指“主题；科目”；当作动词是指“使……服从；提出”；当作形容词是指“服从……的”。在此语境中是当作名词“主题”解释。请看以下例句：

This is a movie on the subject of love.（这是一部以爱情为主题的电影。）

He tried to subject the whole family to his will.（他试图使全家人都服从他的意愿。）

Prices are subject to change.（价格会随情况而变化。）

2 解析重点2 accept

表达“接受”可以用 accept 或者 receive，但是两者有些区别：receive 仅是客观上“收到；接收”，主观上不一定“接受”；accept 则指的是主观上“接受”。请对照以下例句：

I received a bunch of roses yesterday, but I gave it back.（我昨天收到了一束玫瑰花，可是我把它退回去了。）

I'm overjoyed that she accepted my proposal.（她接受了我的求婚，我欣喜若狂。）

高频例句

1. **Thank you very much for your *invitation*[7].**
   非常感谢您的邀请。
2. **I would love to go to the museum with you this weekend.**
   我很乐意这个周末和你一起去博物馆。
3. **I would be *delighted*[8] to come!**
   我将乐意前往！
4. **Having lunch on Saturday together would be really nice!**
   星期六一起吃午饭实在是棒极了！
5. **Should I bring any presents to the party?** 我要带礼物去参加派对吗？
6. **It will be great to see you again!**
   能再次见到你真是太好了！
7. **I am really looking forward to seeing your new house.**
   我很想看看你的新房子。

必背关键单词

1. ***chamber*** [ˈtʃeɪmbə(r)] n. 房间；贸易团体
2. ***commerce*** [ˈkɒmɜːs] n. 商业；贸易
3. ***luncheon*** [ˈlʌntʃən] n. 午宴
4. ***mayor*** [ˈmeə(r)] n. 市长
5. ***status*** [ˈsteɪtəs] n. 地位
6. ***opportunity*** [ˌɒpəˈtjuːnətɪ] n. 机会
7. ***invitation*** [ˌɪnvɪˈteɪʃn] n. 邀请
8. ***delighted*** [dɪˈlaɪtɪd] adj. 开心的；高兴的；欣喜的

11 拒绝邀请

这样写就对了

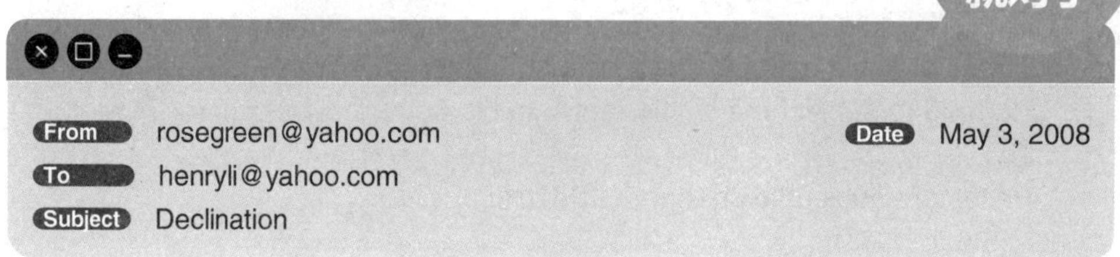
From rosegreen@yahoo.com
To henryli@yahoo.com
Subject Declination
Date May 3, 2008

Dear Henry,

My friend, thank you for your letter of May 2.
Your letter says there will be a ***get-together***[1] on May 6 in Bluestone Park, and you want to invite me to the party with you. ***However***[2], I am ***awfully***[3] sorry to tell you that on that day I am going to see an old friend who's ***seriously***[4] ill, so I am ***afraid***[5] it's ***impossible***[6] for me to go with you then.

I believe that you can find ***another***[7] partner soon.

Yours faithfully,
Rose

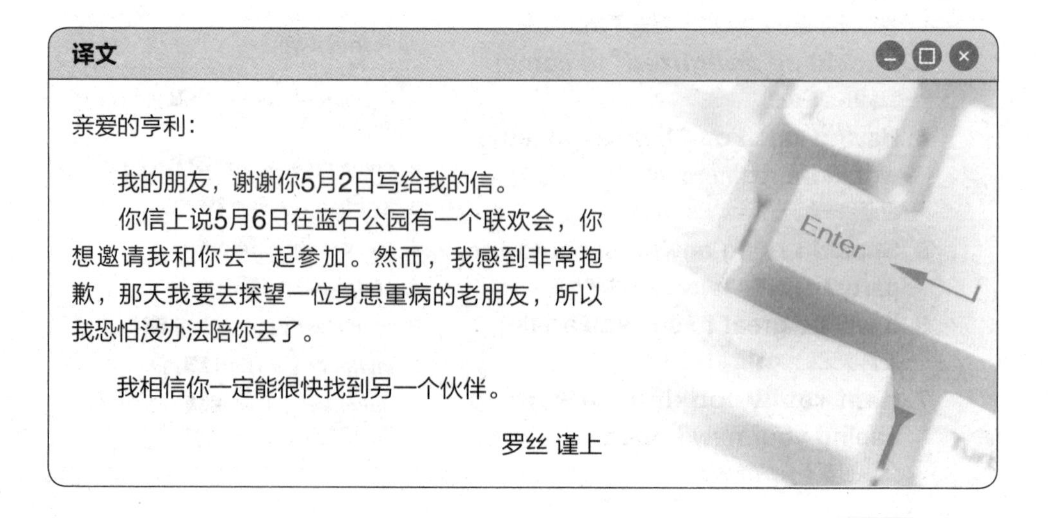
译文

亲爱的亨利：

我的朋友，谢谢你5月2日写给我的信。

你信上说5月6日在蓝石公园有一个联欢会，你想邀请我和你去一起参加。然而，我感到非常抱歉，那天我要去探望一位身患重病的老朋友，所以我恐怕没办法陪你去了。

我相信你一定能很快找到另一个伙伴。

罗丝 谨上

语法重点解析

1 解析重点1 get-together

get-together 指的是非正式的聚会、联欢会，偏口语化。party 则是较为正式的聚会。请看以下例句：

It is a pity that you missed the party.（你未能参加聚会，真是遗憾。）

We're having a little get-together to celebrate David's promotion.（我们为大卫升职举办一个小型庆祝会。）

2 解析重点2 I am awfully sorry to tell you that on that day I am going to see an old friend who's seriously ill.

这句话的意思是“我感到非常抱歉，那天我要去探望一个重病的老朋友”。看得出这是一个委婉拒绝邀请的借口。但是 awfully（非常地；极度地）则表达了深深的歉意，而且因为“要去探望重病的老朋友”而没办法接受邀请，似乎在心理上更容易让对方接受。

1. **I really wish I could go with you, but I promised to help my friend move on Sunday.**
   我真希望能跟你一起去，但是我已经答应星期日去帮朋友搬家了。
2. **If you are free, maybe we could get together sometime next week.**
   如果你有时间的话，我们可以改在下周某个时间聚聚。
3. **Unfortunately, I have a prior engagement this weekend.**
   真可惜，我这周末已经有安排了。
4. **I am sorry to tell you that I have to work *overtime*[8] tomorrow.**
   很抱歉，明天我要加班。
5. **We will find another time to get together soon though.**
   我们尽快另外找个时间聚一聚吧。
6. **I would really like to go, but I have some other plans.**
   我真的很想去，可是我已经有别的安排了。
7. **Unfortunately, I have invited a guest for that day.**
   不巧的是，那天我已经邀请了一位客人。

必背关键单词

1. ***get-together*** [get tə'geðə(r)] n. 聚会；联欢会
2. ***however*** [haʊ'evə(r)] adv. 然而
3. ***awfully*** ['ɔːflɪ] adv. 非常地；极端地
4. ***seriously*** ['sɪərɪəslɪ] adv. 严重地
5. ***afraid*** [ə'freɪd] adj. 担心的；害怕的
6. ***impossible*** [ɪm'pɒsəbl] adj. 不可能的
7. ***another*** [ə'nʌðə(r)] adj. 另一个的
8. ***overtime*** ['əʊvətaɪm] adv. 加班地；超时地

12 邀请出席纪念活动

这样写就对了

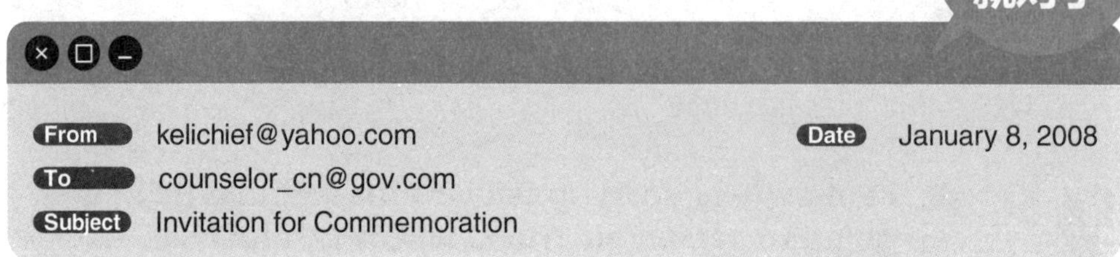
From kelichief@yahoo.com
Date January 8, 2008
To counselor_cn@gov.com
Subject Invitation for Commemoration

Dear Mr. Mayor,

Next Saturday marks the 256th ***anniversary***[1] of the founding of the British Museum. The ***superintendent***[2] of it is planning to ***celebrate***[3] the event with a reception at Claridges Hotel on Saturday, January 17 between 5:00 p.m. and 8:00 p.m. We shall be ***honored***[4] to have you as our ***chief***[5] guest who will **address** the party for about four minutes.

I would appreciate it if you could indicate your *availability*[6] or *inability*[7] to attend the reception before January 13.

Sincerely yours,
Tom Clarke

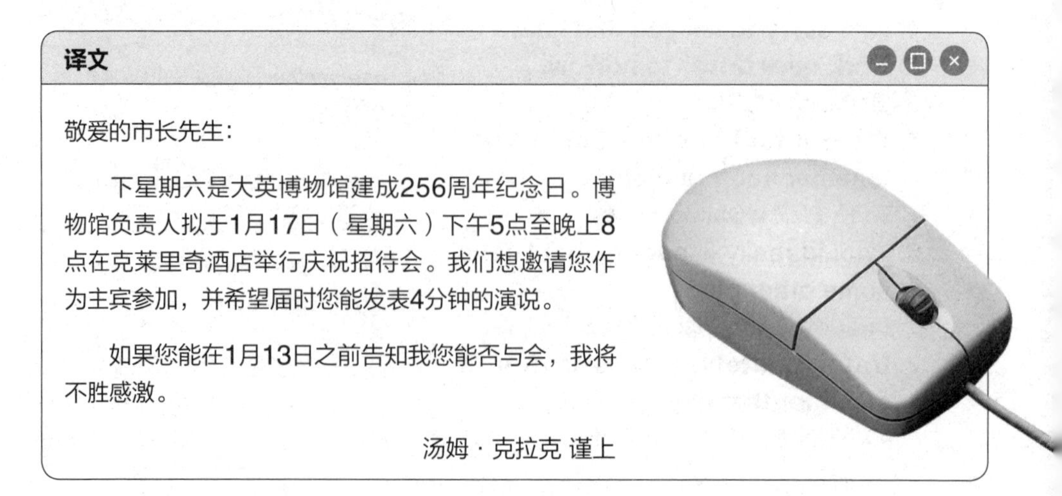
译文

敬爱的市长先生:

下星期六是大英博物馆建成256周年纪念日。博物馆负责人拟于1月17日（星期六）下午5点至晚上8点在克莱里奇酒店举行庆祝招待会。我们想邀请您作为主宾参加，并希望届时您能发表4分钟的演说。

如果您能在1月13日之前告知我您能否与会，我将不胜感激。

汤姆 · 克拉克 谨上

语法重点解析

1 解析重点1 address

address 通常是做名词“地址”用，但在此语境中则做动词“演讲；演说”。请对照以下例句：

Write your address on the back of the envelope.（在信封的背面写上你的地址。）

He addressed the audience in an eloquent speech.（他向听众发表了富有说明力的演说。）

2 解析重点2 I would appreciate it if you could indicate your availability or inability to attend the reception before January 13.

这句话的意思是“如果您能在1月13日之前告知我您能否与会，我将不胜感激”。其中 availability 意为“可利用性”，inability 意为“没办法；不能”。这句话还可以这样表达：I would appreciate it if you could agree to attend the reception before January 13. 这个句子看起来就简洁许多，但如果是非常正式的邀请信，则会选择一些如 availability 或者 inability 等非常正式的用语。

高频例句

1. **I hope you will give me the pleasure of your company on the event.**
   我想邀请贵公司赏光参加我们这次活动。
2. **I would be very pleased if you could come.**
   如果您能参加，我会非常高兴。
3. **We hope you will stay on for the reception following the *ceremony*[8].**
   我们希望您能参加仪式后举行的小型招待会。
4. **Will you do us a favor by joining our party?**
   请参加我们的聚会好吗?
5. **You are cordially invited to the luncheon.**
   我真诚邀请您参加午宴。
6. **I would appreciate it if you could give me your consent.**
   如果您能同意，我将感激不尽。

必背关键单词

1. ***anniversary*** [ˌænɪˈvɜːsərɪ] n. 周年纪念日
2. ***superintendent*** [ˌsuːpərɪnˈtendənt] n.（机关、企业等的）主管；负责人
3. ***celebrate*** [ˈselɪbreɪt] v. 庆祝
4. ***honor*** [ˈɒnə(r)] v. 尊敬；敬意
5. ***chief*** [tʃiːf] adj. 主要的
6. ***availability*** [əˌveɪləˈbɪlətɪ] n. 可利用性；有效性
7. ***inability*** [ˌɪnəˈbɪlətɪ] n. 无能
8. ***ceremony*** [ˈserəmənɪ] n. 典礼；仪式

13 邀请合作

这样写就对了

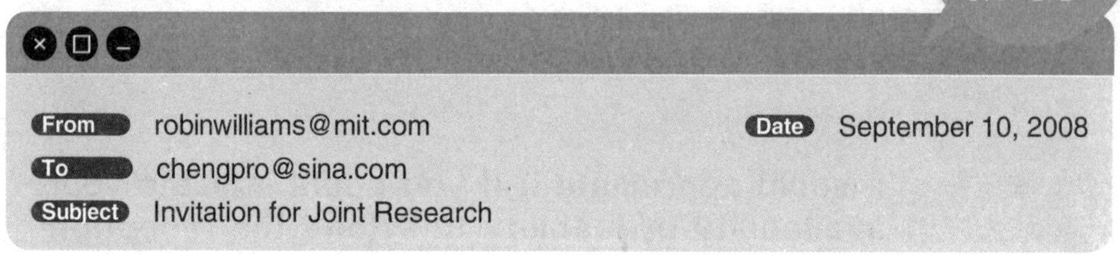

From robinwilliams@mit.com　　Date September 10, 2008
To chengpro@sina.com
Subject Invitation for Joint Research

Dear Professor Cheng,

I am pleased to learn that you have an opportunity to spend a year away from your ***institution***[1] to pursue ***research***[2] in ***Physics***[3], and I'd like to ask you if I could have the honor to invite you to spend that year working in my research ***group***[4] at MIT. I am sure it will be ***beneficial***[5] to both of us.
I am very happy to **cover** all expenses, ***including***[6] **the *costs***[7] **of living expenses, travel costs and research costs**.

I look forward to working with you.

Truly yours,
Robin Williams

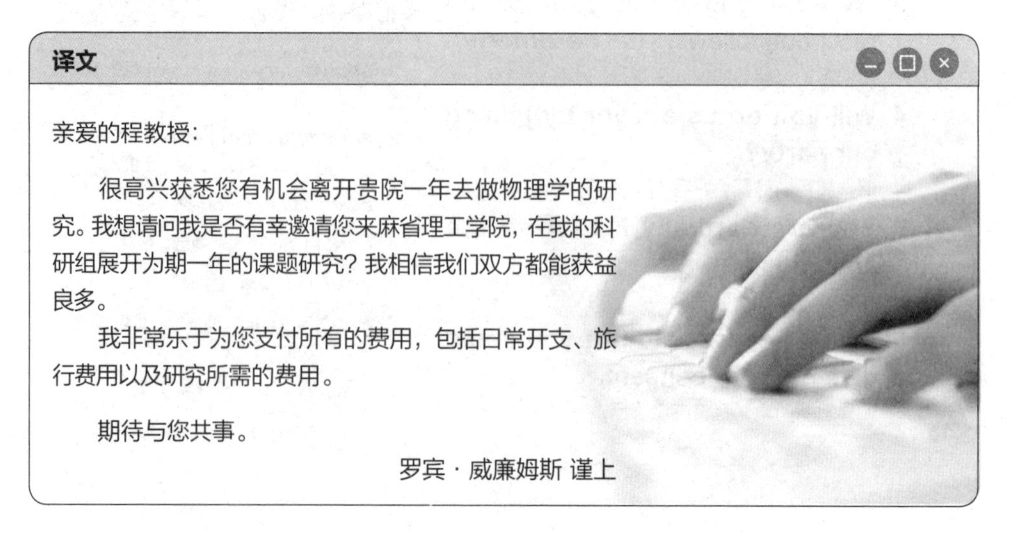

译文

亲爱的程教授：

很高兴获悉您有机会离开贵院一年去做物理学的研究。我想请问我是否有幸邀请您来麻省理工学院，在我的科研组展开为期一年的课题研究？我相信我们双方都能获益良多。

我非常乐于为您支付所有的费用，包括日常开支、旅行费用以及研究所需的费用。

期待与您共事。

罗宾・威廉姆斯 谨上

语法重点解析

1 解析重点1 cover

cover 当作名词是指“封面；盖子；隐蔽处”；当作动词则为“覆盖；涉及；包含”。在此语境是做动词“包含；涵盖”。

I am very happy to cover all expenses. 还可以表达为：

I am happy to pay all your expenses.（我很乐意为您支付所有的费用。）

2 解析重点2 including the costs of living expenses, travel costs and research costs

这句话的意思是“包括日常开支、旅行费用以及研究所需的费用”。including 的意思是“包含；包括”。要注意的是 including 所包含的东西应该置于其之后，如果置于其之前，那么应该用 included。所以此句还可以表达成：the costs of living expenses, travel costs and research costs are included, 请对照以下例句：

They have many pets, including three cats.（他们有很多宠物，包括三只猫。）

Are service charges included?（服务费包含在内吗？）

1. **Would you like to join us in the research?**
   您愿意和我们一起做研究吗？
2. **I would like to invite you to be a visiting professor.**
   我想邀请您做我们的客座教授。
3. **It's my pleasure to work with you.**
   与您共事是我的荣幸。
4. **I am pleased to cover all your expenses.**
   我很乐意为您支付所有的费用。
5. **This cooperation will benefit both of us.**
   此次合作将会使我们双方获益。
6. **I hope that we could do joint research together.**
   我期望我们能一起合作研究。
7. **I believe we will gain more research *achievements*[8].**
   我相信我们能取得更多的研究成果。

必背关键单词 

1. ***institution*** [ˌɪnstɪˈtjuːʃn] n. 机构；学院
2. ***research*** [rɪˈsɜːtʃ] n. & v. 研究
3. ***physics*** [ˈfɪzɪks] n. 物理学
4. ***group*** [gruːp] n. 团体；组
5. ***beneficial*** [ˌbenɪˈfɪʃl] adj. 有益的；有利的
6. ***including*** [ɪnˈkluːdɪŋ] prep. 包括
7. ***cost*** [kɒst] n. 花费；费用
8. ***achievements*** [əˈtʃiːvmənt] n. 完成；成就；成果

14 反客为主的邀请

From hanawei@bisu.com
Date June 24, 2008
To hanyang@yahoo.com
Subject Invitation in Turn

Dear Mr. Yang,

With great ***pleasure***[1] I have received your letter to invite me to have lunch on Saturday, June 28.
I would be very happy to have the opportunity of ***discussing***[2] with you on the ***proposed***[3] visit to Tokyo by your ***Minister***[4] of ***Education***[5]. I would be even happier if you could let me ***host***[6] the lunch for you at 12:00 on that day in the ***International***[7] Hotel. I am sure you **will give me your consent** since you are in Tokyo and I should be the host.

I am looking forward to the pleasure of meeting you **on Saturday at 12:00 in the lobby of the International Hotel**.

With my best regards,
Hanna Wei

译文

亲爱的杨先生：

非常高兴收到您的来信，邀请我参加6月28日（星期六）的午宴。
我非常乐意借此机会与您讨论贵国教育部长拟访问东京的事宜，更乐意于那天中午12点在国际饭店宴请您。既然您来到了东京，我就应尽地主之谊，您一定不会反对吧。

我期待星期六中午12点与您在国际饭店大厅会面。

献上最诚挚的祝福，

汉娜·魏

语法重点解析

1 解析重点1 **give me your consent**

短语 give one's consent 的意思是“答应某人；同意某人”，相当于 consent to sth 或者 agree to do sth。请对照以下例句：

I'd like to know if you will give your consent to his plan?（我想知道你是否同意他的计划？）

Her father reluctantly consented to the marriage.（她父亲勉强地答应了这桩婚事。）

We agreed to their proposal.（我们同意了他们的建议。）

2 解析重点2 **on Saturday at 12:00 in the lobby of the International Hotel**

该句的意思是“星期六中午12点在国际饭店大厅”。事实上，在此之前，书信中就已经提到过午宴的时间是 on Saturday, June 28以及 at 12:00 on that day，但是书信末尾又提了一遍，这只是为了再次强调宴会的准确时间、地点，也再次提醒对方，以免发生误会。

1. I am so pleased to *accept*[8] your invitation.
我非常高兴接受您的邀请。

2. It's my pleasure to have dinner with you.
能与您共进晚餐是我的荣幸。

3. It would be better if you could let me host the lunch for you on April 11.
如果能让我尽地主之谊，在4月11日请您吃午餐就更好了。

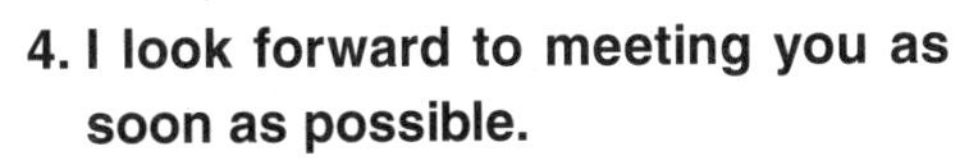

4. I look forward to meeting you as soon as possible.
期待尽快与您会面。

5. There will be an informal reception in the parlor after the meeting.
会议结束后，会客室里将举行一个非正式的招待会。

6. We have sent you the invitation card.
我们已经将邀请函寄给您了。

7. It's my honor to have the opportunity to discuss the matter with you.
能有机会与您讨论此事是我的荣幸。

必背关键单词

1. ***pleasure*** [ˈpleʒə(r)] n. 高兴；乐事
2. ***discuss*** [dɪˈskʌs] v. 讨论
3. ***propose*** [prəˈpəʊz] v. 提议
4. ***minister*** [ˈmɪnɪstə(r)] n. 部长；大臣
5. ***education*** [ˌedʒʊˈkeɪʃn] n. 教育
6. ***host*** [həʊst] n. 主人
7. ***international*** [ˌɪntəˈnæʃnəl] adj. 国际的
8. ***accept*** [əkˈsept] v. 接受

15 取消邀请

这样写就对了

From lindapeal@yahoo.com
To alstonfamily@yahoo.com
Subject Cancelling the Invitation
Date April 2, 2009

Dear Mrs. Alston,

I am very sorry to tell you that I have to cancel the dinner this ***weekend***[1].
I just heard that my mother is seriously ill. My ***husband***[2] and I must go see her at once and we are leaving ***early***[3] tomorrow morning. Therefore, we have to ***recall***[4] our dinner invitation this Saturday, the fourth of April. However, we will ***plan***[5] on a party later on, and you will be cordially invited then.

I am sure that you and Mr. Alston will ***understand***[6] our anxiety, and will ***forgive***[7] this last minute ***change***[8].

Sincerely yours,
Linda Peal

译文

亲爱阿尔斯通夫人：

非常抱歉地通知您，这个周末的晚宴我们不得不取消了。
我刚刚接到我母亲病重的消息，我和我丈夫必须立刻回去探病，我们明天一大早就要出发。所以我们不得不取消本周六，也就是4月4日的晚宴邀请。不过我们打算之后再举办一次聚会，到时我们一定会诚挚地邀请您。

我相信您和阿尔斯通先生一定能理解我们焦急的心情，也一定会原谅我们最后时刻的计划变更。

琳达 · 皮尔 谨上

语法重点解析

1 解析重点1 cancel

表达“取消”可以说 cancel, recall, call off 等，请看以下例句：

That is why we decide to cancel the discussion.（这就是我们决定取消讨论的原因。）

We have to recall the reception for the coming hurricane.（我们不得不因即将到来的飓风取消招待会。）

He phoned me and called the appointment off.（他打电话给我并取消了这次预约。）

2 解析重点2 We are leaving early tomorrow morning.

这句话的意思是“我们明天一大早就要离开”。这个句型其实是用现在进行时表达将要发生的事。所以这个句子相当于 We are going to leave early tomorrow morning. 用法类似的词还有：arrive, come, go, get, have, leave 等，请看以下例句：

We are arriving punctually at four o'clock. 我们将于4点钟准时到达。

Don't rush me. I am coming right now! 别催了，我马上就来了！

1. I am sorry to have to recall the party.
很抱歉，我不得不取消这次的聚会。

2. I feel terribly sorry to tell you that the dinner has been cancelled.
非常抱歉地告诉您晚宴已经取消了。

3. My sister had a car *accident* [9] and I have to look after her in the hospital.
我妹妹出了车祸，我得在医院照顾她。

4. You are cordially invited to our party on next Saturday.
我们诚挚地邀请您参加下周六的聚会。

5. We are leaving tomorrow because of my father's sudden illness.
由于我父亲突患疾病，我们明天就要离开。

6. I express my great regret for this change.
对于此次变动我感到非常抱歉。

1. ***weekend*** [ˌwiːkˈend] n. 周末
2. ***husband*** [ˈhʌzbənd] n. 丈夫
3. ***early*** [ˈɜːlɪ] adv. 早；先前
4. ***recall*** [rɪˈkɔːl] v. 回忆；召回；取消
5. ***plan*** [plæn] v. 计划
6. ***understand*** [ˌʌndəˈstænd] v. 理解
7. ***forgive*** [fəˈɡɪv] v. 原谅
8. ***change*** [tʃeɪndʒ] n. & v. 变化；改变
9. ***accident*** [ˈæksɪdənt] n. 意外；事故

Unit 6 通知 Notice

01 搬迁通知

From tonyt@mar.com　　Date July 30, 2008
To sirs_mar@yahoo.com
Subject Notice: We're Moving

Dear ***Customers***[1],

We are pleased to ***announce***[2] that our Marketing Department will ***move***[3] to Seaside ***Mansion***[4], Room 1704 at 36 Leo ***Street***[5] from July 8, 2008.
Our telephone number ***remains***[6] ***unchanged***[7] and mail should continue to be addressed to the Post Office Box No.31.
Each staff member of our company takes this opportunity to solicit your continued support and attention.

Yours faithfully,
Tony Tsou
Marketing Manager

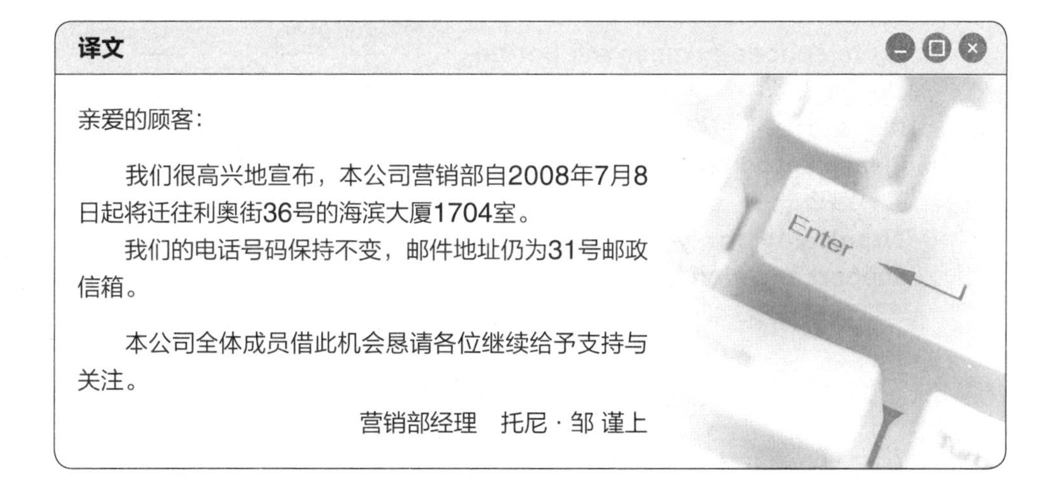
译文

亲爱的顾客：

我们很高兴地宣布，本公司营销部自2008年7月8日起将迁往利奥街36号的海滨大厦1704室。

我们的电话号码保持不变，邮件地址仍为31号邮政信箱。

本公司全体成员借此机会恳请各位继续给予支持与关注。

营销部经理　托尼·邹 谨上

语法重点解析

1 解析重点1 We are pleased to announce that...

We are pleased to announce that... 意思是“我们很高兴地宣布……”。此句型一般作为宣布某事时的开场白使用。宣布某事还可以说 We are happy to announce that... / We are glad to declare that... / We are delighted to make the announcement that...等。请看以下例句：

We are pleased to announce that we have started a business as a printer since this day.（我们很高兴地宣布，即日起我们将开辟印刷业务。）

2 解析重点2 solicit

这个单词在日常英语中很少用到，在书信中，一般表示“恳求；请求”。但是请注意，solicit 还有“拉客；诱惑”的意思，此时它就是贬义词，在一般场合下要谨慎使用。solicit 作“恳求；请求”解时，相当于 ask for, beg 等，请看以下例句：

May I solicit your advice on a matter of some importance?（我可以向您请教一件重要的事吗？）

We ask for the cooperation of all concerned.（我们请求一切相关的合作。）

I beg of you to keep the matter secret.（我请求你对此事保密。）

1. **We are going to move.**
   我们要搬家了。
2. **Please remember our new address.**
   请记下我们的新地址。
3. **My E-mail address *stays*[8] the same.**
   我的电子邮件地址没变。
4. **My telephone number will not be changed.**
   我的电话号码将不会有变动。
5. **Please keep in touch.**
   请保持联络。
6. **Thank you for your continued support and kind help.**
   感谢您继续予以支持和友好的帮助。

必背关键单词

1. ***customer*** [ˈkʌstəmə(r)] n. 顾客；客户
2. ***announce*** [əˈnaʊns] v. 宣布；通告
3. ***move*** [muːv] v. 移动；搬迁
4. ***mansion*** [ˈmænʃn] n. 大厦
5. ***street*** [striːt] n. 街道
6. ***remain*** [rɪˈmeɪn] v. 保持
7. ***unchanged*** [ʌnˈtʃeɪndʒd] adj. 无变化的；未改变的
8. ***stay*** [steɪ] v. 停留；保持

02 电话号码变更通知

这样写就对了

From lincolnburrows@noc.com Date July 10, 2008
To kidney@yahoo.com
Subject Notice: I'm Changing My Phone Number

Dear Mr. Kidney,

I am ***ready***[1] to tell you about my latest change.
My ***office***[2] phone ***number***[3] has changed from 2555-3412 to 2555-7623. You could also ***fax***[4] me at this ***new***[5] number.
Please contact with me at my new number from now on.

I look forward to ***talking***[6] with you as soon as you can.

Yours faithfully,
Lincoln Burrows

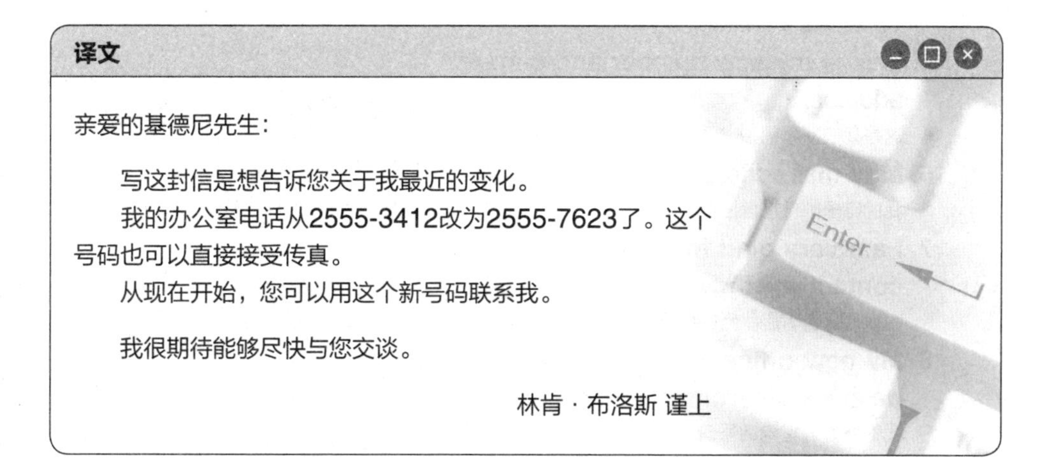

译文

亲爱的基德尼先生：

写这封信是想告诉您关于我最近的变化。
我的办公室电话从2555-3412改为2555-7623了。这个号码也可以直接接受传真。
从现在开始，您可以用这个新号码联系我。

我很期待能够尽快与您交谈。

林肯 · 布洛斯 谨上

语法重点解析

1 解析重点1 latest

latest 在此语意是“最近的；最新的”。另外要注意的是 latest 是形容词 late 的最高级，还可以表示“最后的；最迟的”。因此，一定要根据上下文来判断它的意思。请对照以下例句：

We enclose a copy of our latest price list.（随函寄出我方最新价格表一份。）

Who got up latest this morning?（今天早上谁最晚起床？）

2 解析重点2 from now on

表示“从……开始”可以使用短语 from...on 来表示。如果是“从现在开始”就是 from now on；“从明天开始”就是 from tomorrow on；“从那时开始”就是 from then on。在此类变更通知中，一般都会提到从何时开始的时间点，因此经常使用这个短语。请看以下例句：

We must study hard from now on.（我们必须从现在起努力学习。）

Her music career began from then on.（她的音乐事业是从那时开始的。）

1. I am going to change my phone number.
我要换电话号码了。

2. Please *save*[7] my new cell phone number.
请保存我的新手机号。

3. You can fax me at this new number.
你可以用这个新号码传真给我。

4. I am looking forward to talking with you later.
我期待稍后与您交谈。

5. This is my new number and e-mail address.
这是我的新号码和电子邮件地址。

6. My e-mail address stays unchanged.
我的电子邮件地址没变。

7. I am very glad to tell you my new contact *method*[8].
我很高兴告诉您我新的联系方式。

8. My new office phone number will be available from tomorrow on.
我的新办公室电话号码从明天开始启用。

必背关键单词

1. ***ready*** [ˈredɪ] adj. 准备好的
2. ***office*** [ˈɒfɪs] n. 办公室
3. ***number*** [ˈnʌmbə(r)] n. 号码
4. ***fax*** [fæks] v. 传真
5. ***new*** [njuː] adj. 新的
6. ***talk*** [tɔːk] v. 谈话；会谈
7. ***save*** [seɪv] v. 保存
8. ***method*** [ˈmeθəd] n. 方式；方法

03 职位变更通知

From marksmith@ind.com　Date April 5, 2008
To colepaul@yahoo.com
Subject Notice of Replacement

Dear Mr. Cole,

I would like to ***introduce***[1] you Ms. Sarah Brown. She is ***replacing***[2] me as the ***superintendent***[3] for the ***Import***[4] Department as of April 10.
I am greatly ***grateful***[5] for all your support during my ***tenure***[6]. I hope that you will ***extend***[7] to Ms. Sarah Brown the same ***kindness***[8] and assistance you have ever shown me in the past years.

Thank you again for your long-standing support.

Yours faithfully,
Mark Smith

译文

亲爱的科尔先生：

我想把莎拉·布朗女士介绍给您认识。她将于4月10日开始接任我进口部主管一职。
对于您在我任职期间的支持，我要表示由衷地感谢。我希望您以后也能像过去支持我一样给予莎拉·布朗女士同样的照顾与协助。

再次感谢您长久以来的支持。

马克·史密斯 谨上

语法重点解析

1 解析重点1 **She is replacing me.**

She is replacing me 的意思是“她将接替我的职位”，而不是“她正在接替我的职位”，这是用现在进行时表示将来的用法。同样的表达还可以说成 She is succeeding me 或 She is substituting for me。要注意的是 substitute 一般表示的是因住院或休产假等原因造成工作的暂时变更。类似的表达还有:

She will substitute for me during this time.（这段时间她将接替我的职位。）

She will fill in for me while I am away.（我不在的时候将由她来代理我的职务。）

2 解析重点2 **as of**

as of 这个短语的意思是“从……时起；到……时候为止”。与它同义的短语还有 as from，不过 as from 一般是作为较正式的公文用语使用。请看以下例句:

He was to be Acting Dean as of July.（自7月起，他将担任代理院长。）

Article.59 shall become effective as from the date of promulgation.（条例第五十九条自颁布之日起施行。）

1. **I would like to introduce Miss Chen to you.**
   容我向您介绍一下陈小姐。
2. **He is replacing my *current*[9] position.**
   他将接替我现任的职位。
3. **She will substitute for me while I am away.**
   我不在的时候将由她来接替我。
4. **I will be substituting for her while she is in hospital.**
   在她住院期间将由我来接替她的工作。
5. **Please give her the same care and support.**
   请给予她同样的关照和支持。
6. **Thank you for your support and help during my tenure.**
   感谢您在我任职期间给予的支持和帮助。
7. **I am very pleased to introduce Mr. Lee to you.**
   我非常乐意把李先生介绍给您。
8. **Miss Lin will fill in for me during the time.**
   这段时间将由林小姐来代理我的工作。

必背关键单词  

1. ***introduce*** [ˌɪntrəˈdjuːs] v. 介绍
2. ***replace*** [rɪˈpleɪs] v. 接替；取代
3. ***superintendent*** [ˌsuːpərɪnˈtendənt] n. 主管；负责人
4. ***import*** [ˈɪmpɔːt] n. & v. 进口
5. ***grateful*** [ˈgreɪtfl] adj. 感激的；感谢的
6. ***tenure*** [ˈtenjə(r)] n. 任期
7. ***extend*** [ɪkˈstend] v. 延伸；给予
8. ***kindness*** [ˈkaɪndnəs] n. 仁慈；好意
9. ***current*** [ˈkʌrənt] adj. 目前的

04 暂停营业通知

这样写就对了

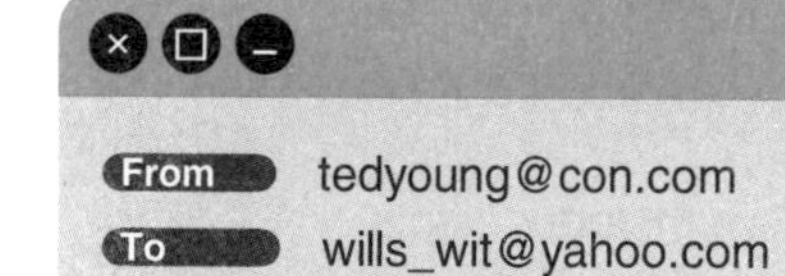

From tedyoung@con.com
To wills_wit@yahoo.com
Subject Notice of Temporary Closure
Date August 5, 2008

Dear Mr. Wills,

I am writing to inform you that our ***store***[1] will be ***closed***[2] ***temporarily***[3] from August 8 to 15, due to ***renovations***[4] to the ***interior***[5] of the building.
We are planning to ***reopen***[6] on August 16, and we will be sure to inform you if there are any changes then.

We are ***deeply***[7] sorry if it may ***cause***[8] you any inconvenience.

Best wishes,
Ted Young

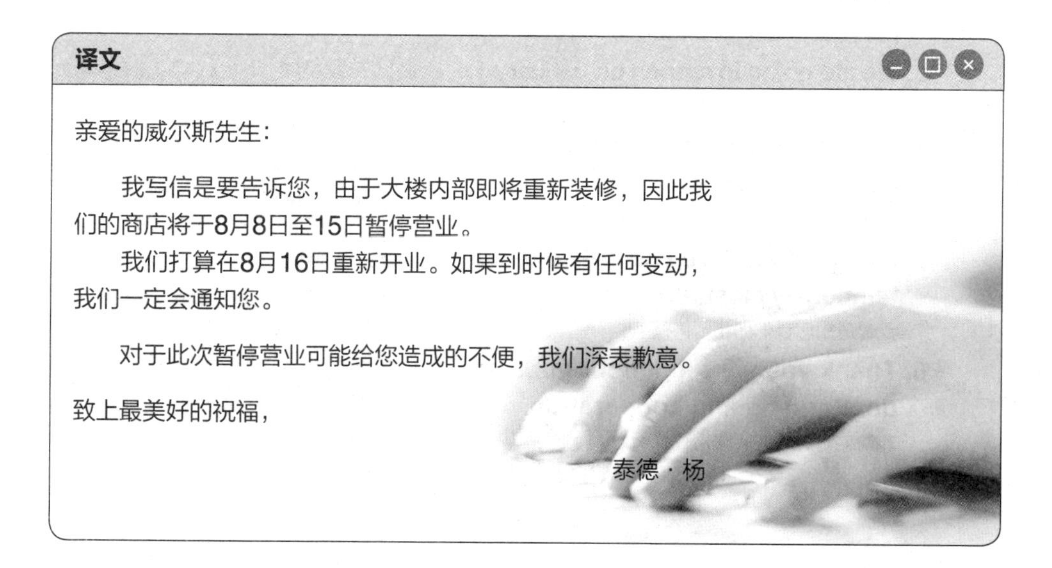

译文

亲爱的威尔斯先生：

我写信是要告诉您，由于大楼内部即将重新装修，因此我们的商店将于8月8日至15日暂停营业。

我们打算在8月16日重新开业。如果到时候有任何变动，我们一定会通知您。

对于此次暂停营业可能给您造成的不便，我们深表歉意。

致上最美好的祝福，

泰德·杨

语法重点解析

1 解析重点1 **due to renovations to the interior of the building**

这句话的意思是"由于大楼内即将进行装修翻新"。renovation是"翻新；整修"的意思；interior 是指"内部的"。内部装潢、重新翻修是商店暂停营业的普遍原因，所以请注意此表达方法。在这里还要注意以下几个表达方法："店面维护"为 for maintenance；"因盘点货存而停业"为 for inventory；如果因为"员工旅行"而停业，则一般用 on holiday。例如：The store will be on holiday from October 1 to 7.（本店将于10月1日至7日因员工旅行而暂停营业。）

2 解析重点2 **We are deeply sorry if it may cause you any inconvenience.**

这句话的意思是"对于可能造成的不便，我们深表歉意"。此句中的 it 指的是商店将 temporary closure"暂停营业"这一事实。副词 deeply 意为"深深地"，注意这里的"深"是表示抱歉程度的"深"，而不是物理意义的深浅。请看以下例句：

The earthquake shook the heart of the Chinese deeply.（这次地震深深地震撼了中国人民的心。）

高频例句

1. **Our shop will be closed from May 1 to 8.**
   本店将于5月1日至8日暂停营业。
2. **We have to close temporarily due to *reconditioning*[9] the machines.**
   由于要修理机器，我们不得不暂停营业。
3. **We are going to reopen on January 1.**
   我们将于1月1日重新开业。
4. **We are sorry for any inconvenience this may cause you.**
   如果给您造成不便，我们深感歉意。
5. **I am going to give you notice of our temporary closure.**
   我要通知您我们即将暂停营业。
6. **Thank you very much for your understanding and cooperation.**
   非常感谢您的理解和合作。

必背关键单词

1. ***store*** [stɔ:(r)] n. 商店
2. ***close*** [kləuz] v. 关闭
3. ***temporarily*** ['temprərəlɪ] adv. 暂时地
4. ***renovation*** [ˌrenə'veɪʃn] n. 革新；翻新；整修
5. ***interior*** [ɪn'tɪərɪə(r)] adj. 内部的；在内的
6. ***reopen*** [ˌri:'əupən] v. 重开；再开
7. ***deeply*** ['di:plɪ] adv. 深深地
8. ***cause*** [kɔ:z] v. 造成；引起
9. ***recondition*** [ˌri:kən'dɪʃn] v. 修理；修复；修补翻新

05 开业通知

这样写就对了

From jmproduct@gzt.com　　Date August 21, 2008
To sirsmadam@yahoo.com
Subject Notice of Opening

Dear Sir or Madam,

We are pleased to inform you that on account of ***rapid***[1] ***increase***[2] in the ***volume***[3] of our trade, we ***decide***[4] to open another sales office for our products here in New York on August 28. We ***employ***[5] a staff of ***consultants***[6] and a well-trained service group, which makes ***routine***[7] checks on all ***equipment***[8] purchased from us.
We would be delighted if you would take full advantage of our services and favorable shopping environment. We fully ***guarantee***[9] the quality of our products.

Yours faithfully,
Jim Green

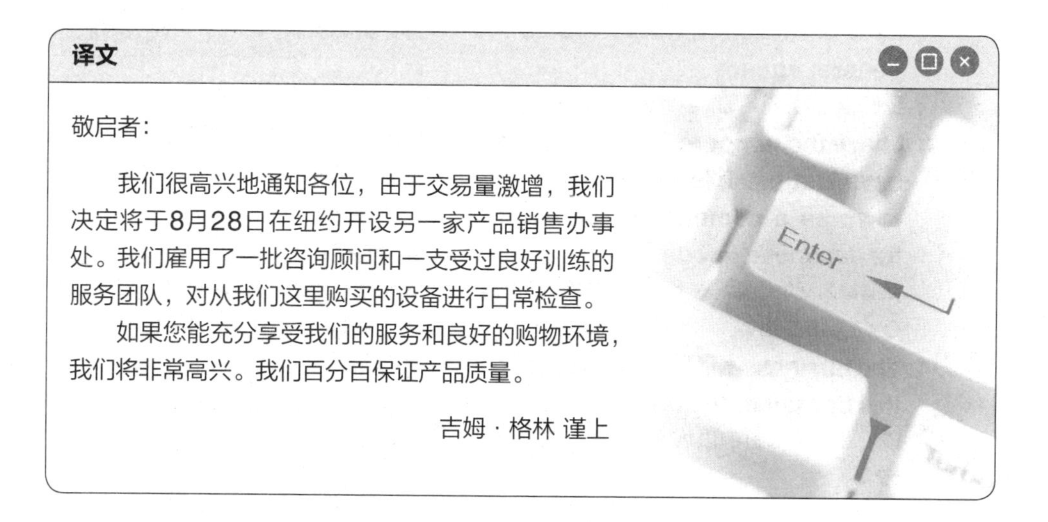

译文

敬启者：

我们很高兴地通知各位，由于交易量激增，我们决定将于8月28日在纽约开设另一家产品销售办事处。我们雇用了一批咨询顾问和一支受过良好训练的服务团队，对从我们这里购买的设备进行日常检查。

如果您能充分享受我们的服务和良好的购物环境，我们将非常高兴。我们百分百保证产品质量。

吉姆 · 格林 谨上

语法重点解析

1 解析重点1 on account of

on account of 是表达原因的短语，意思是“为了……的缘故；因为；由于”。表达原因的短语还有 owing to 等。请看以下例句：

The price dropped greatly on account of large offerings from other sources.（由于其他渠道的大量供货，价格严重下跌。）

Owing to their intransigent attitude, we were unable to reach an agreement.（由于他们态度强硬，我们无法达成协议。）

2 解析重点2 We would be delighted if you would take full advantage of our services and favorable shopping environment.

这句话的意思是“如果您能充分享受我们的服务和良好的购物环境，我们将不胜感激”。新店面开张免不了邀请老顾客前来参加开业庆典，而汉语中开业通知的用词十分的讲究，既要表现诚恳，又要表达清晰。英文开业通知也不例外。随着对外贸易的不断扩大，英文邀请通知也被广泛地使用，其用词习惯和语法表达也十分的考究。

1. **It is our *intention*[10] to confine ourselves to the wholesale business of silk goods.**
   本公司专门经营各种丝制品的批发业务。
2. **We are informing you that we have established ourselves as the general agency.**
   我们已经成为总代理，特此告知。
3. **I have the honor to inform you that I have just established myself in this town as a commission merchant for Japanese goods.**
   我非常荣幸地通知您，我刚刚在本镇开了一家日本商品代销店。
4. **The business will be carried on from this day under the firm of W&G.**
   该项业务从即日起以W&G公司的名义继续经营。

必背关键单词

1. ***rapid*** ['ræpɪd] adj. 快速的
2. ***increase*** [ɪn'kriːs] n. 增长；增加
3. ***volume*** ['vɒljuːm] n. 体积；量
4. ***decide*** [dɪ'saɪd] v. 决定
5. ***employ*** [ɪm'plɔɪ] v. 雇用
6. ***consultant*** [kən'sʌltənt] n. 顾问
7. ***routine*** [ruː'tiːn] adj. 常规的；例行的
8. ***equipment*** [ɪ'kwɪpmənt] n. 设备
9. ***guarantee*** [ˌɡærən'tiː] v. 保证
10. ***intention*** [ɪn'tenʃn] n. 目的；意向

06 营业时间变更通知

这样写就对了

From bjbc@nec.com
To dearclients@bj.com
Subject Notice of Changing Business Hours
Date October 27, 2008

Dear ***Clients***[1],

We are pleased to make an announcement ***hereby***[2].
Effective November 3, 2008, our new ***operating***[3] hours will be from 9:00 a.m. to 9:00 p.m., Monday to Friday.
We sincerely hope that this change of office hours will ***allow***[4] us to ***provide***[5] the ***fullest***[6] and most ***considerate***[7] service for you.

Yours faithfully,
Collins Book Company

译文

尊敬的各位客户：

我们在此很高兴地宣布一项事宜。

自2008年11月3日起，本公司的营业时间将变更为星期一到星期五的上午9点至晚上9点。

我们诚挚地希望此次营业时间的变更能使我们为各位提供最全面、最周到的服务。

柯林斯图书公司 谨上

语法重点解析

1 解析重点1 effective

effective 的意思是“有影响的；有效的”，在此语境下则是指“有效的；生效的”，一般在告示或通知等英文书信中常会见到。经常使用的词组有 become effective，指“开始生效”，该用法经常在法律法规中使用。请看以下例句：

When does the new system become effective?（新制度何时生效？）

These Regulations hereof become effective as of January 1, 2008.（这些条例自2008年1月1日起生效。）

2 解析重点2 We sincerely hope that this change of office hours will allow us to provide the fullest and most considerate service for you.

这句话的意思是“我们诚挚地希望此次营业时间的变更能使我们为各位提供最全面、最周到的服务”。句子看上去非常礼貌客气，这正体现了英文书信写作原则中的 Consideration（体贴）原则，即在英文书信写作过程中，写信者应设身处地为对方设想，采取所谓的 You-Attitude。此句中的 allow us “允许我们”正是 You-Attitude 的体现。

1. **We are very happy to inform you of the change of our office hours.**
   非常高兴通知您关于我们营业时间变更的信息。
2. **Please note our new business hours.**
   请注意我们新的营业时间。
3. **Our new operating hours are *effective*[8] April 4, 2008.**
   我们新的营业时间自2008年4月4日起生效。
4. **Our business hours are 9:30 to 20:30.**
   我们的营业时间是早上9:30到晚上8:30。
5. **Please phone me during business hours.** 请在营业时间给我打电话。
6. **What are the operating hours for your buffet?**
   请问你们自助餐厅的营业时间是什么时候？

必背关键单词

1. ***client*** [ˈklaɪənt] n. 顾客；客户
2. ***hereby*** [ˌhɪəˈbaɪ] adv. 在此；特此
3. ***operate*** [ˈɒpəreɪt] v. 运转；经营
4. ***allow*** [əˈlaʊ] v. 允许
5. ***provide*** [prəˈvaɪd] v. 提供
6. ***full*** [fʊl] adj. 全面的
7. ***considerate*** [kənˈsɪdərət] adj. 体贴的；周到的
8. ***effective*** [ɪˈfektɪv] adj. 有影响的；有效的

07 缴费通知

这样写就对了

From waterco@rwc.com
To resident@huati.com
Subject Bill Notice
Date August 2, 2008

Dear Resident,

This is to ***notify***[1] you that you have to pay RMB 74 water rate in July.
This month you have made use of 20 ton water, and each ton costs RMB 3.70, which reaches the ***total***[2] amount of RMB 74.
In order to ***facilitate***[3] water payment, reduce the ***troubles***[4] of paying bills and improve the ***efficiency***[5], I sincerely ***recommend***[6] users to pay water ***charges***[7] through bank transfer.

Thank you for your understanding and support!

Sincerely yours,
Water Corporation

译文

亲爱的居民:

通知您需缴纳七月份水费74元。

本月您共使用20吨水，每吨单价为3.70元，总计金额为74元。

为了使水费缴纳更为便利，减少缴费麻烦，并提高效率，我们诚挚建议用户通过银行转账的方式缴纳水费。

感谢您的理解和支持!

自来水公司 谨上

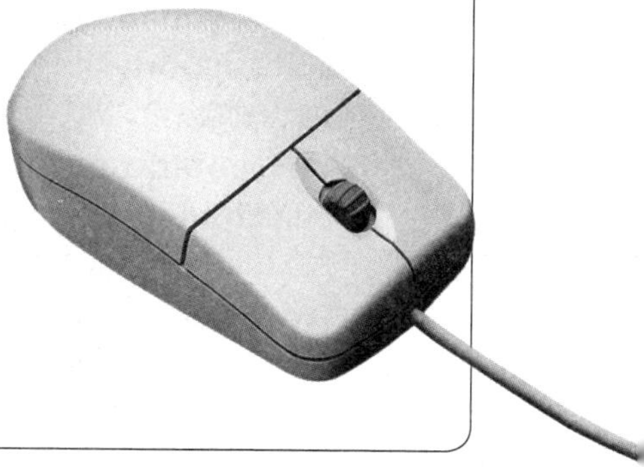

语法重点解析

1 解析重点1 rate

rate 有“比率；等级；价格；费用”的意思，在此语境下是指“费用”。表达“费用”的词还有 charge, fee, expense, cost, fare, tip等。要注意的是，charge 通常指收取的费用，如 stand charge“摊位费”；fee 通常指一些机构收取的费用，如 tuition fee“学费；会费”；expense 和 cost 则指一般的花费；fare 一般指交通费用；tip 特指小费。

2 解析重点2 made use of

短语 made use of 是“利用；使用”的意思。表达“利用；使用”还可以用 use, utilize 等单词。请看以下例句：

She made the best use of her opportunities.（她充分利用了一切机会。）

May I use your knife for a while?（我可以借用一会你的刀吗？）

Scientists are trying to find more efficient way of utilizing solar energy.（科学家们正在寻找利用太阳能的更有效的方法。）

1. **This is to notify you to pay your utility bill.**
   这是通知您需要缴纳水电费。
2. **You used 18 tons of water in total this month.**
   本月您一共使用了18吨的水。
3. **You have to pay RMB 92.5 water *rate*[8].**
   您需缴纳水费92.5元。
4. **Please take the *initiative*[9] to pay the utility bill.**
   请主动缴纳水电费。
5. **Please actively cover all payment.**
   请主动缴纳所有费用。
6. **Thank you for your understanding and support.**
   感谢您的理解和支持。
7. **You could pay the fee through bank transfer for convenience.**
   方便起见，您可以通过银行转账缴费。
8. **We sincerely recommend you to pay the charges through bank transfer.**
   我们诚挚地建议您通过银行转账缴费。

必背关键单词

1. ***notify*** [ˈnəʊtɪfaɪ] v. 通知
2. ***total*** [ˈtəʊtl] adj. 总计的；全部的
3. ***facilitate*** [fəˈsɪlɪteɪt] v. 使便利
4. ***trouble*** [ˈtrʌbl] n. 麻烦；困难
5. ***efficiency*** [ɪˈfɪʃnsɪ] n. 效率
6. ***recommend*** [ˌrekəˈmend] v. 推荐；建议；劝告
7. ***charge*** [tʃɑːdʒ] n. 费用
8. ***rate*** [reɪt] n. 比率；等级；价格
9. ***initiative*** [ɪˈnɪʃətɪv] n. 主动性；主动权

08 盘点通知

From oriental@olc.com
To cusup@21c.com
Subject Notice of Stocktaking
Date December 15, 2008

Dear customers and suppliers,

Thank you very much for your kind support and cooperation!
Please be informed that our ***annual***[1] ***stocktaking***[2] will be held from December 28 to 31, 2008. During the period, all of our delivering and receiving operations will be stopped. Therefore, we would not arrange any goods ***delivery***[3] to all of your ***warehouses***[4] ***except***[5] special requests by your Purchasing Department before December 22, 2008. We will **resume** the normal delivering and receiving ***operations***[6] from January 2, 2009.

We also ***apologize***[7] for any inconvenience caused.

Yours faithfully,
Oriental Logistics Co.

译文

亲爱的客户及供应商：

非常感谢各位对我们的支持与配合！
谨此通知，本公司将于2008年12月28日至12月31日进行年度存货盘点，在此期间我们将会暂停所有收发货业务。因此若有特殊情况，请贵采购部预先于本月22日前告知，本公司会另行安排。本公司将从在2009年1月2日恢复正常营运。

不便之处敬请谅解。

东方物流公司 谨上

语法重点解析

1 解析重点1 annual stocktaking

annual stocktaking 的意思是“年度存货盘点”，也可以说 annual inventory。annual 意为“每年的；年度的”；stocktaking 意为“存货盘点”。所谓盘点，是指定期或临时对库存商品的实际数量进行清查、清点的作业，以便掌握货物的流动情况，对仓库现有物品的实际数量与库存账上记录的数量相核对，以便准确地掌握库存数量。英文表达“盘点”还可用：make an inventory of, check, draw up an inventory, take stock 等。

2 解析重点2 resume

resume 这个单词有名词和动词两种词性，意思分别为“简历；履历”和“再继续；重新开始”。在此语境下做动词使用。请对照以下例句：

Please send a detailed resume to our company.（请将详细的简历寄至我们公司。）

The failure of the strike enabled the company to resume normal bus services.（罢工的失败使公司恢复了正常的公交车运营。）

1. **We are going to make an *inventory*[8] from December 25 to 30.**
   我们将于12月25日至30日进行存货盘点。
2. **The warehouse is closed for the annual stocktaking.**
   仓库关闭，以进行年度存货盘点。
3. **Our food store takes stock every week.**
   我们这家食品店每周都盘点存货。
4. **Stocktaking today, business as usual tomorrow.**
   今日盘点，明日照常营业。
5. **I want an inventory of all items in the warehouse.**
   我想要一张仓库所有货品的存货清单。
6. **We make an inventory of all accessories every month.**
   我们每月对各类配件进行盘点。

必背关键单词

1. ***annual*** [ˈænjʊəl] adj. 每年的；年度的
2. ***stocktaking*** [ˈstɒkteɪkɪŋ] n. 存货盘点
3. ***delivery*** [dɪˈlɪvərɪ] n. 递送；交付
4. ***warehouse*** [ˈweəhaʊs] n. 仓库
5. ***except*** [ɪkˈsept] prep. 除了……之外
6. ***operation*** [ˌɒpəˈreɪʃn] n. 运转；经营
7. ***apologize*** [əˈpɒlədʒaɪz] v. 道歉
8. ***inventory*** [ˈɪnvəntrɪ] n. 存货清单

09 求职录用通知

这样写就对了

From xinyi@adv.com　Date November 25, 2008
To mikakitty@yahoo.com
Subject Notice of Employment

Dear Miss Mika,

In view of your ***interview***[1] on November 23, it is a great pleasure to inform you that you have been approved by the ***Board***[2] of Directors, and the Personnel Department has decided to appoint you as the ***secretary***[3] to ***General***[4] Manager, commencing from December 1, 2008.
We will send you a ***Notification***[5] of an Offer later on. In addition, we will arrange a time to sign the ***employment***[6] contract with you. If you have any questions, please do not ***hesitate***[7] to contact me.

We are looking forward to having you here with us!

Yours sincerely,
Novelty Co. Ltd.

译文

亲爱的米卡小姐：

鉴于您在11月23日的面试，我非常高兴地通知您，您已经通过了董事会的批准，人事部已经决定自2008年12月1日起，聘用您作为总经理秘书。

稍后我们会将录用通知寄送给您。另外，我们还会安排时间与您签订雇用合约。如果您还有任何问题，尽请与我联系。

期待您的加入！

新意有限公司 谨上

语法重点解析

1 解析重点1 in view of

短语 in view of 的意思是“鉴于；考虑到”，正式书信中常用到此短语。表达相同意思的短语还有 in consideration of，in the light of 等。请看以下例句：

In view of the facts, it seems useless to continue.（鉴于这些事实，继续下去似乎是无用的。）

In consideration of our friendship, I forgave him.（考虑到我们的友谊，我原谅了他。）

In the light of these changes, we must revise our plan.（鉴于这些变化，我们必须修改我们的计划。）

2 解析重点2 appoint

appoint 的意思是“任命；委派；指定”。职场上表达“任命；任免”的单词还有 designate, nominate 等。请看以下例句：

He was appointed as the sales manager.（他被任命为业务经理。）

The chairman has designated her as his successor.（主席已指定她作为他的继任者。）

The board nominated him as the new director.（董事会指定他为新任董事。）

1. I am pleased to inform you that you are recruited.

我非常高兴地通知您，您被录用了。

2. It's my pleasure to tell you that you have passed the interview.

我非常荣幸地通知您，您通过了面试。

3. The Personnel Department has decided to *appoint* [8] you as sales assistant.

人事部已经决定任命你为销售助理。

4. We are ready to sign the contract with you.

我们正准备与您签合约。

5. We will arrange a time to sign the contract of labor with you.

我们将安排时间与您签订劳动合同。

6. Welcome to *join* [9] our team!

欢迎加入我们的团队！

必背关键单词

1. ***interview*** [ˈɪntəvjuː] n. 面试；面谈
2. ***board*** [bɔːd] n. 板；董事会
3. ***secretary*** [ˈsekrətrɪ] n. 秘书
4. ***general*** [ˈdʒenrəl] adj. 总的；一般的；普通的
5. ***notification*** [ˌnəʊtɪfɪˈkeɪʃn] n. 通知；告示
6. ***employment*** [ɪmˈplɔɪmənt] n. 工作
7. ***hesitate*** [ˈhezɪteɪt] v. 犹豫；迟疑
8. ***appoint*** [əˈpɔɪnt] v. 任命；指定
9. ***join*** [dʒɔɪn] v. 加入

10 节假日通知

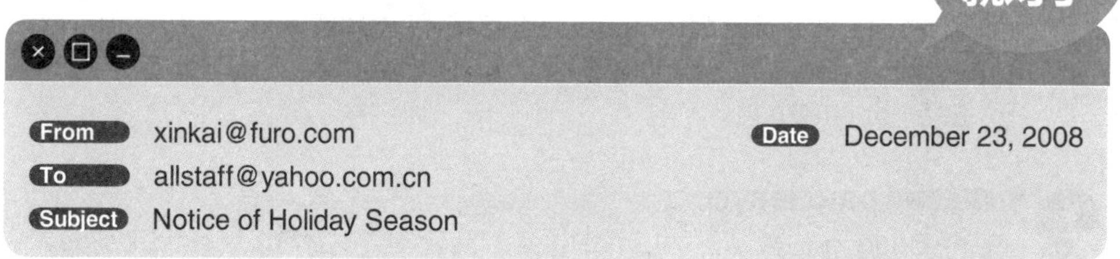

From xinkai@furo.com
Date December 23, 2008
To allstaff@yahoo.com.cn
Subject Notice of Holiday Season

Dear all,

I am very pleased to announce the good ***news*** [1]!

To express our ***appreciation*** [2] for your hard work this year and ***enable*** [3] everyone to ***spend*** [4] the holidays with friends and family, we have decided to close the office from December 25 to January 2, **inclusive**. All ***personnel*** [5] will be given **paid *leave*** [6] during this period.

Wishing all of you a happy and ***healthy*** [7] holiday season!

Sincerely yours,
Administrative Department

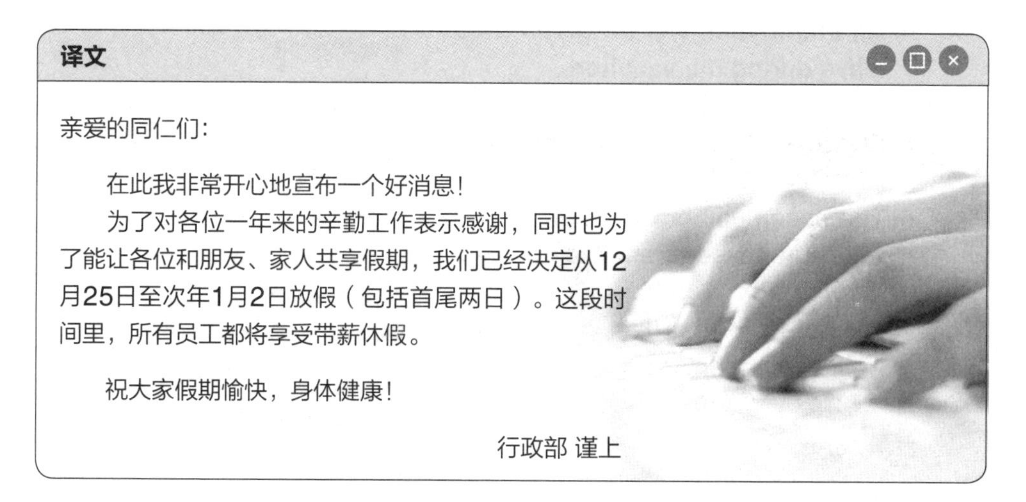

译文

亲爱的同仁们：

在此我非常开心地宣布一个好消息！

为了对各位一年来的辛勤工作表示感谢，同时也为了能让各位和朋友、家人共享假期，我们已经决定从12月25日至次年1月2日放假（包括首尾两日）。这段时间里，所有员工都将享受带薪休假。

祝大家假期愉快，身体健康！

行政部 谨上

语法重点解析

1 解析重点1 inclusive

形容词 inclusive 的意思是“包含的；包括……在内的”，表达此意思的短语有 be inclusive of。请对照以下例句：

A calendar year is from January 1 to December 31 inclusive.（日历的一年是从1月1日起至12月31日，包含首尾两日。）

The monthly rent is US$200 inclusive of all utility fees.（月租金总共两百美元，包括一切水电费在内。）

2 解析重点2 paid leave

paid leave 的意思是“付工资的休假，带薪假期”，也可以用 paid holiday 表示。表达假期的短语还包括 maternity leave with full pay / paid maternity leave（带薪产假），private affair leave（事假），unpaid leave（无薪假），sick leave（病假）， maternity leave（产假），paternity leave（陪产假）等。

高频例句

1. **We are happy to announce that all staff will be given two additional vacation days.**
   我们非常高兴地向大家宣布，全体员工均将享受到两天额外的假期。
2. **The total number of vacation days will still be determined by each employee's length of service.**
   休假总天数将仍按照每位员工的工作年限而定。
3. **The vacation is to express our appreciation for your hard work.**
   此次休假是为了感谢各位的辛勤工作。
4. **All employees will be given paid leave during the vacation.**
   所有员工都能够享受带薪假期。
5. **Please enjoy your holiday *seasons*[8].**
   请尽情享受你们的假期。
6. **The base number of vacation days will be increased from 7 to 10 days.**
   基本假期将从7天增加至10天。
7. **We wish you all a wonderful holiday season.**
   祝福大家都有个美好的假期。

必背关键单词

1. ***news*** [nju:z] n. 新闻；消息
2. ***appreciation*** [əˌpri:ʃɪˈeɪʃn] n. 感谢；感激
3. ***enable*** [ɪˈneɪbl] v. 使能够
4. ***spend*** [spend] v. 度过
5. ***personnel*** [ˌpɜ:səˈnel] n. 人员；职员
6. ***leave*** [li:v] n. 放假；休假
7. ***healthy*** [ˈhelθɪ] adj. 健康的
8. ***season*** [ˈsi:zn] n. 季节

11 裁员通知

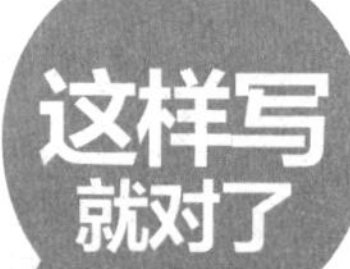

Dear Ben,

We had been hoping that during this difficult ***period***[1] of ***reorganization***[2] we could keep all of our ***employees***[3] with the company. Unfortunately, this is not the ***case***[4].
It is with regret, therefore, that we have to inform you that we will be *unable*[5] to *utilize*[6] your services anymore. We have been pleased with the ***qualities***[7] you have ***exhibited***[8] during your tenure of employment, and will be sorry to lose you.

We wish you a promising future!

Yours truly,
ABC Co.

译文

亲爱的本：

我们一直希望能够在此次重组的困难时期留下公司的全体员工，不幸的是，事实并不是这样。

因此，公司不得不遗憾地通知你，我们无法再继续雇用你。我们一直很满意你在受聘期间展现出的才能，并为失去你这样的员工而感到遗憾。

祝你前程似锦！

ABC公司 谨上

语法重点解析

1 解析重点1 We had been hoping that...

We had been hoping that...是“我们一直希望……”的意思。had been doing 是过去完成进行时态，表示过去正在进行的动作或状态，一直持续到过去的某个时间点。请看以下例句：

Everybody knew what he had been doing all those years.（大家都知道那些年他在做什么。）

The child had been missing for a week.（那个孩子已经失踪一个星期了。）

2 解析重点2 It is with regret, therefore, that we have to inform you that we will be unable to utilize your services anymore.

这句话的意思是“因此，公司不得不遗憾地通知您，我们无法再继续雇用您”，表达得非常委婉，完全没有生硬突兀之感。with regret，have to，unable 都含蓄地表达了公司辞退员工是出于无奈。这句话如果说成 We must inform you that you are fired. 则会显得太过直接、不近人情。所以在辞退员工时一定要顾及到对方的感受，在措辞上力求委婉、诚恳。

1. **I am afraid that I have to tell you a bad news.**
   我恐怕不得不告诉你一个坏消息。
2. **We regret to inform you that your employment with the *firm*[9] shall be terminated.**
   我们很遗憾地通知您，公司将解除对您的雇用。
3. **I regret having to tell you that your service will have to be terminated.**
   我很遗憾地告诉您，我们将解除对您的雇用。
4. **We are really sorry to see you leave actually.**
   事实上，我们真的不愿看到你离开。
5. **Please arrange for the return of company property in your possession.**
   请安排归还您所使用的公司物品。
6. **Again, we regret that this action is necessary.**
   我们再一次对这一必要举动表示遗憾。

必背关键单词 

1. ***period*** [ˈpɪərɪəd] n. 时期；（一段）时间
2. ***reorganization*** [rɪˌɔːgənaɪˈzeɪʃn] n. 重组；重新安排
3. ***employee*** [ɪmˈplɔɪiː] n. 员工
4. ***case*** [keɪs] n. 情形；情况；案例
5. ***unable*** [ʌnˈeɪbl] adj. 不能的；不会的
6. ***utilize*** [ˈjuːtəlaɪz] v. 利用
7. ***quality*** [ˈkwɒlətɪ] n. 品质；才能
8. ***exhibit*** [ɪgˈzɪbɪt] v. 展现；显示
9. ***firm*** [fɜːm] n. 公司

12 人事变动通知

From adam_smith@idea.com
To colleague_i@idea.com
Subject Notice of Changing Leadership
Date June 24, 2008

Dear Colleagues,

We are very pleased to announce, **effective the same day**, the ***appointment*** [1] of Julia Turner as ***Chief*** [2] Information ***Officer*** [3] of IDEA Company. Ms. Turner **possesses** 12 years of ***experience*** [4] in the information ***technology*** [5] sector and her ***extensive*** [6] knowledge and experience will be an ***invaluable*** [7] ***asset*** [8] to our company.

Please join us in welcoming Ms. Turner to IDEA!

Sincerely yours,
Adam Smith

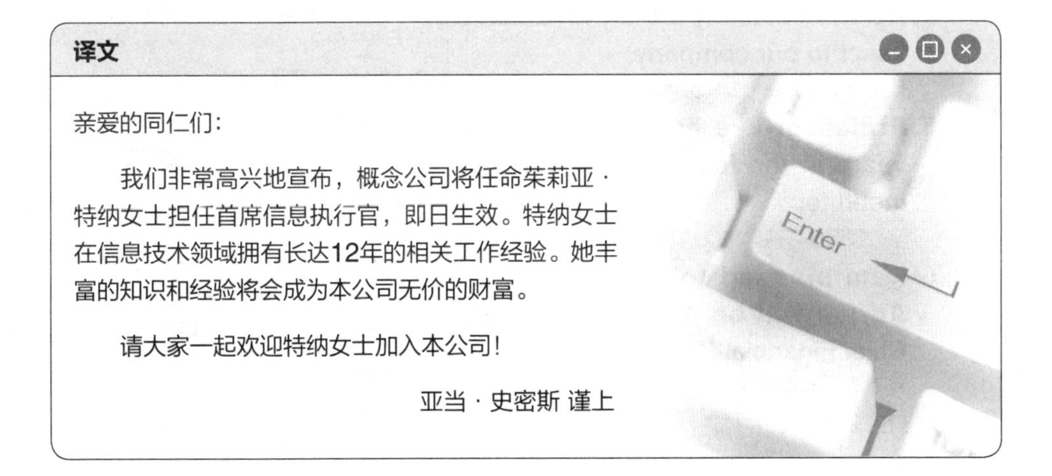

译文

亲爱的同仁们：

我们非常高兴地宣布，概念公司将任命茱莉亚·特纳女士担任首席信息执行官，即日生效。特纳女士在信息技术领域拥有长达12年的相关工作经验。她丰富的知识和经验将会成为本公司无价的财富。

请大家一起欢迎特纳女士加入本公司！

亚当·史密斯 谨上

1 解析重点1 effective the same day

effective the same day 的意思是“当日生效”。effective 意为“有效的；生效的”；the same day 意为“当日”。表达“当日”还可以使用 that very day；表达“立即生效”则可用短语 effective immediately。请看以下例句：

Visa-free status for the four countries is effective immediately.（这四个国家的免签证规定立即生效。）

2 解析重点2 possess

possess 的意思是“拥有；具有；占据”，相当于 have，但是 have 为普通用词，possess 则相对正式。请对照以下例句：

Do you have friends there?（你在那儿有朋友吗？）

All possess something that must be loved.（所有人都有令人喜爱的地方。）

高频例句

1. **It is with great pleasure for me to announce an appointment.**
   我非常荣幸地向大家宣布一个任命通知。
2. **We are pleased to announce the appointment of Mr. Keller as our new general manager.**
   我们很高兴宣布任命凯勒先生为我们的新任总经理。
3. **He has broad knowledge and *rich*[9] experience.**
   他拥有渊博的知识和丰富的经验。
4. **He will certainly be an invaluable asset to our company.**
   他一定会成为我们公司宝贵的财富。
5. **Professor Lee has served for 10 years in the faculty of Commerce Institute.**
   李教授已经在商学院任教10年了。
6. **I am pleased to announce that Adrian will be appointed as our Chief Financial Officer.**
   我很高兴地宣布阿德里安将被任命为我们的首席财务官。

必背关键单词

1. ***appointment*** [əˈpɔɪntmənt] n. 任命；约会
2. ***chief*** [tʃiːf] adj. 主要的；首席的
3. ***officer*** [ˈɒfɪsə(r)] n. 官员；执行官
4. ***experience*** [ɪkˈspɪərɪəns] n. 经验；经历；阅历
5. ***technology*** [tekˈnɒlədʒɪ] n. 技术
6. ***extensive*** [ɪkˈstensɪv] adj. 广泛的；广阔的
7. ***invaluable*** [ɪnˈvæljʊəbl] adj. 无价的
8. ***asset*** [ˈæset] n. 资产；有用的东西
9. ***rich*** [rɪtʃ] adj. 丰富的

13 公司破产通知

From courtsjz@org.com Date August 30, 2008
To employees@shjz.com
Subject Notice of Bankruptcy

All employees:

According to the ***application*** [1] of the ***creditor*** [2], the Court, in accordance with the law, has declared the ***debtor*** [3] to pay off their debts and set up a ***liquidation*** [4] team to take over all assets.
All workers are now required to strictly comply with the law, to protect the business ***property*** [5], everyone shall not illegally deal with business books, ***writs*** [6], materials, seal and license, and **shall not** ***conceal*** [7] or divide the business property.
The Enterprise's legal representative shall not be absent from duty and ***implement*** [8] any acts to ***holdback*** [9] the liquidation before the end of **bankruptcy proceedings**.

People's Court

译文

全体职工：

根据债权人的申请，本院已依法裁定宣告债务人需偿还债务，并成立清算组接管所有资产。

现要求全体职工严格遵守法律规定，保护好企业财产，任何人不得非法处理企业账册、文书、资料、印章和证照，不得隐匿、分摊企业财产。

企业的法定代表人在破产程序结束前不得擅离职守，或实施妨碍破产清算的行为。

人民法院

语法重点解析

1 解析重点1 shall not

在法律英语中，shall not不能解释为“不应该”，而应该翻译为“不得”。所以在法律英语中的 shall 表示“应当”，而不是“应该”或“将要”。请看以下例句：

Employees shall not deal with business books illegally.（员工不得非法处理企业账册。）

Criminal responsibility shall be borne for intentional crimes.（故意犯罪应当负刑事责任。）

2 解析重点2 bankruptcy proceedings

bankruptcy proceedings 就是经济领域里常见的“破产程序”。破产程序是指对资不抵债的企业进行破产处理的司法程序，亦写作 bankruptcy procedure。bankruptcy court（破产法庭）在收到 bankruptcy petition（破产申请）并决定立案后还会发布 bankruptcy notice（破产公告）。proceeding 在这里表示“程序；进程”，如 institute / take / start legal proceedings against...（对……开始 / 着手法律手续），条件成熟后便可进入 hearings and written proceeding（开庭和书面审理程序）。

1. **Government declared this morning that ABC Group was going into *bankruptcy*[10] proceedings.**
   市政府今天上午宣称，ABC集团将进入破产程序。
2. **Will the factories close down because of bankruptcy?**
   工厂是否会因为破产而关闭？
3. **This hastened the bankruptcy of the peasants' handicraft industries.**
   这加速了农民手工业的破产。
4. **All workers are now required to strictly comply with the law.**
   全体职工现在都必须严格遵守法律规定。
5. **The liquidation team will take over all assets.**
   破产小组将会接管所有资产。

必背关键单词 

1. ***application*** [ˌæplɪˈkeɪʃn] n. 申请
2. ***creditor*** [ˈkredɪtə(r)] n. 债权人
3. ***debtor*** [ˈdetə(r)] n. 债务人
4. ***liquidation*** [ˌlɪkwɪˈdeɪʃn] n. 清算
5. ***property*** [ˈprɒpətɪ] n. 财产
6. ***writ*** [rɪt] n. 令状；文书
7. ***conceal*** [kənˈsiːl] v. 隐藏；隐匿
8. ***implement*** [ˈɪmplɪment] v. 实施
9. ***holdback*** [ˈhəʊldbæk] v. 妨害
10. ***bankruptcy*** [ˈbæŋkrʌptsɪ] n. 破产

14 公司停业通知

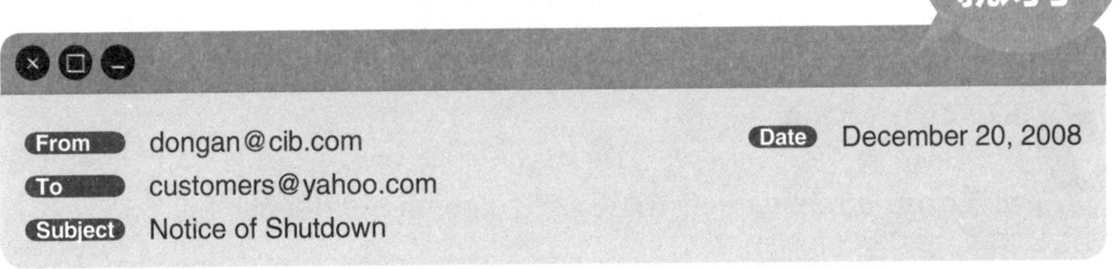

Dear ***Customers***[1],

We are going to close down for some time.
Our year-end holidays have been ***scheduled***[2] as ***follows***[3]:
The office will ***close***[4] on December 25, 2008 and ***reopen***[5] on January 4, 2009.
We thank you for your support over the past year and we wish you all a joyous holiday season!

Yours faithfully,
Liberty Co., Ltd.

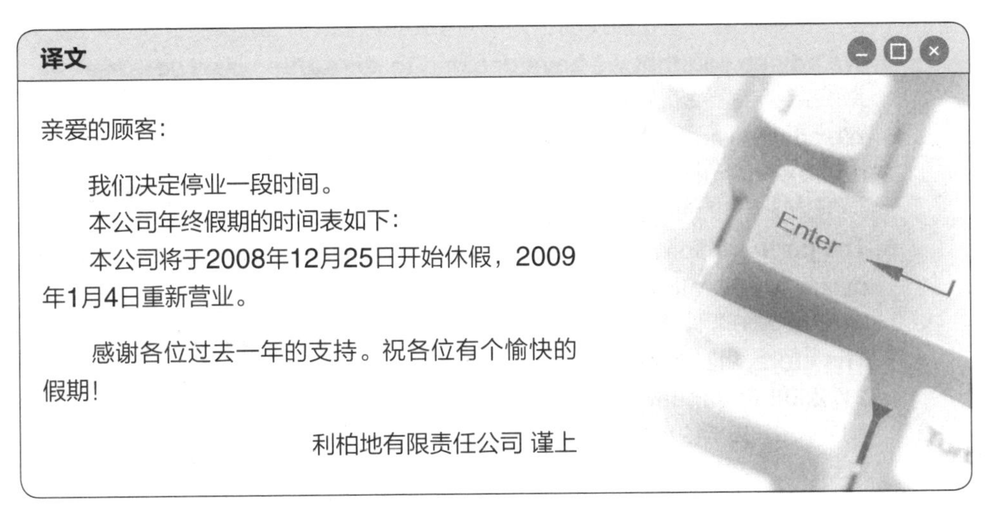

译文

亲爱的顾客：

我们决定停业一段时间。
本公司年终假期的时间表如下：
本公司将于2008年12月25日开始休假，2009年1月4日重新营业。

感谢各位过去一年的支持。祝各位有个愉快的假期！

利柏地有限责任公司 谨上

语法重点解析

1 解析重点1 customer

表达“顾客；客户；购买者”可以用buyer, customer, shopper, client 等。与顾客有关的短语包括 secure customers / draw customers（招揽顾客），regular customer（老顾客），Customers First（顾客至上），customer service（客户服务）等。

2 解析重点2 close

close 有“关闭”的意思，还有“停业；倒闭”的意思。表达“关闭”还可以使用 close down, turn off 等短语。但是 turn off 一般指关闭水龙头、煤气、电源等。请对照以下例句：

The firm has decided to close down its Chicago branch.（公司已决定关闭某芝加哥的分公司。）

Do you mind if I turn off the light?（你介意我把灯关掉吗？）

1. **We have decided not to continue our business on and after September 1.**
   我们已经决定自9月1日起停止营业。
2. **Notice is hereby given that our corporation has been *discontinued*[6] by agreement.**
   兹通知本公司已协议终止营业。
3. **As a member of the company, I decide to drop out.**
   作为公司合伙人之一，我决定退出。
4. **We advise you that we have decided to *dissolve*[7] *partnership*[8].**
   我们已经决定解除合伙关系，特此通知。
5. **We advise you that we have decided not to continue our partnership.**
   我们已经决定停止合伙经营，谨此通知。
6. **The partnership will be discontinued owing to the retirement of Mr. Cole.**
   由于科尔先生的退休，合作关系将中断。
7. **The store will close from December 21, 2008 to January 2, 2009.**
   本店将于2008年12月21日至2009年1月2日停止营业。

必背关键单词

1. ***customer*** [ˈkʌstəmə(r)] n. 顾客；客户
2. ***schedule*** [ˈʃedjuːl] v. 安排；计划
3. ***follow*** [ˈfɒləʊ] n. 跟随；追随
4. ***close*** [kləʊz] v. 关闭
5. ***reopen*** [ˌriːˈəʊpən] v. 重开
6. ***discontinue*** [ˌdɪskənˈtɪnjuː] v. 中断
7. ***dissolve*** [dɪˈzɒlv] v. 解散
8. ***partnership*** [ˈpɑːtnəʃɪp] n. 合伙；合作关系

15 商品出货通知

这样写就对了

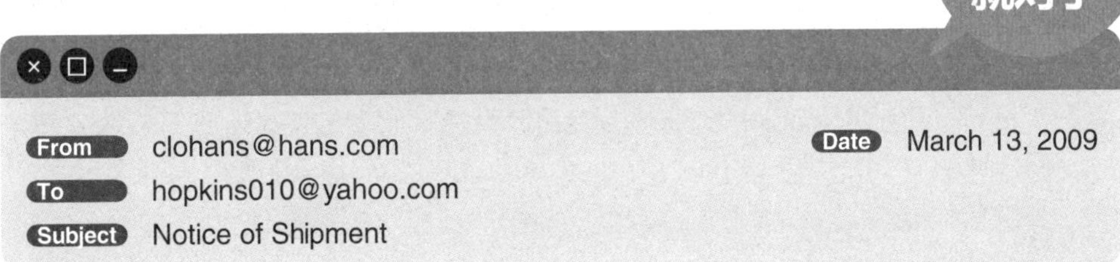
From clohans@hans.com
To hopkins010@yahoo.com
Subject Notice of Shipment
Date March 13, 2009

Dear Mr. Hopkins,

This is as a ***notification***[1] of ***shipment***[2].
We have shipped your ***order***[3] No. 3108 as of March 12, 2009, and we also sent you the ***relevant***[4] **shipping documents** by fax. Please ***check***[5].
You ***should***[6] receive the **goods** you ordered by March 20, 2009 if there is no ***accident***[7].

Please let us know when they arrive. Thank you very much!

Yours truly,
HANS Company

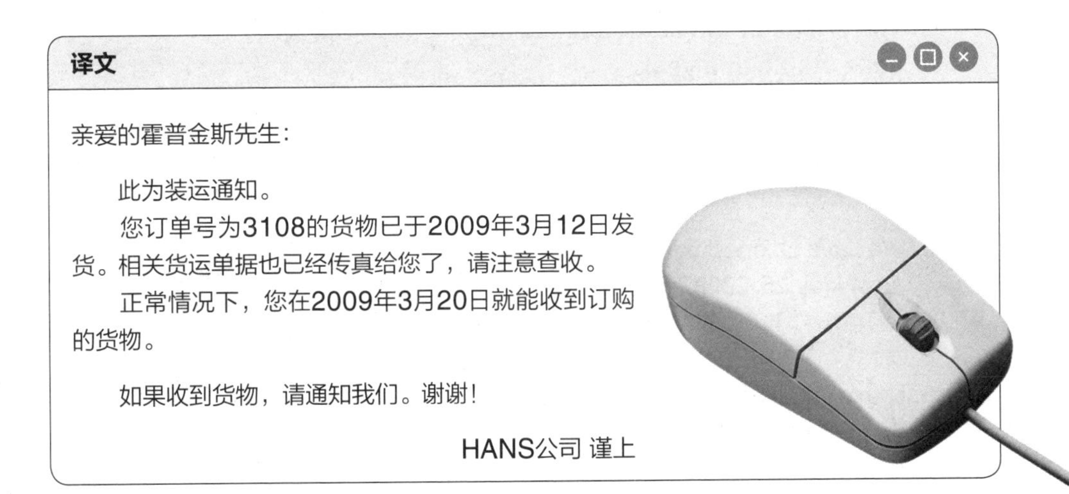
译文

亲爱的霍普金斯先生:

此为装运通知。
您订单号为3108的货物已于2009年3月12日发货。相关货运单据也已经传真给您了，请注意查收。
正常情况下，您在2009年3月20日就能收到订购的货物。

如果收到货物，请通知我们。谢谢!

HANS公司 谨上

语法重点解析

1 解析重点1 shipping documents

短语 shipping document 的意思是“货运清单”。表达相关出货文件的短语还有 bill of lading（提单），shipping invoice（装运发票），shipping order（运货单）等。请看以下例句：

I have brought a set of the duplicate of our shipping document.（我带来了我方一整份货运清单的副本。）

The bill of lading shows an issuing date.（提单表明了签发日期。）

Could you please fax me shipping documents?（请您将货运清单传真给我好吗？）

2 解析重点2 goods

除了可以用 goods 表达“货物”外，还可以用 cargo / freight 表示“（船装载的）货物”，个人订购的货品可以用 merchandise（商品），product（产品），commodity（日用品），item（物品）等表示。请看以下例句：

The customs impound the whole cargo.（海关扣押了全部货物。）

Branded merchandise is the one bearing a standard brand name.（品牌商品是具有标准商标名称的商品。）

1. **I have received a bill of lading for forty *bales*[8] of cotton by that vessel.**
   我已经收到了该船运送的40大包棉花的提单。
2. **Did you note the damage on the bill of lading?**
   你把损坏情况在提单上注明了吗?
3. **The prices of the commodities are quite stable this year.**
   今年的物价相当稳定。
4. **The bill of lading should be marked as "freight prepaid".**
   提单上应该注明“运费预付”的字样。
5. **We have shipped your order No. 3110 as of May 25, 2008.**
   您订单编号3110的货物已于2008年5月25日发货。
6. **I have faxed you the relevant shipping documents.**
   我已经将相关的货运清单传真给您了。

必背关键单词

1. ***notification*** [ˌnəʊtɪfɪˈkeɪʃn] n. 通知；通告
2. ***shipment*** [ˈʃɪpmənt] n. 装运；装载的货物
3. ***order*** [ˈɔːdə(r)] n. 订单
4. ***relevant*** [ˈreləvənt] adj. 相关的
5. ***check*** [tʃek] v. 核查；检查
6. ***should*** [ʃʊd] aux. 应该
7. ***accident*** [ˈæksɪdənt] n. 事故；意外
8. ***bale*** [beɪl] n. 大包；大捆

16 样品寄送通知

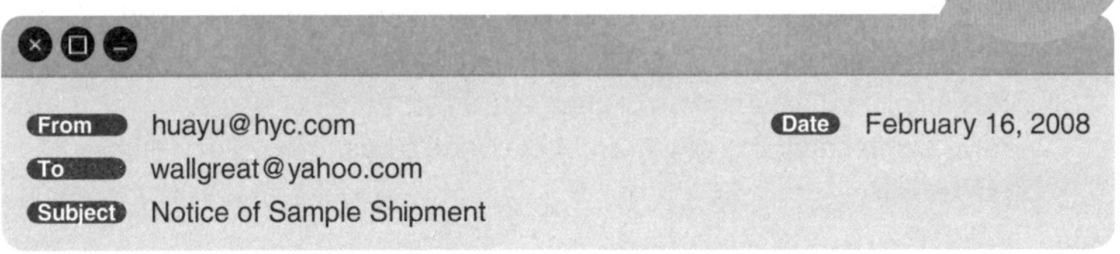

From huayu@hyc.com
Date February 16, 2008
To wallgreat@yahoo.com
Subject Notice of Sample Shipment

Dear Mr. Wall,

I am writing to inform you that the ***samples***[1] you ***requested***[2] were shipped today by ***Federal***[3] ***Express***[4].
I have ***attached***[5] a price list and ***color***[6] ***swatches***[7], too. Please inform me immediately when they arrive. Thanks a lot!

I am looking forward to your feedback.

Sincerely yours,
HY Company

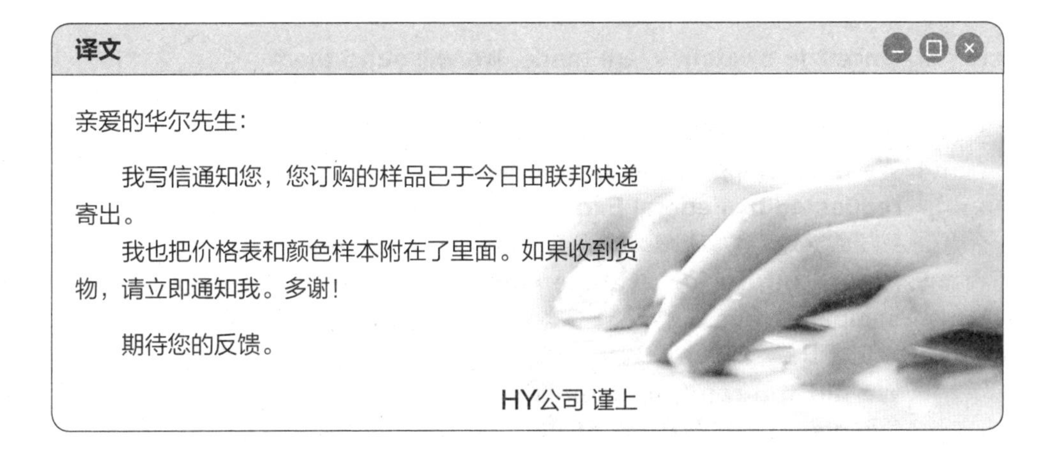

译文

亲爱的华尔先生：

我写信通知您，您订购的样品已于今日由联邦快递寄出。
我也把价格表和颜色样本附在了里面。如果收到货物，请立即通知我。多谢！

期待您的反馈。

HY公司 谨上

语法重点解析

1 解析重点1 swatch

“样品”可以用 sample 和 swatch 表示。但是 swatch 仅做名词使用，多用于衣物、纺织品等，而 sample 既可以作为名词，又可以当作动词，解释为“取样；品尝”。请看以下例句：

Once the swatch is made, we will send you at once.（一旦样品制作好，我们会立刻寄给您。）

The sample will be sent free of charge.（样品会免费寄送。）

I have sampled all the cakes and I like Jane's best.（我尝了所有的蛋糕，我最喜欢简做的。）

2 解析重点2 feedback

feedback 是名词，意思是“反馈”。注意，如果拆成 feed back，则做动词词组，表示“反馈；反作用”。请对照以下例句：

The more feedback we get from viewers, the better.（从观众那儿得到的反馈越多越好。）

The sales clerks feed back information to the firm about its sales.（销售员们把销售情况反馈给公司。）

1. Samples will be present to you immediately upon your request.
只要您有需要，我们会立刻寄样品给您。

2. The distributors looked with favor on your sample.
您（发来）的样品得到了经销商们的青睐。

3. Once the swatches are made, we will send them to you at once.
样品一旦制作好，我们会马上寄给您。

4. We have shipped the sample you requested by Federal Express.
我们已经通过联邦快递把您要的样品寄给您了。

5. Please inform us at once when samples arrive.
如果收到样品，请马上通知我们。

6. We did the random sampling of five pieces per *batch*[8] of 100.
我们在每批的100件货物中随机抽样5件。

必背关键单词 

1. ***sample*** [ˈsɑːmpl] n. 样品；标本
2. ***request*** [rɪˈkwest] v. 要求
3. ***federal*** [ˈfedərəl] adj. 联邦的
4. ***express*** [ɪkˈspres] n. 快递；快车
5. ***attach*** [əˈtætʃ] v. 附上；贴上
6. ***color*** [ˈkʌlə(r)] n. 颜色
7. ***swatch*** [swɒtʃ] n. 样品；样本
8. ***batch*** [bætʃ] n. 一批；一炉

17 订购商品通知

这样写就对了

From ouyam@mall.com **Date** December 26, 2008
To membersoy@yahoo.com
Subject Notice of Ordering Commodities

Dear members,

First of all, we wish everyone a Happy New Year in ***advance***[1]!
We are pleased to inform you that you could order any ***commodities***[2] **as usual** during the holiday. You just need to ***choose***[3] what you want on the ***website***[4] of our ***mall***[5], and write down your ***detailed***[6] address, and then we will provide the service of ***cash***[7] **on delivery**.

Please enjoy your shopping!

Sincerely yours,
Asian Mall

译文

亲爱的会员朋友：

首先，我们预祝大家新年快乐！
我们很高兴通知各位会员朋友，大家可以在假日正常订购商品。您只需要在我们商场的网站上选择您想要的商品，然后写明您的具体地址，之后我们会提供货到付款服务。

祝您购物愉快！

亚洲购物中心 谨上

语法重点解析

1 解析重点1 as usual

短语 as usual 的意思是“照例；像往常一样”。run true to form（一如既往）也可以表达相同的意思。请看以下例句：

After supper, Jim dived into his work as usual.（晚饭后，吉姆一如往常地投入到工作中。）

Run true to form, we will support our new and old customers.（我们将一如既往地支持我们的新老客户。）

2 解析重点2 cash on delivery

cash on delivery 的意思是“货到付款”。现在流行的新型购物方式主要有：Order by Phone（电话订购），Order Online（网络购物），Order by Fax（传真订购）和 Order by Mail（邮寄订购）等。而这些购物的付款方式主要有：cash on delivery（货到付款），payment by post（邮寄付款）和 bank transfer（银行转账）等。请看以下例句：

Could we pay cash on delivery?（我们能货到付款吗？）

The common international export transaction is via bank transfer.（常见的国际出口交易是通过银行转账的。）

1. **I bought a new computer using cash on delivery.**
   我以货到付款的方式买了台新电脑。
2. **We insist on payment in cash on delivery without allowing any discount.**
   我们坚持货到付款，没有任何折扣。
3. **Don't hesitate! Pick up the phone and order!**
   别再犹豫了，拿起电话订购吧！
4. **We accept Cash on Delivery only up to this moment. Sorry for any inconvenience caused.**
   我们目前只接受货到付款，不便之处，敬请原谅。
5. **We are a high-volume discount mail-order *corporation*[8].**
   我们是大规模折扣邮购公司。
6. **The firm will do business for the Christmas holidays as usual.**
   这家公司在圣诞节照常营业。

必背关键单词

1. ***advance*** [ədˈvɑːns] n. 预先
2. ***commodity*** [kəˈmɒdətɪ] n. 商品；日用品
3. ***choose*** [tʃuːz] v. 选择
4. ***website*** [ˈwebsaɪt] n. 网站
5. ***mall*** [mɔːl] n. 商场；购物中心
6. ***detailed*** [ˈdiːteɪld] adj. 详细的
7. ***cash*** [kæʃ] n. 现金
8. ***corporation*** [ˌkɔːpəˈreɪʃn] n. 公司；企业

18 确认商品订购通知

From cuteclo@ccc.com
To burns@yahoo.com
Subject Notice of Receipt
Date September 13, 2008

Dear Mr. Burns,

I am writing to ***inform***[1] you that we ***received***[2] the ***shipment***[3].
On September 12 we received your shipment of ***waistcoats***[4], invoice No. 91-37457.
We are very ***gratified***[5] that they were delivered so ***fast***[6]. Actually, they are selling ***quite***[7] well. We will keep you informed if we need ***extra***[8] orders.

Yours truly,
Coulter Clothing Co.

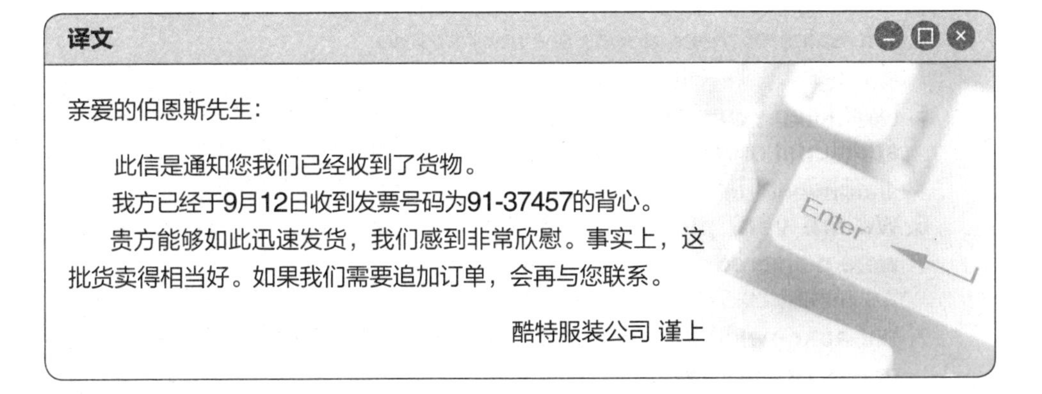
译文

亲爱的伯恩斯先生：

此信是通知您我们已经收到了货物。
我方已经于9月12日收到发票号码为91-37457的背心。
贵方能够如此迅速发货，我们感到非常欣慰。事实上，这批货卖得相当好。如果我们需要追加订单，会再与您联系。

酷特服装公司 谨上

语法重点解析

1 解析重点1 receive

receive 做动词时表示“接受；收到”。表达“接受”含义还可以用 accept, take 等。但是 receive 表示客观上被动地接受；accept 表示主动而且高兴地接受；take 所表示的接受包含着他人赠予的意味。请对照以下例句：

She has received his present, but she will not accept it.（她收到了他的礼物，但她是不会接受的。）

He takes anything he is given.（给他什么他都收。）

2 解析重点2 They are selling quite well.

这句话的意思是“它们卖得非常好”，sell well 即为“卖得好”。关于销售状况可以用以下方式表达：sales are booming（销售量激增），sales are good / strong（销售量不错），sales are going up（销售量上升），sales are fair（销量尚可），sales are flat（销量平平），sales are going down（销量下降），sales are sluggish（销售停滞），sales are bad / poor（销路不畅）。

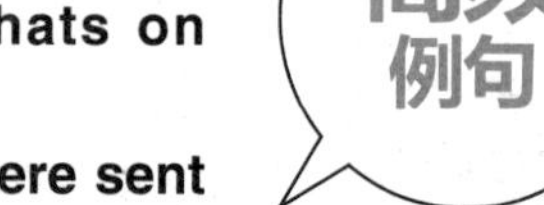

1. **I am pleased to receive your goods on time.**
   我很高兴准时收到了贵方寄送的货物。
2. **We have received your shipment of hats on August 10.** 我们已于8月10日收到了贵方寄送的帽子。
3. **We are very *satisfied*[9] that the goods were sent so quickly.**
   我们非常满意发货发得如此之快。
4. **The sales of these goods are very strong.**
   这些货物的销量非常不错。
5. **I will keep you informed regarding additional orders we may have.**
   如果还需要追加订货，我会与你们联系。
6. **We are very pleased that they were delivered so fast.**
   货物运送如此之快令我们感到很高兴。
7. **We acknowledge receipt of your goods of the 15th.**
   告知贵方我方已收到你方15号发来的货物。
8. **Please stay in touch in case we need to make an additional order.**
   请保持联络，以便我们追加订单。

必背关键单词

1. ***inform*** [ɪnˈfɔːm] v. 通知
2. ***receive*** [rɪˈsiːv] v. 收到
3. ***shipment*** [ˈʃɪpmənt] n. 装运；装载的货物
4. ***waistcoat*** [ˈweɪskəʊt] n. 背心；马甲
5. ***gratify*** [ˈɡrætɪfaɪ] v. 使满足；使高兴
6. ***fast*** [fɑːst] adv. 快地；迅速地
7. ***quite*** [kwaɪt] adv. 很；十分
8. ***extra*** [ˈekstrə] adj. 额外的
9. ***satisfied*** [ˈsætɪsfaɪd] adj. 满意的

19 商品缺货通知

这样写就对了

From horizon@hebc.com　**Date** February 16, 2008
To subscriber@hebc.com
Subject Notice: Out of Stock

Dear ***subscribers***[1],

Due[2] to ***booming***[3] sales, we are so sorry to inform you that the book *Harry Potter VII* you ordered has been **out of stock**.
However[4], we will ***replenish***[5] **stock** at once. If you still wish to order it three days later, please let us know if you would like to have it delivered by ***airmail***[6] or by ***surface***[7] mail.

We are looking forward to hearing from you soon.

Sincerely yours,
Horizon English Book Co.

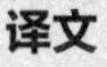

译文

亲爱的订购客户:

由于销售量激增，很抱歉通知您，您订购的《哈利波特7》已经没货了。

不过我们会马上进货。如果3天后您还想要订购，请告知我们您想要用航空邮寄还是平邮。

期待您能尽快回信。

地平线英语图书公司 谨上

语法重点解析

1 解析重点1 **out of stock**

短语 out of stock 的意思是“无现货的；脱销的”，意思相当于 sell out（卖完）。而表达“有货；有库存”则用 in stock。请看以下例句：

The book is out of stock.（此书没有现货。）

This edition of the dictionary is sold out.（这个版本的词典已卖完了。）

Do you have any grey pullovers in stock?（你们的灰色套头毛衣有现货吗？）

2 解析重点2 **replenish stock**

短语 replenish stock 的意思是“进货”，replenish 意为“补充”，stock 意为“库存；存货”。表达“进货”还可以用 purchase of merchandise, lay in a stock of merchandise 等。请看以下例句：

We must replenish our stock of coal.（我们必须补充煤的储备。）

Mom is preparing to lay in a stock of merchandise for New Year's Day.（妈妈正在准备置办年货。）

1. **The blue shirts are out of *stock*[8].**
   这种蓝衬衫已经没货了。
2. **The item you ordered is out of stock.**
   你订购的产品目前缺货。
3. **So many children have bought toy rockets that the store is now out of stock.**
   太多的孩子买了玩具火箭，以致于现在商店都没货了。

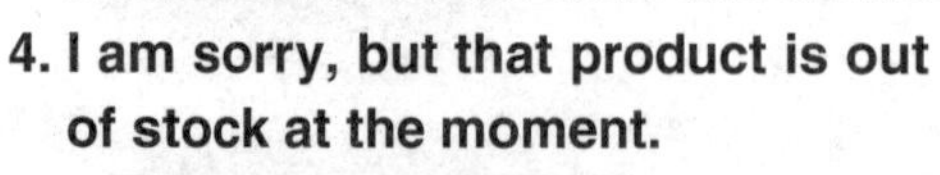

4. **I am sorry, but that product is out of stock at the moment.**
   很抱歉，该商品目前已没有存货了。
5. **As a matter of fact, we have run out of stock for a few weeks.**
   事实上，我们已经缺货几个星期了。
6. **The dictionary you asked for is out of stock.**
   你要买的词典现在没货了。
7. **In reply to your recent inquiry, the book you mentioned is not in stock.**
   您近日询问的图书没有现货，谨此回复。

必背关键单词

1. ***subscriber*** [səbˈskraɪbə(r)] n. 订购者
2. ***due*** [djuː] adj. 到期的；应归于的
3. ***booming*** [ˈbuːmɪŋ] adj. 兴旺的；繁荣的
4. ***however*** [haʊˈevə(r)] adv. 然而；不过
5. ***replenish*** [rɪˈplenɪʃ] v. 补充
6. ***airmail*** [ˈeəmeɪl] n. 航空邮件
7. ***surface*** [ˈsɜːfɪs] n. 表面；地面；水面
8. ***stock*** [stɒk] n. 存货；库存

20 付款确认通知

这样写就对了

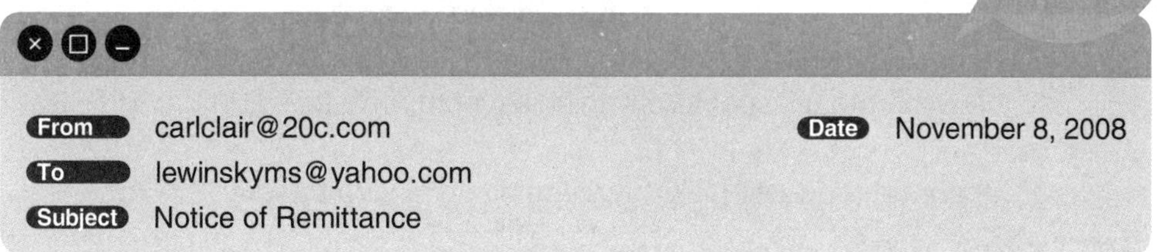
From carlclair@20c.com
To lewinskyms@yahoo.com
Subject Notice of Remittance
Date November 8, 2008

Dear Ms. Lewinsky,

On November 7, 2008, the ***amount***[1] of 5,700.00 US ***dollars***[2] as payment for your invoice No. 25167842 were **transferred** into your account. Please kindly ***check***[3].

I have also faxed a ***copy***[4] of the ***remittance***[5] ***slip***[6] **for your** ***reference***[7].

Sincerely yours,
Carl Clair

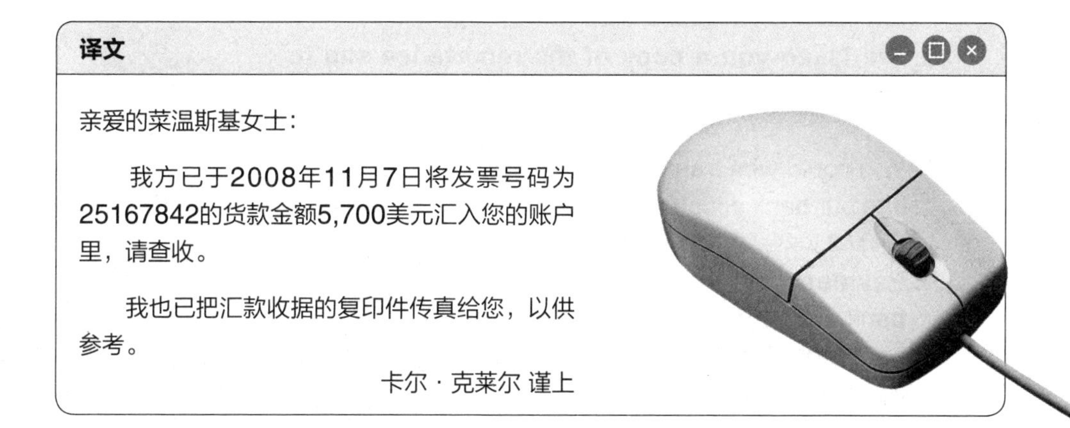
译文

亲爱的莱温斯基女士：

我方已于2008年11月7日将发票号码为25167842的货款金额5,700美元汇入您的账户里，请查收。

我也已把汇款收据的复印件传真给您，以供参考。

卡尔·克莱尔 谨上

语法重点解析

1 解析重点1 transfer

transfer 的基本意思是“转移”，既有名词词性，又有动词词性，在此语境下是做动词，表示“转移；汇款”。表达“汇款”可以用 remit 或 send。注意以下几个有关“汇款”的短语：wire transfer（电汇），transfer / wire to one's bank account（汇入某人的银行账户）。请看以下例句：

Please remit the full cost to our bank account.（请将所有款项汇入我们的银行账户。）

All withdrawal requests must be processed via wire transfer.（要求所有提款必须以电汇的形式支付。）

2 解析重点2 for your reference

短语 for reference 的意思是“参考；备案”。与“参考”相关的短语还有：for reference only（仅供参考）；reference price（参考价格）；reference documents（参考文献）；reference data / reference material（参考资料）。请看以下例句：

Please keep the original receipt for reference.（请保留收据正本以便参考。）

The sample is for reference only.（样品仅供参考。）

1. **We expect to hear from you at once with a remittance.**
   我们希望能尽快收到您的来函与汇款。
2. **I've faxed you a copy of the remittance slip for reference.**
   我已将汇款通知单的复印件传真给您以供参考。
3. **You could wire transfer the payment into our bank account.**
   你可以将款项电汇至我们的银行账户。
4. **Payment by direct deposit into bank account will usually take five additional working days more than by check mailing.**
   相较于支票支付，直接存款需要多5个工作日的处理时间。

必背关键单词

1. ***amount*** [əˈmaʊnt] n. 数量；总额
2. ***dollar*** [ˈdɒlə(r)] n. 美元
3. ***check*** [tʃek] v. 检查
4. ***copy*** [ˈkɒpɪ] n. 拷贝；副本
5. ***remittance*** [rɪˈmɪtns] n. 汇款
6. ***slip*** [slɪp] n. 失足；纸条；纸片
7. ***reference*** [ˈrefrəns] n. 参考；参照

21 入账金额不足通知

这样写就对了

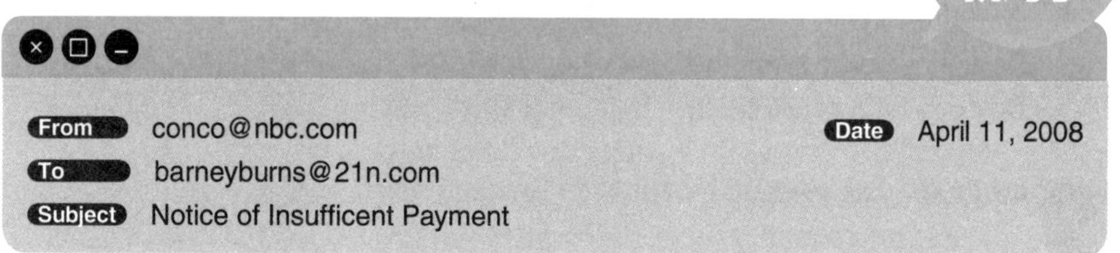

From conco@nbc.com　Date April 11, 2008
To barneyburns@21n.com
Subject Notice of Insufficent Payment

Dear Mr. Burns,

We have received your ***telegraphic***[1] ***transfer***[2] of US$3,000.00 on April 10, 2008. However, I am afraid you have neglected to add the cost of ***freight***[3] that was ***indicated***[4] on the ***invoice***[5] we faxed to you.
We would therefore like to ask you to transfer an additional US$150.00 so that we can ship the order to you.
On receipt of the full ***payment***[6], we will immediately ***forward***[7] the goods.

Best regards,
NBC Corporation

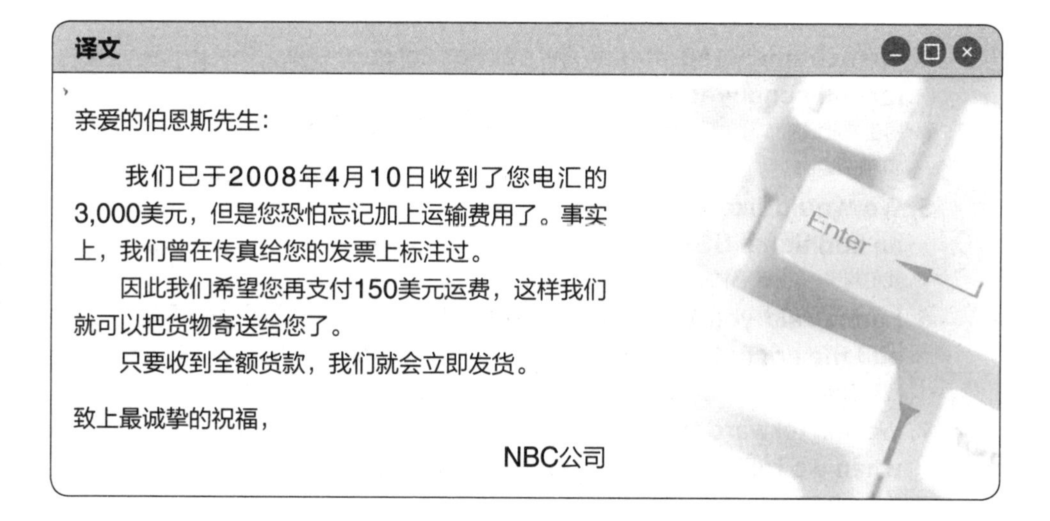

译文

亲爱的伯恩斯先生：

我们已于2008年4月10日收到了您电汇的3,000美元，但是您恐怕忘记加上运输费用了。事实上，我们曾在传真给您的发票上标注过。

因此我们希望您再支付150美元运费，这样我们就可以把货物寄送给您了。

只要收到全额货款，我们就会立即发货。

致上最诚挚的祝福，

NBC公司

语法重点解析

1 解析重点1 neglect to

neglect 有“忽视；忽略；遗漏”的意思。neglect to add... 表示“忘记加上……”，也可以用 forget to add... 来表达同样的意思。此外，表达“省略；遗漏”还可以用 omit。

请看以下例句：

Do not neglect to lock the door when you leave.（你走的时候别忘了锁门。）

Do not omit a single detail.（不要漏掉任何细节。）

2 解析重点2 on receipt of

短语 on receipt of 的意思是“一收到……就……”，同样的意思还可以用 upon receipt of 来表达。请看以下例句：

We will remit the amount of invoice on receipt of the invoice of the shipment.（一收到出货发票，我们就会立即汇款。）

The total amount must be paid in full upon receipt of the documents.（全部款项在收到单据后必须全额付清。）

高频例句

1. **Quality is something we never *neglect*[8].**
   我们一直十分重视品质。
2. **Upon receipt of your L/C we will immediately ship your order.**
   一接到贵公司的信用证，我们将立即发货。
3. **On receipt of your check, we shall ship the goods immediately.**
   一收到贵方的支票后，我们会立即发货。
4. **We extremely regret that we omitted to quote you the price you recently enquired.**
   对贵方近日的询价，我们因疏忽未能报价，对此感到非常抱歉。
5. **We would like to ask you to pay an additional US$125.00.**
   我们想请您付清额外的125美元。
6. **I am afraid you have forgotten to add the cost of freight.**
   我想您恐怕忘记加上运费了。
7. **We will forward the goods at once when we receive the full payment.**
   收到全额款项后，我们会立即出货。

必背关键单词

1. ***telegraphic*** [ˌtelɪˈgræfɪk] adj. 电报的；电信的
2. ***transfer*** [trænsˈfɜː(r)] n. 转移；迁移；转账
3. ***freight*** [freɪt] n. 货运；运费
4. ***indicate*** [ˈɪndɪkeɪt] v. 指示；表明
5. ***invoice*** [ˈɪnvɔɪs] n. 发票；发货单
6. ***payment*** [ˈpeɪmənt] n. 支付；付款
7. ***forward*** [ˈfɔːwəd] v. 运送；转寄
8. ***neglect*** [nɪˈglekt] v. 忽略

Unit 7 业务开发维护
Business Establishing & Maintaining

01 开发业务

From "Hugh Jackman" (hjackman@tdn.com) **Date** Fri., March 13, 2009

To "Ben Affleck" (baffleck@aso.com)

Subject Developing a Business Relationship

Dear Mr. Affleck,

I learned from your ***message***[1] by ***accident***[2] on the Internet that your company needs car components.

We are a car parts ***manufacturer***[3]. I am sure you'll be interested in our products. We do hope there is an opportunity for us to ***collaborate***[4] with each other in the coming days.

An early reply will be appreciated!

Yours sincerely,

Hugh Jackman

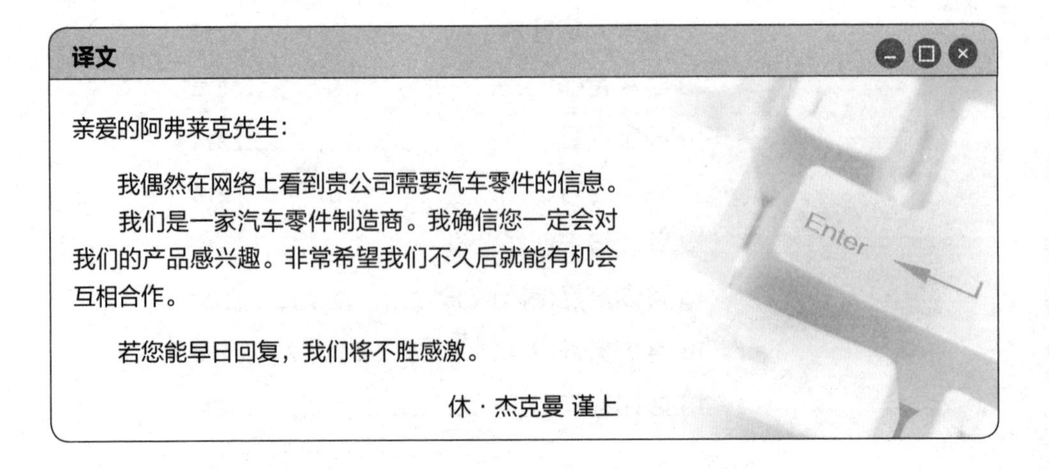

译文

亲爱的阿弗莱克先生：

我偶然在网络上看到贵公司需要汽车零件的信息。

我们是一家汽车零件制造商。我确信您一定会对我们的产品感兴趣。非常希望我们不久后就能有机会互相合作。

若您能早日回复，我们将不胜感激。

休·杰克曼 谨上

语法重点解析

1 解析重点1 I learned from your message by accident on the Internet that...

一般我们在开发业务的时候，都是自己主动去找客户、挖掘客户。有时候在网络上会偶然发现潜在客户的信息。通常表达“意外地”意思时，我们会选择用 suddenly 来表达。其实，这里如果我们使用 by accident 则更合适。suddenly 多指事情在突然之间、没有防备的情况下发生。例如：the lights went off suddenly（灯突然熄灭了），而 by accident 这个短语才表示“事情在偶然、无意间发生”的意思。

2 解析重点2 We do hope there is an opportunity for us to collaborate with each other in the coming days.

发现潜在客户之后，关键就是要表达想与其合作的意愿。因此我们必须在邮件中提出这一愿望，而且，要表达得强烈一点，让对方看到我们的诚意。上面邮件内文中的句子：We do hope there is an opportunity for us to collaborate with each other in the coming days（很希望我们不久能有机会合作），就是使用了助动词 do 加上动词原形的结构来表达强烈的情感，强调了渴望合作的意愿。

高频例句

1. **I'm sure our products will *attract*[5] you.**
   我确信您一定会对我们的产品感兴趣。
2. **We manufacture a wide *range*[6] of products.**
   我们生产的产品范围广泛。
3. **Please don't hesitate to call us whenever you want.**
   请在需要的情况下随时向我们致电。
4. **I shall be grateful if you will favor me with an early reply.**
   若能早日回复则不胜感激。
5. **We produce five *different*[7] new items recently.**
   我们最近生产了5种不同的新产品。
6. **We are willing to *explore*[8] every possibility for new business.**
   我们愿意探索各种可能性，以开拓新的业务。

必背关键单词

1. ***message*** [ˈmesɪdʒ] n. 信息
2. ***accident*** [ˈæksɪdənt] n. 机遇；偶然；事故
3. ***manufacturer*** [ˌmænjʊˈfæktʃərə(r)] n. 制造者
4. ***collaborate*** [kəˈlæbəreɪt] v. 协力；合作
5. ***attract*** [əˈtrækt] v. 吸引；引起……的注意
6. ***range*** [reɪndʒ] n. 范围
7. ***different*** [ˈdɪfrənt] adj. 不同的
8. ***explore*** [ɪkˈsplɔː(r)] v. 探索；探究；仔细查看

02 拓展业务

这样写就对了

From "Topher Grace" (tgrace@aol.com) Date Sat, March 14, 2009
To "Timmy Cruise" (tcruise@bbc.com)
Subject Hoping to Establish New Business Relations

Dear Mr. Cruise,

We have obtained your company information from china.alibaba.com, so we are e-mailing you to ***enquire***[1] whether you would be willing to ***establish***[2] new business ***relations***[3] with us.
We have been a manufacturer of deluxe ***toiletries***[4] for many years. Now we plan to ***extend***[5] our range. If the prices of your products are competitive, we would expect to transact a ***significant***[6] ***volume***[7] of business.

An early reply will be obliged!

Yours sincerely,
Topher Grace

译文

亲爱的克鲁斯先生：

我们从阿里巴巴中文网站上得到了贵公司的信息。我们发送此封邮件给您，是想询问您是否愿意与我们建立新的业务关系。
多年来，我们一直致力于高级化妆品的生产。现在，我们计划着拓展我们的业务范围。如果你们能给我们有竞争力的产品价格，我们期待能达成大规模交易。

如您能早日回复，则不胜感激！

托弗 · 格雷斯 谨上

语法重点解析

1 解析重点1 **We are e-mailing you to enquire whether...**

拓展业务是在已有的业务基础上发展新业务。比如，企业现有产品种类若比较局限，就可以考虑从别的企业引进新产品，使产品多样化，提供给客户更多的选择，从而创造更多的卖点。写信询问企业产品，当然就是要告知别人自己的意向。We are e-mailing you to enquire whether...（我们发送此封邮件给您，是想询问……）中的 whether 在这里是表示后面是可选择的内容，如此问话显得语气更加委婉。

2 解析重点2 **Now we plan to extend our range.**

上面的邮件说到想跟别的企业合作，这是企业自身寻求新的发展契机的结果。想买进新产品，拓展业务，才会想到和别的企业合作。因此，在自报家门之后，邮件中又写到：Now we plan to extend our range（现在，我们想拓展我们的业务范围），这样，对方就明白（Clearness）事情的来龙去脉了。

1. **Would you like to establish business relations with us?**
   贵公司愿意和我们建立业务关系吗？
2. **We have been an importer of shoes for many years.**
   我们做鞋子进口生意已经很多年了。
3. **We are interested in *expanding*[8] our business.**
   我们对拓展自身业务很感兴趣。
4. **I would appreciate your catalogues and quotations.**
   若您能提供商品目录和报价单，我将不胜感激。
5. **We want to extend our business to FPD.**
   我们想将业务范围扩展到平板显示器。
6. **We write you with a view to establish trade relations.**
   我们写这封信是想要和贵方建立贸易关系。
7. **We are one of the largest importers of electric goods in this country.**
   我们是国内最大的电器进口商之一。

必背关键单词 

1. ***enquire*** [ɪnˈkwaɪə(r)] v. 打听；询问
2. ***establish*** [ɪˈstæblɪʃ] v. 建立
3. ***relation*** [rɪˈleɪʃn] n. 关系
4. ***toiletry*** [ˈtɔɪlɪtrɪ] n. 化妆品
5. ***extend*** [ɪkˈstend] v. 延长；扩展
6. ***significant*** [sɪɡˈnɪfɪkənt] adj. 有意义的
7. ***volume*** [ˈvɒljuːm] n. 卷；册；音量；容积
8. ***expand*** [ɪkˈspænd] v. 扩展；展开

03 介绍新产品

From "Paul Walker" (pwalker@cct.com) Date Sun., March 15, 2009
To "Robert Keating" (rkeating@nyn.com)
Subject Introducing New Products

Dear Mr. Keating,

We are pleased to inform you that we have just ***marketed***[1] our new products. We believe that you will find our new products more ***competitive***[2] both in quality and prices. **They should get a very good *reception*[3] in your market. Please let us know if you would like to take the *matter*[4] further.**

Look forward to hearing from you.

Yours sincerely,
Paul Walker

译文

亲爱的基廷先生：

很高兴通知贵公司，我们刚刚推出了新产品。

我们相信您会发现我们的新产品在品质和价格上都更有竞争力，必定能得到贵公司客户的认可。如果您感兴趣，敬请告知。

敬候佳音。

保罗 · 沃克 谨上

语法重点解析

1 解析重点1 They should get a very good reception in your market.

向客户介绍新产品时，我们一般要说自己的产品好卖、有市场，这样别人才会考虑是否要跟我们做生意。通常我们形容某一个产品好卖时会说：Our product will sell well in your market. 其实，我们也可以像上述邮件内文中一样用：They should get a very good reception in your market（新产品必定能得到贵公司顾客的认可），其中get a good reception 表达的意思是“（产品在市场上）受欢迎”，但是，和 sell well 相比，更让人有耳目一新的感觉。

2 解析重点2 Please let us know if you would like to take the matter further.

在介绍产品后，我们通常会想知道对方是否有购买的意愿，而一般我们常用：Please let us know if you are interested in our products（如果您对我们的产品感兴趣，敬请告知），而在上面的邮件中，则用了 ...if you would like to take the matter further（如感兴趣……），变换一个说法会使对方更有新鲜感。

高频例句

1. It covers the latest ***designs***[5] which are now available from stock.
   这款涵盖了最新设计的产品现在有货。
2. I think you have to acknowledge that this ***feature***[6] will appeal to the many users.
   我想您不得不承认这一特点将吸引不少用户。
3. One of our main ***strengths***[7] is the quality of our products.
   我们的主要优势之一是我们产品的品质。
4. There is a good market opportunity for ***snazzy***[8] products.
   时髦的产品在市场上有很大的商机。
5. After going through our SWOT process, I think we're in good shape.
   经过态势分析过程后，我认为我们的经营状况很好。

必背关键单词 

1. ***market*** ['mɑ:kɪt] v. （在市场上）销售
2. ***competitive*** [kəm'petətɪv] adj. 有竞争力的
3. ***reception*** [rɪ'sepʃn] n. 接受
4. ***matter*** ['mætə(r)] n. 问题；事件
5. ***design*** [dɪ'zaɪn] n. 设计
6. ***feature*** ['fi:tʃə(r)] n. 特征；特色
7. ***strength*** [streŋθ] n. 长处；优点
8. ***snazzy*** ['snæzɪ] adj. 新潮的；时髦的；漂亮的

04 介绍附加服务

From "Jimmy Diamond" (jdiamond@hot.com) Date Mon, March 16, 2009
To "Jim Block" (jblock@apt.com)
Subject Introducing Additional Service

Dear Mr. Block,

We are pleased to inform you we are offering ***maintenance***[1] ***service***[2] in China's market from now on.
Purchasers[3] can ***enjoy***[4] three years of free ***warranty***[5] service and a life-long maintenance in local service centers. If you have any questions, please ***dial***[6] our Customer Service Hotline at 2999-6666.

Yours sincerely,
Jimmy Diamond

译文

亲爱的布洛克先生:

很高兴通知您，自即日起，我们公司将提供中国市场的维修服务。
购买者可到当地服务中心享受产品3年免费保修及终身维修的服务。如果您有任何疑问，请拨打我们的客服热线：2999-6666。

吉米 · 戴蒙德 谨上

语法重点解析

1 解析重点1 Purchasers can enjoy three years of free warranty service and a life-long maintenance in local service centers.

向客户介绍附加服务的时候，尽量要做到详细和具体（Concreteness）。一般来说，厂商会承诺购买者在购买产品之后，可以享受几年免费保修以及是否提供终身维修。free warranty service（免费保修服务）是厂商提供的免费服务，而 a life-long maintenance（终身维修服务）就可能会需要收取适当的费用。这些都是需要说清楚的，以免购买者产生误解，引发不必要的误会。

2 解析重点2 If you have any questions, please dial our Customer Service Hotline at 2999-6666.

虽然我们已经告知顾客或用户可以到当地服务中心享受服务，但是最好还要告知客服专线。有时候顾客或用户根本不了解当地是否有服务中心，或是服务中心的具体位置，这时要是有客服专线，就方便多了。

1. **We offer a personal service to our customers.**
   我们为顾客提供个人服务。
2. **The computer comes with a year's *guarantee*[7].**
   这台电脑保修期为一年。
3. **We offer a free *backup*[8] service to customers.**
   我们为顾客提供免费支援服务。
4. **Please allow me to make a presentation of the services we can offer.**
   请允许我介绍我们公司所能够提供的各项服务。
5. **We commit to providing professional after-sales service to our *customers*[9].**
   我们承诺为用户提供专业的售后服务。

必背关键单词 

1. ***maintenance*** [ˈmeɪntənəns] n. 维修
2. ***service*** [ˈsɜːvɪs] n. 服务
3. ***purchaser*** [ˈpɜːtʃəsə(r)] n. 购买者
4. ***enjoy*** [ɪnˈdʒɔɪ] v. 享受；欣赏
5. ***warranty*** [ˈwɒrəntɪ] n. 担保；保证
6. ***dial*** [ˈdaɪəl] v. 拨（电话）
7. ***guarantee*** [ˌgærənˈtiː] n. 保证；担保；保证书；抵押品
8. ***backup*** [ˈbækʌp] adj. 替代的；备用的；候补的
9. ***customer*** [ˈkʌstəmə(r)] n. 顾客

05 恢复业务关系

这样写就对了

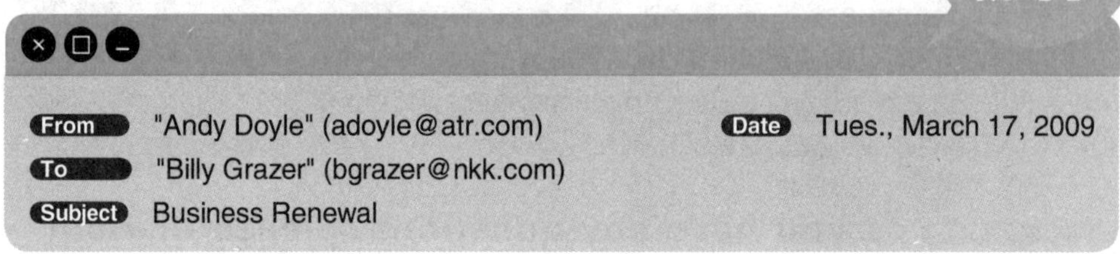
From "Andy Doyle" (adoyle@atr.com)　Date Tues., March 17, 2009
To "Billy Grazer" (bgrazer@nkk.com)
Subject Business Renewal

Dear Mr. Grazer,

We ***understand***[1] from our trade ***contacts***[2] that you have reestablished your business in London.
We would like to extend our ***congratulations***[3] and offer our very best ***wishes***[4] for your continued success. Since our last trade, our lines have changed a lot. The catalogue is enclosed for your reference.

Looking forward to hearing from you.

Yours sincerely,
Andy Doyle

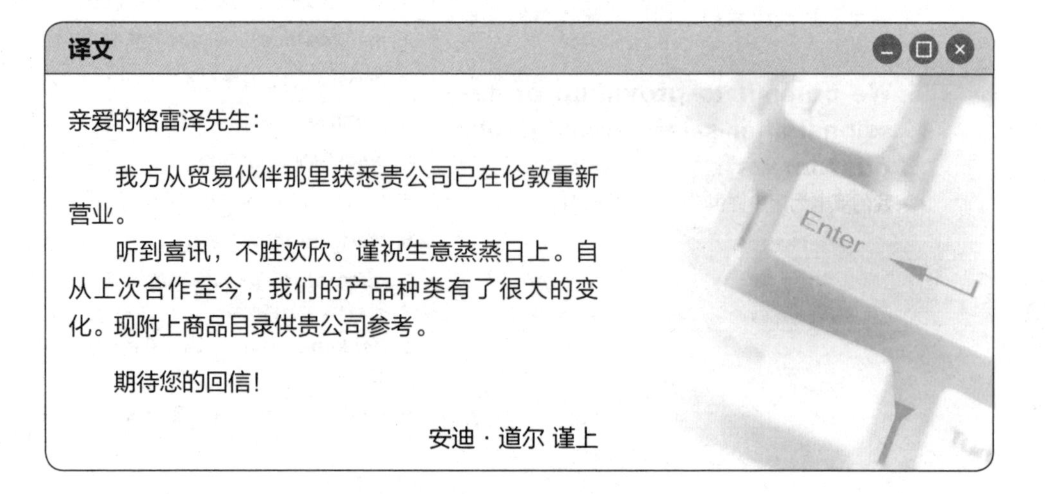
译文

亲爱的格雷泽先生：

　　我方从贸易伙伴那里获悉贵公司已在伦敦重新营业。

　　听到喜讯，不胜欢欣。谨祝生意蒸蒸日上。自从上次合作至今，我们的产品种类有了很大的变化。现附上商品目录供贵公司参考。

　　期待您的回信！

安迪 · 道尔 谨上

语法重点解析

1

解析重点1 **We would like to extend our congratulations and offer our very best wishes for your continued success.**

对方重新营业了，对他们和我们来说都是一件好事。曾经的合作伙伴可以再次寻求合作的机会，当然要向对方好好祝贺一番：We would like to extend our congratulations（听到喜讯，不胜欢欣）；同时，也不忘祝贺对方今后生意兴隆：We offer our very best wishes for your continued success（谨祝生意蒸蒸日上）。

2

解析重点2 **Since our last trade, our lines have changed a lot.**

向对方祝贺了一番后，我们还是得回到重点，那就是，告知对方自己公司产品发展的情况，以寻求新的合作机会。Since our last trade, our lines have changed a lot（自从上次合作至今，我们产品种类变化很大），line 在这里指“（产品的）种类”，上面这个句子说明在对方停业的那段时间，自己公司的产品发生的变化，为接下来附上新目录，寻求新合作做铺垫。

高频例句

1. **A *booklet*[5] including a *general*[6] introduction of business is enclosed for your reference.**
   随函附上包含公司业务概况的小册子供您参考。
2. **We have had considerable transactions with your corporation for the past five years.**
   我们在过去的5年中曾与贵公司有着非常频繁的贸易往来。
3. **Should you wish to receive samples for closer *inspection*[7], we would be very happy to forward them.**
   如贵方需仔细查看样品，我方非常乐意提供。
4. **Your company is once again trading successfully in your region.**
   贵公司在贵地区的贸易再次取得成功。

必背关键单词 

1. ***understand*** [ˌʌndəˈstænd] v. 了解；明白；领会
2. ***contact*** [ˈkɒntækt] n. 接触；亲近
3. ***congratulation*** [kənˌɡrætʃʊˈleɪʃn] n. 祝贺；庆贺
4. ***wish*** [wɪʃ] n. 愿望；希望
5. ***booklet*** [ˈbʊklət] n. 小册子
6. ***general*** [ˈdʒenrəl] adj. 普遍的；一般的
7. ***inspection*** [ɪnˈspekʃn] n. 检查；调查

06 巩固业务关系

From "Brandon Rodd" (brodd@coi.com) Date Wed., March 18, 2009
To "Williams Smith" (wsmith@opt.com)
Subject To Consolidate Our Business Relationship

Dear Mr. Smith,

In your last letter, **you asked whether we could give you a 5% *discount*[1]**. Since we have had a ***close***[2] business relationship with your company all these years, we decide to offer you such a price, **though this product is in great *demand*[3] and the *supply*[4] is limited**.

Your early reply will be greatly appreciated.

Yours sincerely,
Brandon Rodd

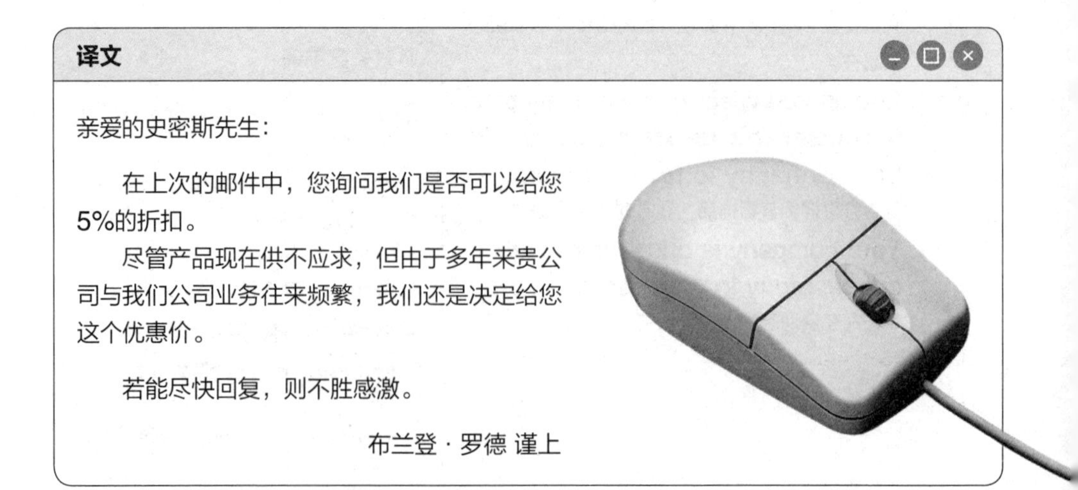

译文

亲爱的史密斯先生：

在上次的邮件中，您询问我们是否可以给您5%的折扣。

尽管产品现在供不应求，但由于多年来贵公司与我们公司业务往来频繁，我们还是决定给您这个优惠价。

若能尽快回复，则不胜感激。

布兰登·罗德 谨上

语法重点解析

1 解析重点1 You asked whether we could give you a 5% discount.

在巩固业务关系的时候，当我们的客户提出特殊的要求，我们经过考虑，认为可以满足对方时，就要适当满足客户，以巩固双方的业务关系。You asked whether we could give you a 5% discount（您曾询问我们是否可以给您打个95折）这句话表明客户提出了希望给他打95折的要求。以这句话开头，可以说明这封信的主旨，同时表明我们在积极解决和回复客户的问题。

2 解析重点2 ...though this product is in great demand and the supply is limited.

由于业务往来频繁的客户提出了打折这一要求，所以邮件中，我们考虑到彼此的业务关系，答应会满足他的这一要求：We decide to offer you such a price（我们还是决定给你这个优惠价）。但是，我们还是要让他明白我们确确实实给了他一个不小的优惠，也给他施加一点小小的压力。例如，邮件中就说：This product is in great demand and the supply is limited（产品现在供不应求）。

高频例句

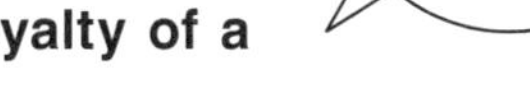

1. **We have built up a *solid*[5] *connection*[6] with your company.**
   我们现已和贵公司建立起了稳固的业务关系。
2. **The letter is intended to secure the loyalty of a satisfied customer.**
   此信是想获得一位满意的客户对我们的忠诚。
3. **We must do everything possible to *consolidate*[7] our established relations with the firms.**
   我们应尽一切可能巩固和我们有业务往来的公司之间的关系。
4. **For the past five years, we have done a lot of *trade*[8] with you.**
   在过去的5年中，我方与你有着非常频繁的贸易往来。
5. **The *normal*[9] price is US$50, but now I can give you a 5% discount.**
   正常价是50美元，但现在我可以给您打95折。

必背关键单词

1. ***discount*** [ˈdɪskaʊnt] n. 折扣
2. ***close*** [kləʊz] adj. 亲密的；亲近的
3. ***demand*** [dɪˈmɑːnd] n. 要求
4. ***supply*** [səˈplaɪ] n. 供应；补给
5. ***solid*** [ˈsɒlɪd] adj. 结实的；稳固的
6. ***connection*** [kəˈnekʃn] n. 联系；连接
7. ***consolidate*** [kənˈsɒlɪdeɪt] v. （使）巩固；（使）加强
8. ***trade*** [treɪd] n. 贸易
9. ***normal*** [ˈnɔːml] adj. 正常的

07 加深业务联系

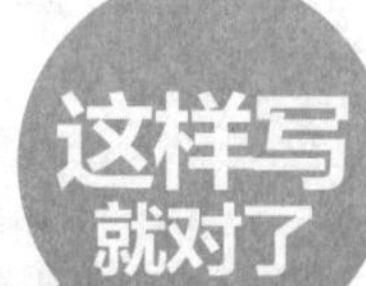

From "Simon Grimes" (sgrimes@cot.com) Date Thurs., March 19, 2009
To "Ross Scott" (rscott@rim.com)
Subject To Further Our Trade Links

Dear Mr. Scott,

Thank you for your cooperation with our business in the recent years.
Now we are ***keen***[1] to ***enlarge***[2] our trade in various kinds of ***electric***[3] ***equipments***[4], but unfortunately we do not have enough circulating ***funds***[5]. Please don't hesitate to call us if there is any possibility of cooperation between us.

Your early reply will be greatly appreciated.

Yours sincerely,
Simon Grimes

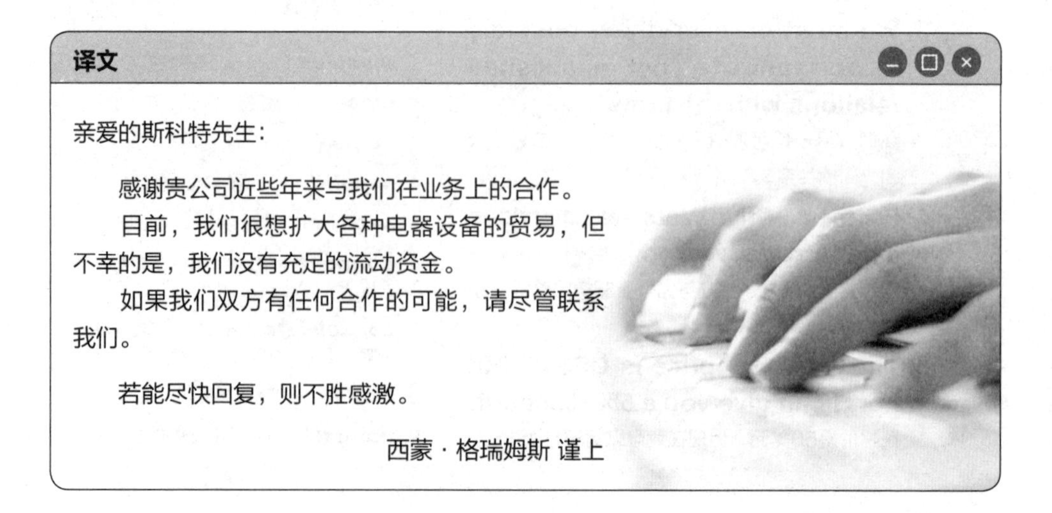

译文

亲爱的斯科特先生：

感谢贵公司近些年来与我们在业务上的合作。
目前，我们很想扩大各种电器设备的贸易，但不幸的是，我们没有充足的流动资金。
如果我们双方有任何合作的可能，请尽管联系我们。

若能尽快回复，则不胜感激。

西蒙·格瑞姆斯 谨上

语法重点解析

1 解析重点1 **We are keen to enlarge our trade in various kinds of electric equipments.**

当我们表达“非常希望做某事”时，最先出现在脑海中里的词语，就是 want, wish, hope等。这些是使用频率很高的词语，但是它们往往表达不出强烈的情感，需要加上副词修饰才行。而我们上述的邮件中，用到了 be keen to...（热衷于……），也可以理解成“很想……；热切希望……”的意思。比那些常用的词语，是不是更为贴切、生动呢？

2 解析重点2 **But unfortunately we do not have enough circulating funds.**

这句话很清楚地让对方了解了目前公司的状况。..., but unfortunately...（……，但是，不幸的是……）把前后两个句子连接起来，使表达更连贯，逻辑性更强。

高频例句

1. **We have vast potential for cooperation between us.**
   我们之间有着巨大的合作潜力。
2. **Our company had *grown*[6] rapidly in the recent years.**
   最近几年我们公司发展迅速。
3. **Over the last ten years, our company has been developing steadily.**
   最近10年，我们公司发展稳定。
4. **It is possible for us to cooperate with each other in this field.**
   我们在这个领域进行合作是有可能的。
5. **We should *grasp*[7] the opportunity to *expand*[8] our business cooperation.**
   我们应把握机会，扩展彼此的业务合作。
6. **Our business has been expanding *rapidly*[9] these years.**
   这些年我们的生意迅速扩大。

必背关键单词

1. ***keen*** [kiːn] adj. 热心的；渴望的
2. ***enlarge*** [ɪnˈlɑːdʒ] v. 扩大
3. ***electric*** [ɪˈlektrɪk] adj. 电的
4. ***equipment*** [ɪˈkwɪpmənt] n. 装备；设备
5. ***fund*** [fʌnd] n. 资金；基金
6. ***grow*** [grəʊ] v. 生长；增加；发展
7. ***grasp*** [grɑːsp] v. 掌握；抓牢
8. ***expand*** [ɪkˈspænd] v. 扩展
9. ***rapidly*** [ˈræpɪdlɪ] adv. 快速地

08 请求介绍客户

这样写就对了

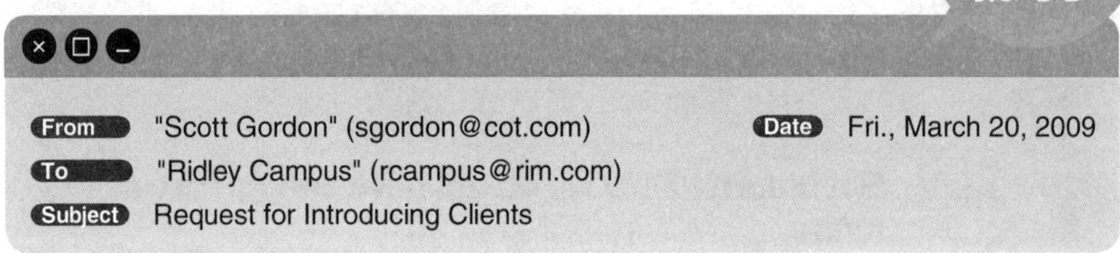

From "Scott Gordon" (sgordon@cot.com)　Date Fri., March 20, 2009
To "Ridley Campus" (rcampus@rim.com)
Subject Request for Introducing Clients

Dear Mr. Campus,

Thank you for your cooperation for our business.
We would like to explore the new potential market and ***increase***[1] the ***export***[2] of ***textiles***[3].
Therefore we shall appreciate it very much if you could kindly ***introduce***[4] us to some of the most ***capable***[5] ***importers***[6] who are interested in them.

Your early reply will be greatly appreciated.

Yours sincerely,
Scott Gordon

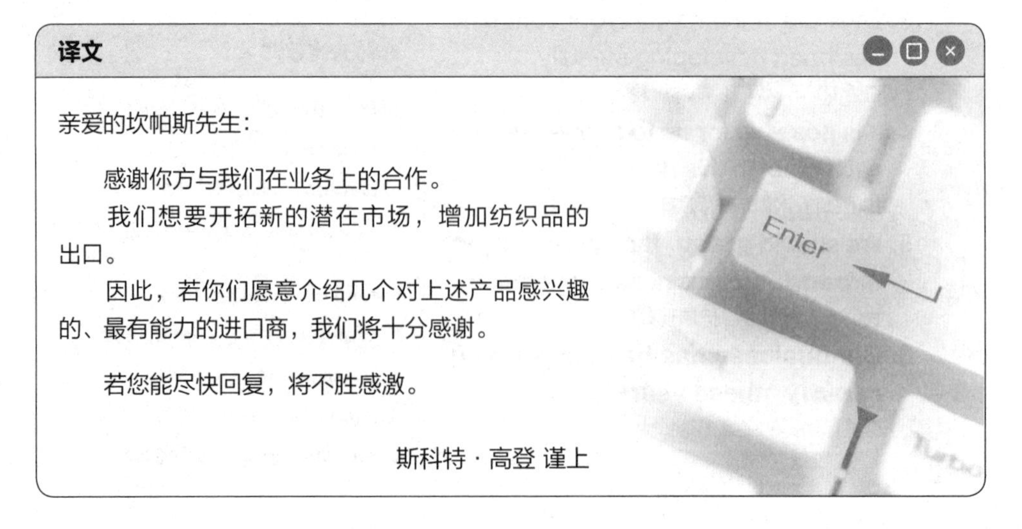

译文

亲爱的坎帕斯先生：

感谢你方与我们在业务上的合作。

我们想要开拓新的潜在市场，增加纺织品的出口。

因此，若你们愿意介绍几个对上述产品感兴趣的、最有能力的进口商，我们将十分感谢。

若您能尽快回复，将不胜感激。

斯科特 · 高登 谨上

语法重点解析

1

解析重点1 **We would like to explore the new potential market and increase the export of textiles.**

一般用英语来表达“开发市场”这个意思的时候，我们会用 develop the market，而在上述的邮件中，用的却是 explore the market（开拓市场）。在这篇请求介绍客户的邮件中，后者更为合适。因为 explore 有“探究；探索”的意思，强调发现、寻找（新市场）。它更加清楚（Clearness）、准确地说明了这封邮件的真正用意。

2

解析重点2 **...if you could kindly introduce us to some of the most capable importers.**

请求认识的一方介绍第三方，这里的“介绍”，用的是 introduce。present 也有“介绍”的意思，但尤指把他人介绍给地位更高的人。当然，我们这里还可以使用另外一个单词，那就是 recommend（推荐），例如：

..., if you will kindly recommend some of the most capable importers to us.（如果您能介绍几个能力最强的进口商给我们，……）

高频例句

1. **We are planning to expand our trade.**
   我们计划扩大我们的贸易。
2. **We are looking for an *investor*[7] who can invest in our project.**
   我们正在寻找能够投资我们项目的投资商。
3. **The main *purpose*[8] is to find a partner to set up a joint venture.**
   主要目的是想寻找一个建立合资企业的合作伙伴。
4. **We hope to expand in international petroleum market.**
   我们期望拓宽国际石油市场。
5. **We'll continue to expand the market *access*[9].**
   我们将继续扩大市场准入的范围。

必背关键单词 

1. ***increase*** [ɪnˈkriːs] v. 增加
2. ***export*** [ˈekspɔːt] n. 出口物；出口
3. ***textile*** [ˈtekstaɪl] n. 纺织
4. ***introduce*** [ˌɪntrəˈdjuːs] v. 介绍；引进
5. ***capable*** [ˈkeɪpəbl] adj. 有能力的
6. ***importer*** [ɪmˈpɔːtə(r)] n. 进口商
7. ***investor*** [ɪnˈvestə(r)] n. 投资商
8. ***purpose*** [ˈpɜːpəs] n. 目的；意图
9. ***access*** [ˈækses] n. 接近；进入

09 寻求合作

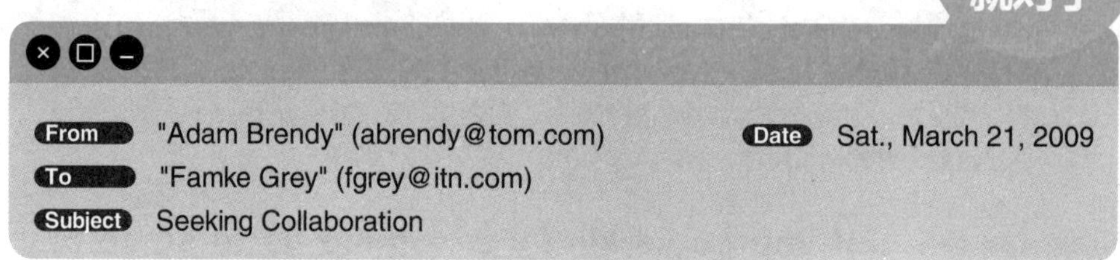
From "Adam Brendy" (abrendy@tom.com) Date Sat., March 21, 2009
To "Famke Grey" (fgrey@itn.com)
Subject Seeking Collaboration

Dear Mr. Grey,

We have obtained your address from the *Time* and are writing to you to ***seek***[1] collaboration.
We are very well connected with all the major ***dealers***[2] of electronic products here, and feel confident that we can sell large quantities of them if you can give us a ***special***[3] offer.

Your early reply will be greatly appreciated.

Yours sincerely,
Adam Brendy

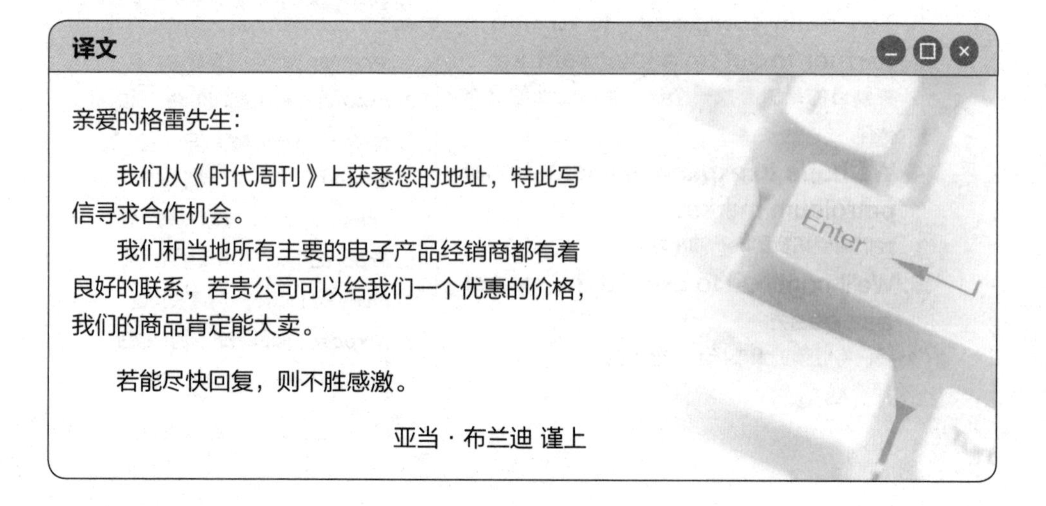
译文

亲爱的格雷先生：

我们从《时代周刊》上获悉您的地址，特此写信寻求合作机会。

我们和当地所有主要的电子产品经销商都有着良好的联系，若贵公司可以给我们一个优惠的价格，我们的商品肯定能大卖。

若能尽快回复，则不胜感激。

亚当 · 布兰迪 谨上

语法重点解析

1 解析重点1 We are very well connected with all the major dealers of electronic products here.

寻求合作的时候，最重要的是让对方明白我们雄厚的实力和强大的业务网络。拥有良好的生意人脉和渠道，是产品销售的一个非常有利的条件。We are very well connected with all the major dealers of electronic products here（我们和当地所有的电子产品大经销商都有着密切的联系），对方听了这句话，肯定会对我们寻求合作的意愿加以考虑。

2 解析重点2 We feel confident that we can sell large quantities of them if you can give us a special offer.

贸易往来的时候，往往需要的是肯定的语气，才能让别人觉得你给出的是积极有效的答复，让他们愿意与你合作。feel confident...的意思是“对……满怀信心；确信……”。而后面紧跟着的 if 从句，附上了前提条件，显得整个意思表达很清楚、合理。

1. **Please let us have all *necessary*[4] information *regarding*[5] your products.**
   请告知我们有关你们产品的所有必要信息。
2. **We are looking for a *partner*[6] who can supply us with such goods.**
   我们正在寻找可以给我们提供这种货物的合作伙伴。
3. **Please inform us what special offer you can give us based on a quantity of 600 tons.**
   请告知在600吨订购量的基础上，贵公司能给予什么样的优惠价格。
4. **Can you give us a special offer for the *purpose*[7] of introducing your product to our market?**
   为了把你们的产品引入我们的市场，可否给我们一个优惠报价？
5. **We plan to *showcase*[8] our products and seek collaboration opportunities.**
   我们计划把我们的产品展示出来，以寻求合作机会。

必背关键单词

1. ***seek*** [siːk] v. 寻找
2. ***dealer*** [ˈdiːlə(r)] n. 商人；经销商
3. ***special*** [ˈspeʃl] adj. 专门的；特别的
4. ***necessary*** [ˈnesəsərɪ] adj. 必要的；不可缺少的
5. ***regarding*** [rɪˈɡɑːdɪŋ] prep. 关于；就……而论
6. ***partner*** [ˈpɑːtnə(r)] n. 伙伴
7. ***purpose*** [ˈpɜːpəs] n. 目的；意图
8. ***showcase*** [ˈʃəʊkeɪs] v. 使展现；使亮相

10 肯定回复

这样写就对了

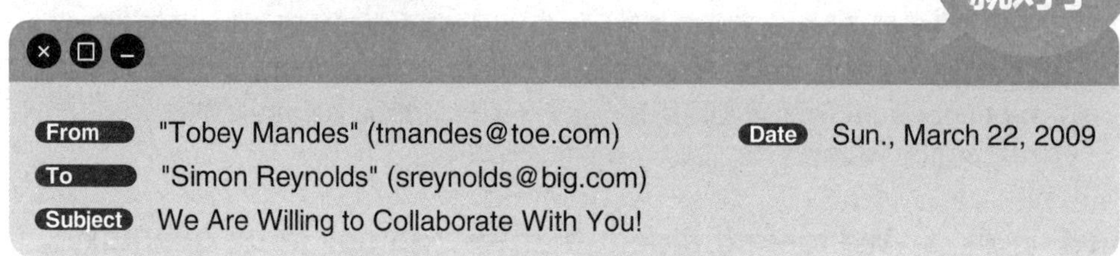

From "Tobey Mandes" (tmandes@toe.com) Date Sun., March 22, 2009
To "Simon Reynolds" (sreynolds@big.com)
Subject We Are Willing to Collaborate With You!

Dear Mr. Reynolds,

Thank you for your E-mail of March 21. **We shall be *glad*[1] to *enter*[2] into business relations with your company**.
Complying[3] **with your request, we are sending you our latest catalogue and price list**. If you find business possible, please e-mail us.

Your early reply will be greatly appreciated.

Yours sincerely,
Tobey Mandes

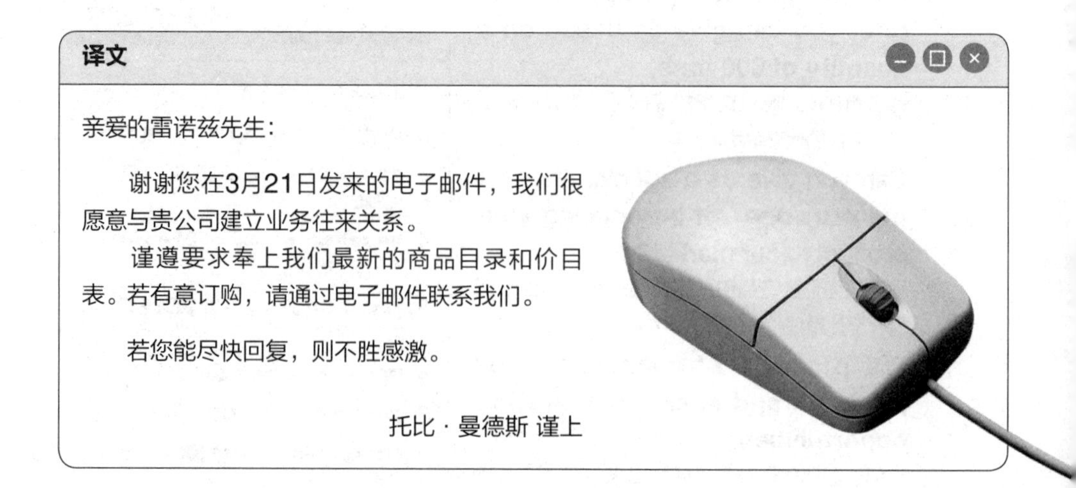

译文

亲爱的雷诺兹先生:

谢谢您在3月21日发来的电子邮件，我们很愿意与贵公司建立业务往来关系。

谨遵要求奉上我们最新的商品目录和价目表。若有意订购，请通过电子邮件联系我们。

若您能尽快回复，则不胜感激。

托比 · 曼德斯 谨上

语法重点解析

1 解析重点1 We shall be glad to enter into business relations with your company.

对于对方发出的建立业务关系的请求，不管我们是否接受，都要十分礼貌并且及时地做出回复。当我们也很想跟对方建立业务关系的时候，我们可以像上述邮件那样说：We shall be glad to enter into business relations with your company（我们很愿意与贵公司建立业务往来），shall 在这里表达了希望合作的强烈意愿。

2 解析重点2 Complying with your request, we are sending you our latest catalogue and price list.

一般来说，对方发邮件请求建立业务关系，都会在内文中咨询产品的目录、报价单之类的资料以供参考。那么，在回复过程中，我们可以说，Complying with your request, we are sending you our latest catalogue and price list（谨遵要求，奉上我们最新的商品目录和价目表），这里的 complying with your request 还可以换成 in compliance with your request。这样，就可以显示出我们非常尊重对方。

1. We are willing to ***collaborate***[4] with you in the line of ***processing***[5] materials.
   我们愿与贵方就材料加工业务进行合作。
2. We are ***ready***[6] to enter into friendly co-operation with you.
   我们愿意和你们进行友好合作。
3. 
   We would like to collaborate with you in this work.
   我们愿意在这项工作上与你们合作。
4. If this proposal is ***acceptable***[7], please let us know so that we can discuss details.
   假若贵方愿意接受我们的建议，请通知我们，以便我们进一步商讨合作的细节。
5. We are willing to collaborate with you and if necessary, we can make some ***concessions***[8].
   我们愿意和贵公司合作，必要的话，我们还可以做些让步。

必背关键单词

1. ***glad*** [glæd] adj. 高兴的
2. ***enter*** [ˈentə(r)] v. 加入；参加
3. ***comply*** [kəmˈplaɪ] v. 遵从；依从；服从
4. ***collaborate*** [kəˈlæbəreɪt] v. 合作
5. ***process*** [ˈprəʊses] v. 加工；处理
6. ***ready*** [ˈredɪ] adj. 做好准备的
7. ***acceptable*** [əkˈseptəbl] adj. 可接受的
8. ***concession*** [kənˈseʃn] n. 让步；妥协

11 婉拒对方

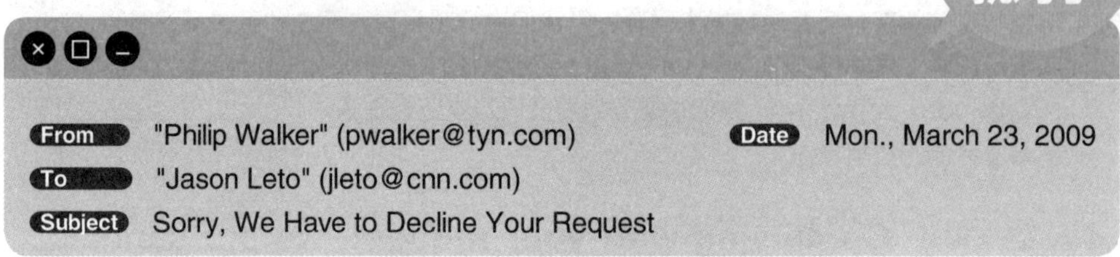

From "Philip Walker" (pwalker@tyn.com)
Date Mon., March 23, 2009
To "Jason Leto" (jleto@cnn.com)
Subject Sorry, We Have to Decline Your Request

Dear Mr. Leto,

Thank you for your e-mail of March 22.
We are sorry to inform you that the products, though with high ***content***[1] of ***science***[2] and ***technology***[3], are not well ***calculated***[4] for our market.
We are ***hoping***[5] to work with you next time. Please keep in ***touch***[6] for more business.

Looking forward to hearing from you.

Yours sincerely,
Philip Walker

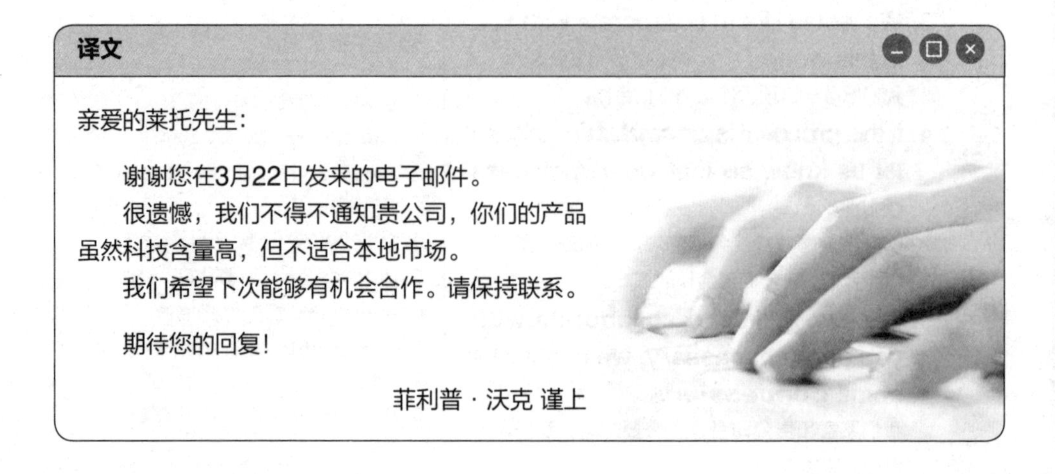

译文

亲爱的莱托先生:

谢谢您在3月22日发来的电子邮件。

很遗憾，我们不得不通知贵公司，你们的产品虽然科技含量高，但不适合本地市场。

我们希望下次能够有机会合作。请保持联系。

期待您的回复!

菲利普 · 沃克 谨上

语法重点解析

1

解析重点1 **The products, though with high content of science and technology, are not well calculated for our market.**

婉拒对方贸易请求的时候，我们首先还是要肯定对方产品的优点，再适时地说出拒绝的理由。对方产品的优点是 the products with high content of science and technology（产品高科技含量高），而拒绝的理由则是 the products are not well calculated for our market（产品不适合本地市场）。though 在这里起到了很好的衔接作用，而且很简洁（Conciseness）地表明了说话者的否定意味。

2

解析重点2 **We are hoping to collaborate with you next time.**

尽管这次无法合作，但是也许下次有合作机会。企业要维持长久的贸易关系，才能持续发展，立于不败之地，因此说话要留有余地。We are hoping to collaborate with you next time.（希望下一次有机会合作），这句话实际上也是对对方的一种鼓励。

1. **Will you please let us know other goods suitable for the market?**
   请告知其他适合市场的商品好吗?
2. **We hope we can collaborate with you in the future.**
   希望我们将来有机会合作。
3. **The market will not stand a high-priced line.**
   高价商品并不适合本地市场。
4. **We very much regret that we have to *decline*[7] your request.**
   非常遗憾，我们不得不拒绝您的请求。
5. **These goods don't fit the *local*[8] *market*[9] very much.**
   这批货物非常不适合当地市场。
6. **On account of difference in taste, the *design*[10] doesn't suit this market.**
   由于品味不同，这个设计不适合这一市场。

必背关键单词  

1. ***content*** [ˈkɒntent] n. 内容；目录
2. ***science*** [ˈsaɪəns] n. 科学
3. ***technology*** [tekˈnɒlədʒɪ] n. 科技（总称）；工艺学
4. ***calculate*** [ˈkælkjʊleɪt] v. 计算
5. ***hope*** [həʊp] v. 期望
6. ***touch*** [tʌtʃ] n. 接触；碰；触摸
7. ***decline*** [dɪˈklaɪn] v. 拒绝
8. ***local*** [ˈləʊkl] adj. 当地的
9. ***market*** [ˈmɑːkɪt] n. 市场
10. ***design*** [dɪˈzaɪn] n. 设计

12 再次寻求业务合作

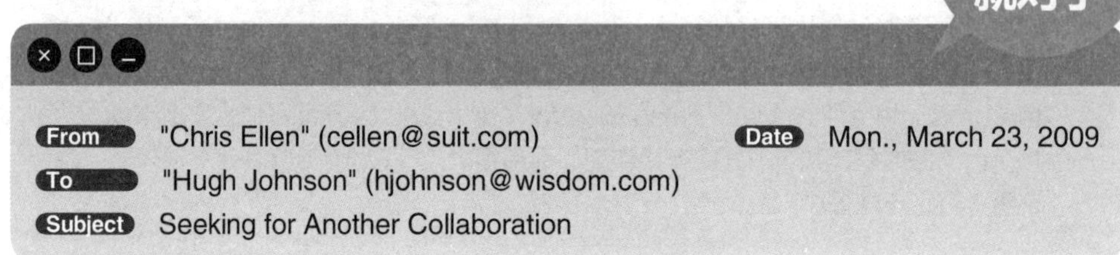

From "Chris Ellen" (cellen@suit.com) Date Mon., March 23, 2009
To "Hugh Johnson" (hjohnson@wisdom.com)
Subject Seeking for Another Collaboration

Dear Mr. Johnson,

Since our last ***conversation***[1], two years have passed.
We very much regret we *lost*[2] our last trade opportunity, but we are so happy that we ***still***[3] have an opportunity ***ahead***[4] of us now. **We have been extending the scope of our products in the past years**. I am sure some items would be of great interest to you.
The catalogue and all necessary information for your reference are ***enclosed***[5].

Looking forward to hearing from you.

Yours sincerely,
Chris Ellen

译文

亲爱的强森先生：

自从上次联系后，一晃两年过去了。

对于上次未能合作一事，我们深表遗憾，但是很高兴我们现在仍有合作的机会。过去的几年中，我们一直在扩大产品范围。相信一些产品将会让您颇感兴趣。

随函附上目录和所有必要的资料，供您参考。

期待您的回复！

克里斯 · 艾伦 谨上

语法重点解析

1 解析重点1 We very much regret we lost our last trade opportunity.

一般我们在口语中表示遗憾或是可惜，可以说：What a pity！而在书写电子邮件或是其他商务信件的时候，我们需要尽量使用比较书面化或是正式一些的词语来表达，这样才能显示出我们的庄重。We very much regret we lost our last trade opportunity（我们对上次不能合作一事感到深深的遗憾），这里 very much regret 加上从句，表达的也是遗憾、可惜，但更正式、严谨。

2 解析重点2 We have been extending the scope of our products in the past years.

企业经过几年发展，再次联系客户的时候，最好要适当地描述一下这些年来的变化和发展，以便对方了解企业目前的发展状况以及企业的新动向。这个句子使用了现在完成时，表示从过去某时（in the past years）一直持续到现在的动作，并且还将持续下去。

1. **It is a thousand *pities*[6] that we missed the chance.**
   错过了那次机会真是太可惜了。
2. **It's a pity we *missed*[7] the opportunity to collaborate last time.**
   上次错过了合作机会，真是可惜。
3. **We really missed a great opportunity last time.**
   我们上次的确是失去了一个很好的机会。
4. **Could we try to collaborate with each other again this time?**
   我们这次可否尝试着再合作一次呢？
5. **We could seek opportunities to collaborate with each other again.**
   我们可以再次寻找相互合作的机会。
6. **I hope we can do business together, and look forward to hearing from you soon.**
   我希望我们有机会合作，并期望尽快收到回信。
7. **We should make the best of this *valuable*[8] opportunity.**
   我们应该善加利用这个宝贵的机会。

必背关键单词

1. ***conversation*** [ˌkɒnvəˈseɪʃn] n. 交谈；谈话
2. ***lose*** [luːz] v. 遗失；失去
3. ***still*** [stɪl] adv. 仍然
4. ***ahead*** [əˈhed] adv. 在前面；预先
5. ***enclose*** [ɪnˈkləʊz] v. 把……装入信封；附带
6. ***pity*** [ˈpɪtɪ] n. 可惜之事；憾事
7. ***miss*** [mɪs] v. 漏掉；错过
8. ***valuable*** [ˈvæljʊəbl] adj. 宝贵的

13 咨询产品使用情况

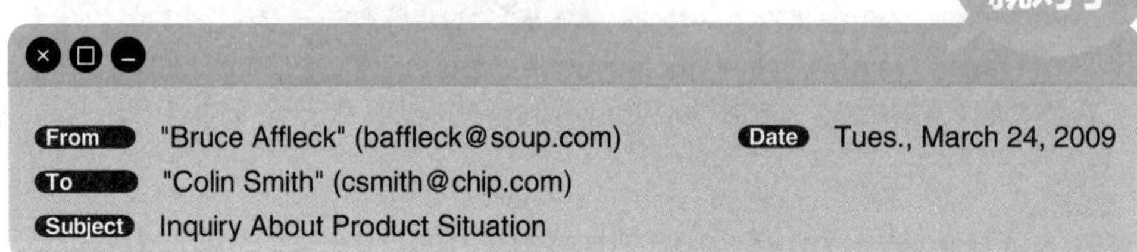
From "Bruce Affleck" (baffleck@soup.com) Date Tues., March 24, 2009
To "Colin Smith" (csmith@chip.com)
Subject Inquiry About Product Situation

Dear Mr. Smith,

You purchased a HP ***personal***[1] computer in our store in Houston last year. Thank you for choosing our ***brand***[2].
Now we would like to know whether our product is in a good ***state***[3]. Please fill out the following ***questionnaire***[4] about its service ***condition***[5] and e-mail us so that we can improve its technology.

Looking forward to hearing from you.

Yours sincerely,
Bruce Affleck

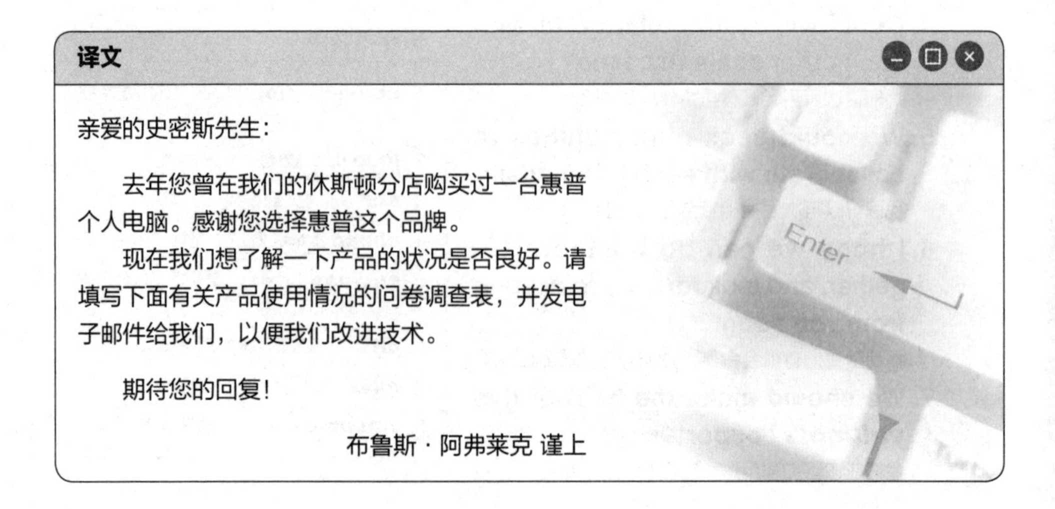
译文

亲爱的史密斯先生：

去年您曾在我们的休斯顿分店购买过一台惠普个人电脑。感谢您选择惠普这个品牌。

现在我们想了解一下产品的状况是否良好。请填写下面有关产品使用情况的问卷调查表，并发电子邮件给我们，以便我们改进技术。

期待您的回复！

布鲁斯·阿弗莱克 谨上

语法重点解析

1 解析重点1 We would like to know whether our product is in a good state.

在回访客户使用情况的时候，我们一般会询问产品使用情况好不好？有没有什么问题？那么，该如何表达呢？我们可以这样说：We would like to know whether our product is in a good state（我们想向您了解一下产品是否使用良好），be in a good state 的意思就是“状况良好”，whether 含有“是否……”的意思，表明两种情况都有可能存在。

2 解析重点2 Please fill out the following questionnaire about its service condition.

一般企业想了解一些情况，会选择使用问卷调查的形式，我们有一个单词专门用来表示“问卷调查”，那就是 questionnaire。Fill out the following questionnaire（请填写以下问卷调查表）。由于近年来网络快速发展，使得邮件问卷调查变得很普遍。这个调查有方便、快捷的优点。但是，由于垃圾邮件的增多，人们也开始反感此类行为，因此，我们也要尽量慎重地向自己的用户发送邮件，以免打扰用户。

高频例句

1. **We still need to *try*[6] and improve on our technology.**
   我们仍需尝试改进技术。
2. **It will *take*[7] you only a few minutes to fill out the questionnaire.**
   填写这份问卷，只需花上您几分钟的时间。
3. **Thanks for your *patience*[8] and *understanding*[9].**
   谢谢您的耐心和理解！
4. **You bought the clothes from our franchised store.**
   您曾在我们的专卖店购买过衣服。
5. **If you have any questions or problems after buying our products, please let us know.**
   如果你在购买我们的产品后有任何的疑问或问题，请与我们联系！

必背关键单词

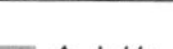

1. ***personal*** [ˈpɜːsənl] adj. 个人的
2. ***brand*** [brænd] n. 品牌
3. ***state*** [steɪt] n. 状态；情形；州；国家
4. ***questionnaire*** [ˌkwestʃəˈneə(r)] n. 问卷；调查表
5. ***condition*** [kənˈdɪʃn] n. 条件；情况
6. ***try*** [traɪ] v. 尝试
7. ***take*** [teɪk] v. 拿；取；花费
8. ***patience*** [ˈpeɪʃns] n. 耐心
9. ***understanding*** [ˌʌndəˈstændɪŋ] n. 理解

14 维护老客户

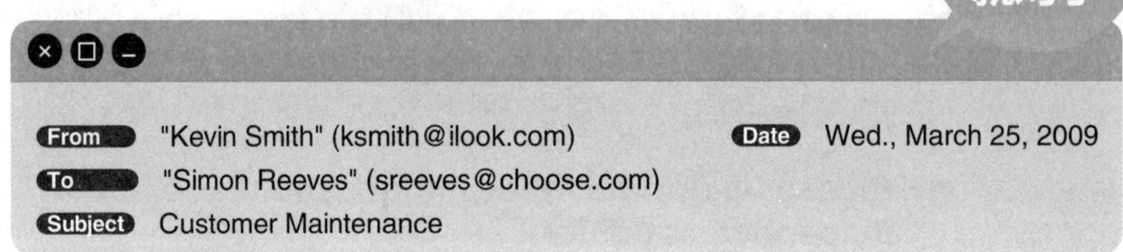

From "Kevin Smith" (ksmith@ilook.com) Date Wed., March 25, 2009
To "Simon Reeves" (sreeves@choose.com)
Subject Customer Maintenance

Dear Mr. Reeves,

We are most gratified that you have, for several years, included a ***selection***[1] of our products in your order catalogues.
We are pleased to inform you our latest product is available now. The new ***machine***[2] vastly ***exceeds***[3] the old one in performance. If you, our old customer, are interested in it, we can offer you a 10% ***discount***[4].

Looking forward to hearing from you.

Yours sincerely,
Kevin Smith

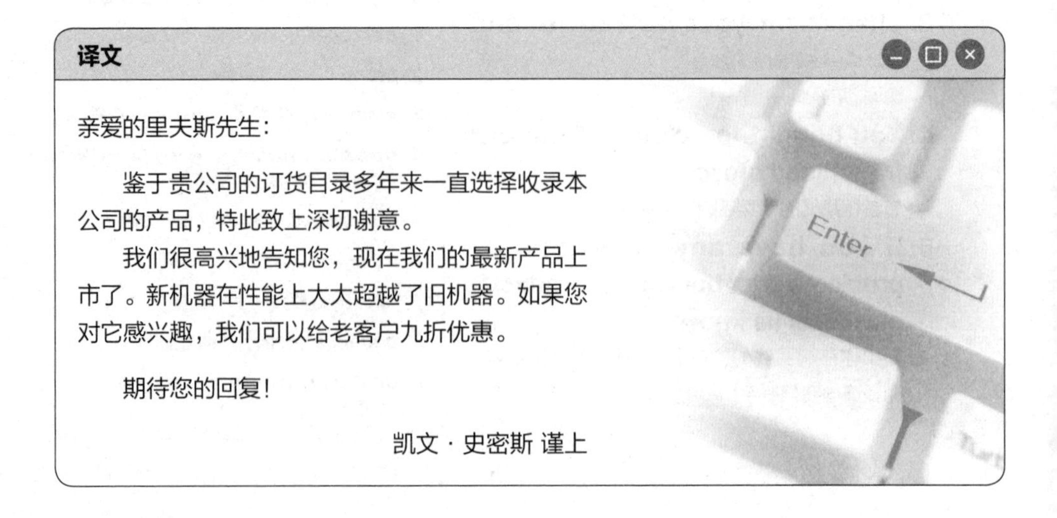

译文

亲爱的里夫斯先生：

鉴于贵公司的订货目录多年来一直选择收录本公司的产品，特此致上深切谢意。
我们很高兴地告知您，现在我们的最新产品上市了。新机器在性能上大大超越了旧机器。如果您对它感兴趣，我们可以给老客户九折优惠。

期待您的回复！

凯文 · 史密斯 谨上

语法重点解析

1 解析重点1 We are most gratified that...

由于邮件内容是针对客户或合作伙伴，所以我们的措辞一般要客气和正式。表达“感谢、感激某事”时，可以用 we are gratified that...，但是，当我们在 be 动词之后再加上 most，感激的程度则更深。邮件内文中的句子 We are most gratified that... 的意思就是“我们非常感谢……”。

2 解析重点2 The new machine vastly exceeds the old one in performance.

当我们在向客户介绍新产品时，肯定要说明新产品和旧产品之间的差别，那么，表示新产品优于旧产品，该如何表达呢?

一般，我们会说：The new machine is better than the old one（新机器比旧机器好）。但是，如果我们用 exceed 这个单词，不仅可以准确表达 be better than 的意思，也可以使句子又简洁（Clearness）了不少。

1. Always with pleasure at your service.
我们随时愿意为您效劳。

2. I'd like to introduce you to our new product *line*[5].
我想向您介绍我们新的产品线。

3. We are glad to have the opportunity to introduce to you our *newly*[6] developed products.
很高兴能有此机会向贵公司介绍我们新开发的产品。

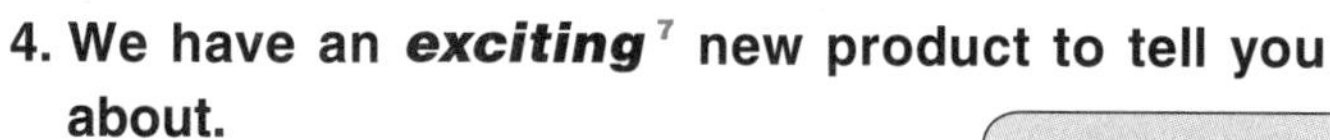

4. We have an *exciting*[7] new product to tell you about.
我们要向大家介绍一款令人兴奋的新产品。

5. It is ever so nice of you to give us support.
非常感谢您对我们的支持。

6. We are ready and eager to serve you.
我们恭候您，并竭诚为您服务。

7. We hope we may be favored with your *commands*[8] which shall at all times have our *utmost*[9] attention.
竭诚欢迎贵公司前来订货，对此，我方将随时予以极大的关注。

必背关键单词

1. *selection* [sɪˈlekʃn] n. 选择
2. *machine* [məˈʃiːn] n. 机器；机械
3. *exceed* [ɪkˈsiːd] v. 胜过；超过
4. *discount* [ˈdɪskaʊnt] n. 折扣
5. *line* [laɪn] n. （产品的）种类；生产线
6. *newly* [ˈnjuːlɪ] adv. 新近；最近
7. *exciting* [ɪkˈsaɪtɪŋ] adj. 使人兴奋的
8. *command* [kəˈmɑːnd] n. 命令
9. *utmost* [ˈʌtməʊst] adj. 极度的；最大的

15 感谢客户

From "Topher Krave" (tkrave@search.com)
Date Thurs., March 26, 2009
To "John Miller" (wmiller@look.com)
Subject Letter of Thanks

Dear Mr. Miller,

Thank you for your support for our company in the last years. We really appreciate your cooperation, and we hope we can ***continue***[1] our ***good***[2] business ***relationship***[3] and interactivity in the ***future***[4].

If we can be of service to you again, please let us know.

Yours sincerely,
Topher Krave

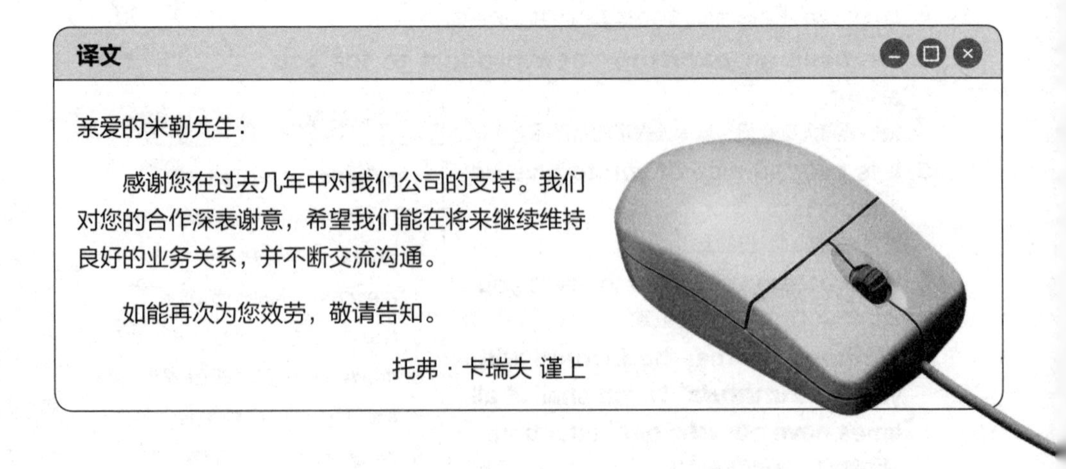

译文

亲爱的米勒先生：

感谢您在过去几年中对我们公司的支持。我们对您的合作深表谢意，希望我们能在将来继续维持良好的业务关系，并不断交流沟通。

如能再次为您效劳，敬请告知。

托弗 · 卡瑞夫 谨上

语法重点解析

1 解析重点1 **We hope we can continue our good business relationship and interactivity in the future.**

我们在答谢客户的时候，还要表达自己希望维系双方贸易关系的意愿。We hope we can continue our good business relationship and interactivity in the future（希望我们能在将来继续维持良好的业务关系，并不断交流沟通），continue our business relationship（继续维持这种业务关系）中的 continue 这个词用的很恰当，是“继续”的意思。

2 解析重点2 **If we can be of service to you again, please let us know.**

一般我们在写邮件内文的结尾时，总要表达自己愿意随时为客户服务的意愿。例如，上面邮件中的 If we can be of service to you again, please let us know（如能再为您效劳，敬请告知）。还有更多类似的说法：

We are always pleased to serve you at any time.（我们很高兴随时为您服务。）

We're always at your service.（我们随时为您服务。）

1. **We thank our clients for their interest in our products and welcome your inquiry and orders with us.**
   本公司衷心感谢客户对我们产品的关注，并热情欢迎您的洽询订货。
2. **Thank you for your support and we will continue to do our best to provide everyone quality service.**
   感谢客户们的支持，我们会继续努力为大家提供优质服务。
3. **We thank our new and old clients for choosing our products.**
   特此感谢新老客户选购我公司的产品。
4. **We thank customers both at home and *abroad*[5] for their *trust*[6] and *firm*[7] *support*[8].**
   感谢国内外新老客户的信赖和鼎力支持。

必背关键单词 

1. ***continue*** [kən'tɪnjuː] v. 继续；持续
2. ***good*** [gʊd] adj. 好的；优良的
3. ***relationship*** [rɪ'leɪʃnʃɪp] n. 关系
4. ***future*** ['fjuːtʃə(r)] n. 未来；将来
5. ***abroad*** [ə'brɔːd] adv. 在国外；到国外
6. ***trust*** [trʌst] n. 信任
7. ***firm*** [fɜːm] adj. 坚定的；坚固的
8. ***support*** [sə'pɔːt] n. 支持

Unit 8 询问 Inquiry

01 咨询商品信息

From "John Hancock" (jhancock@apl.com)　**Date** Mon., February 23, 2009
To "Peter Smith" (psmith@cri.com)
Subject Asking for Product Information

Dear Mr. Smith,

We learned from the ***advertisement***[1] that your company produces ***electronic***[2] products of high ***quality***[3].
We are going to order more of them for we find that they are in a great ***demand***[4] in the local shops. **Is it possible for you to send us a detailed *catalogue***[5] or any ***material***[6] about your products in terms of price, specification and payment method?

Looking forward to hearing from you!

Yours sincerely,
John Hancock

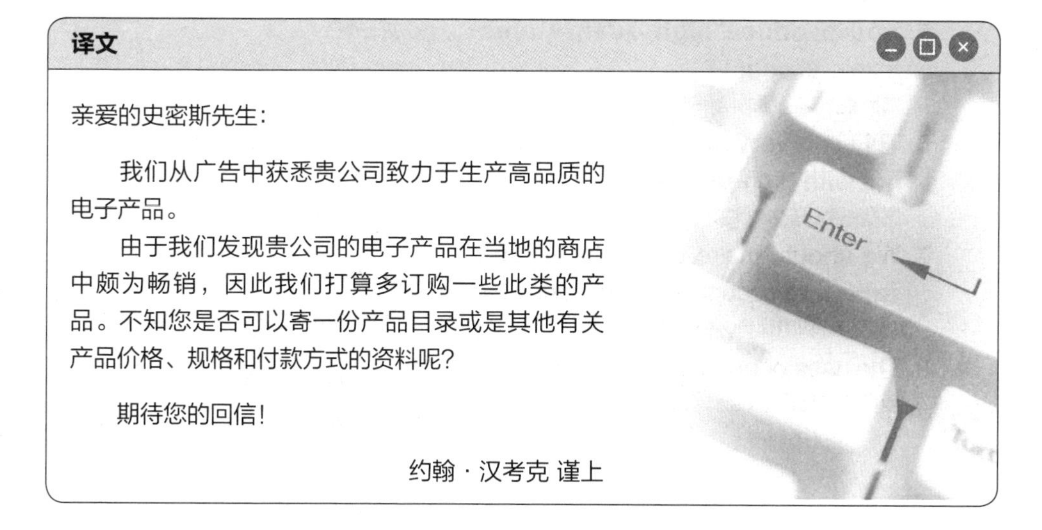

译文

亲爱的史密斯先生:

我们从广告中获悉贵公司致力于生产高品质的电子产品。

由于我们发现贵公司的电子产品在当地的商店中颇为畅销，因此我们打算多订购一些此类的产品。不知您是否可以寄一份产品目录或是其他有关产品价格、规格和付款方式的资料呢?

期待您的回信!

约翰·汉考克 谨上

语法重点解析

1 解析重点1 We are going to order more.

一般我们表达想要进更多的货，我们会说：We want to order more. 这个句子在意思表达上没有问题，不过如果我们把 want to order 改成 be going to order，也许会显得更加有诚意一些。因为 be going to order 表示事先经过考虑后，打算要做某件事情。这样一来，就可以看出我们是经过充分考虑的。

2 解析重点2 Is it possible for you to send us a catalogue?

想要表达“您能寄给我们一份产品目录吗？”我们可以说：Can you give us a catalogue of your products? 这个句子在意思表达上没有问题。Can you...?（你能……？）已经是比较礼貌的问法了，但如果我们使用 Is it possible to do sth（不知可否……？）来询问同样的问题，则会显得更加委婉和礼貌。它充分体现了英文书信7C原则中的 Courtesy（礼貌）原则。

高频例句

1. **I am writing to ask for *information*[7] about your products.**
   我写信是想咨询贵公司的产品信息。
2. **Can you inform us of your products in detail?**
   你能告知我们你们产品的详细信息吗?
3. **Your products are well received locally.**
   你们的产品在当地很畅销。
4. **We would appreciate it if you would send us your catalogue.**
   如能寄来一份商品目录，我们将不胜感激。
5. **You produce high quality electronic products.**
   你们生产高品质的电子产品。
6. **This is a very challenging industry, but with high *potential*[8].**
   这是一个极具挑战性和发展前景的产业。
7. **We should make use of its *advantages*[9] to occupy the market quickly.**
   我们应该利用优势，快速占领市场。
8. **The uses of our products are various.**
   我们的产品用途广泛。

必背关键单词

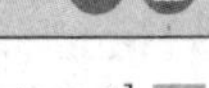

1. ***advertisement*** [əd'vɜ:tɪsmənt] n. 广告；宣传
2. ***electronic*** [ɪˌlek'trɒnɪk] adj. 电子的；电子化的
3. ***quality*** ['kwɒlətɪ] n. 品质；质量
4. ***demand*** [dɪ'mɑ:nd] n. 需求；需要
5. ***catalogue*** ['kætəlɒg] n. 目录；目录册；览表
6. ***material*** [mə'tɪərɪəl] n. 资料；材料
7. ***information*** [ˌɪnfə'meɪʃn] n. 消息；资料；情报
8. ***potential*** [pə'tenʃl] n. 潜力；潜能
9. ***advantage*** [əd'vɑ:ntɪdʒ] n. 有利条件；有利因素；优势

02 咨询交货日期

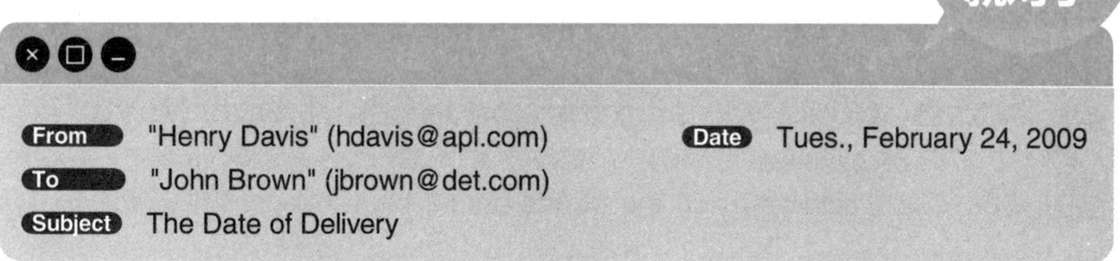

From "Henry Davis" (hdavis@apl.com)　Date Tues., February 24, 2009
To "John Brown" (jbrown@det.com)
Subject The Date of Delivery

Dear Mr. Brown,

Would you please ***inform***[1] us how long it usually takes for you to make ***delivery***[2] of our order of May 20 (Order No. 728) for car ***components***[3]?
Could you ***ship***[4] the ***goods***[5] before early May?
Moreover, please mail the ***invoice***[6] of this order to our company.

Looking forward to hearing from you soon.

Yours sincerely,
Henry Davis

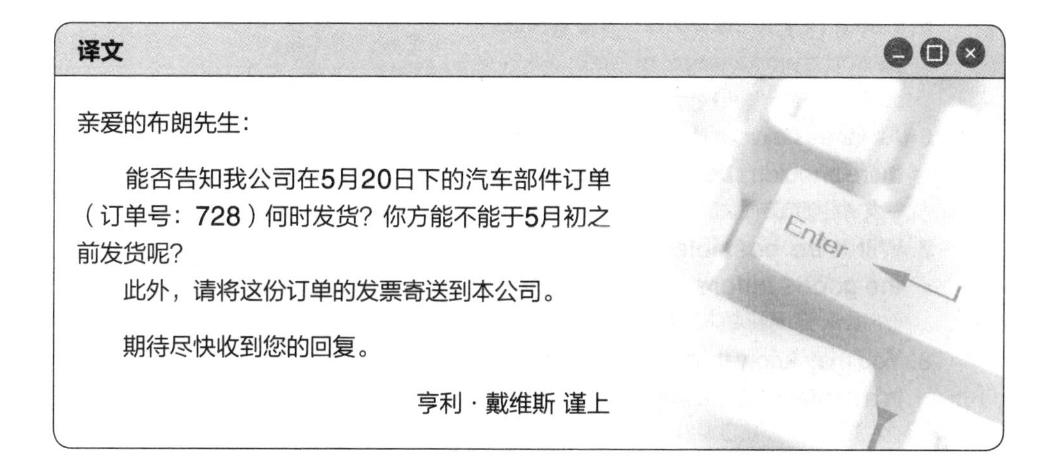

译文

亲爱的布朗先生：

能否告知我公司在5月20日下的汽车部件订单（订单号：728）何时发货？你方能不能于5月初之前发货呢？

此外，请将这份订单的发票寄送到本公司。

期待尽快收到您的回复。

亨利·戴维斯 谨上

1 解析重点1 Would you please inform us...?

当我们在询问对方相关交货日期事宜的时候，我们往往会说：Would you please inform us...?（能否请您告知我们……？）或者 Please inform us...（请告知我们……）。很明显，前者更加礼貌（Courtesy）和客气，隐含着商量的意味。

2 解析重点2 Could you ship the goods before early May?

在询问交货日期时，身为客户，我们可能还会有一些特殊要求，希望对方能够满足。例如 Could you ship the goods before early May?（你们能不能于五月初之前发货呢？）这里也可以使用 Couldn't you...? 但感觉语气比较强硬，通常在货物严重延误时才使用。

高频例句

1. **You mustn't let us down on delivery *dates*[7].**
   贵方不能在发货日期上让我方失望。
2. **We can live with the other terms, *except*[8] the delivery date.**
   我们可以同意交货日期以外的其他条件。

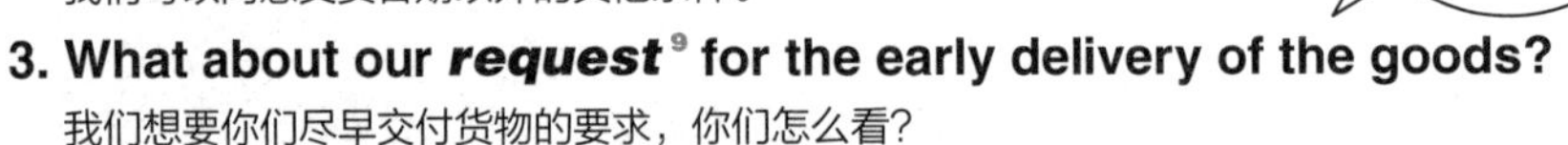

3. **What about our *request*[9] for the early delivery of the goods?**
   我们想要你们尽早交付货物的要求，你们怎么看？
4. **When will you deliver the products to us?**
   你们什么时候可以发货给我们？
5. **Couldn't you *extend*[10] the delivery period by one week or so?**
   贵方不能将交货期再延长一个星期左右吗？
6. **As far as delivery dates are concerned, there shouldn't be any problems.**
   就发货日期这方面，应该没有什么问题了。
7. **Will it be possible for you to ship the goods before early September?**
   你们能否于9月初之前发货呢？
8. **You may know that the time of delivery is a matter of great importance.**
   你们知道的，交货日期事关重大。

必背关键单词

1. ***inform*** [ɪnˈfɔːm] v. 通知
2. ***delivery*** [dɪˈlɪvərɪ] n. 递送；送交
3. ***component*** [kəmˈpəʊnənt] n. 部件；零件
4. ***ship*** [ʃɪp] v. 运送；装船
5. ***goods*** [gʊdz] n. 商品；货物
6. ***invoice*** [ˈɪnvɔɪs] n. 发票
7. ***date*** [deɪt] n. 日期
8. ***except*** [ɪkˈsept] prep. 除……外
9. ***request*** [rɪˈkwest] n. 要求；请求
10. ***extend*** [ɪkˈstend] v. 延长

03 咨询交易条件

这样写就对了

From "Tom Gordon" (tgordon@qdr.com)
Date Wed., February 25, 2009
To "Kevin Williams" (kwilliams@aot.com)
Subject The Terms and Conditions of Business

Dear Mr. Williams,

Thanks for your call last week. To ***confirm***[1] our ***conversation***[2], we'd like to ***inquire***[3] about your trading terms and conditions for providing related ***equipments***[4] and services.
We also want to know how long it will take you to finish this ***project***[5] if we decide to let you do this job.
Please offer us a ***quotation***[6] ***specifying***[7] terms and conditions of business and a work schedule.

Looking forward to hearing from you!

Yours sincerely,
Tom Gordon

译文

亲爱的威廉姆斯先生：

感谢您上周的来电。我想就此确认一下我们电话中所谈及的交易条件，以便能提供相关设备及服务。

此外，我还想知道，如果将该项目交给你们，需要多长时间才能完成。

最后，还要麻烦您为我们提供一份有具体交易条件的报价单以及一份工作日程表。

期待您的回复！

汤姆 · 高登 谨上

语法重点解析

1 解析重点1 To confirm our conversation, we'd like to...

通常，为了说明自己发送邮件的目的，我们需要用上“为了……”这一说法，这时往往会想到用 in order to... 或 for the purpose of...这两个短语，虽然在意思和语法上都没有问题，但是我们可以发现上述的邮件内文中，仅仅只用了一个介词 to 就表达了跟上面这两个短语同样的意思，体现了简洁（Conciseness）这一写作原则，使意思一目了然，同时也避免了句子头重脚轻。

2 解析重点2 If we decide to let you do this job...

当我们打算把业务交给另一方来做的时候，我们往往会说：If we decide to let you do this job...，而询问对方需要多久才能完成交办事项则用：How long will it take you to finish this project?

高频例句

1. **What are the terms and conditions on this trade cooperation?**
   这次贸易合作的条件是什么？
2. **I shall be glad if you will send me your price list, and *state*[8] your best term.**
   如果您能寄给我价格表，并告知最好的交易条件，我将不胜欣喜。
3. **The business is booked when the terms and *conditions*[9] are agreed upon.**
   当这些条款和条件均为双方所接受时，交易达成。
4. **We shall thank you for letting us know your trade terms and forwarding us samples and other helpful literature.**
   感谢贵公司告知交易条件并赠以样品和其他有用的资料。
5. **We'll go on to the other terms and conditions this time.**
   这次让我们来探讨一下其他条件和情况。
6. **All the terms and conditions shall be clearly stated in the quotation.**
   所有条件和情况都应在报价单中清楚列明。

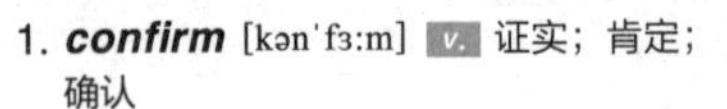

必背关键单词

1. ***confirm*** [kən'fɜ:m] v. 证实；肯定；确认
2. ***conversation*** [ˌkɒnvə'seɪʃn] n. 交谈；谈话；会话
3. ***inquire*** [ɪn'kwaɪə(r)] v. 打听；询问
4. ***equipment*** [ɪ'kwɪpmənt] n. 设备；装备
5. ***project*** ['prɒdʒekt] n. 项目；计划；方案；课题
6. ***quotation*** [kwəʊ'teɪʃn] n. 报价；行情；引用
7. ***specify*** ['spesɪfaɪ] v. 详述
8. ***state*** [steɪt] v. 陈述；声明
9. ***condition*** [kən'dɪʃn] n. 情况；条件

04 咨询库存状况

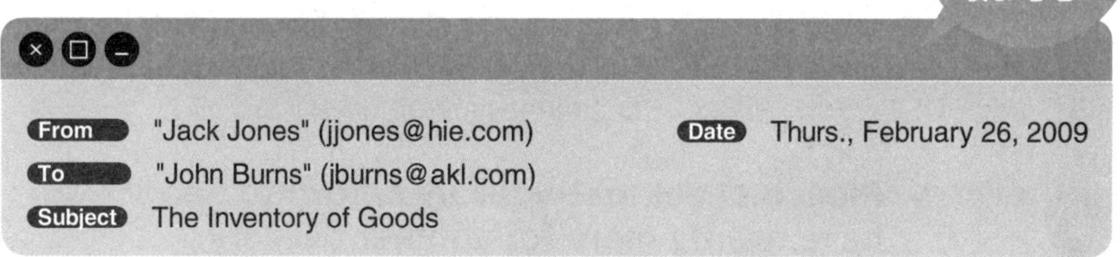
From "Jack Jones" (jjones@hie.com) Date Thurs., February 26, 2009
To "John Burns" (jburns@akl.com)
Subject The Inventory of Goods

Dear Mr. Burns,

Because your ***camera***[1], DSC-T700, sells well here with its high quality and favorable price, we have decided to ***purchase***[2] more.
Please ***check***[3] out your ***inventory***[4] to see if you have twenty more for another delivery.

Looking forward to hearing from you soon!

Yours sincerely,
Jack Jones

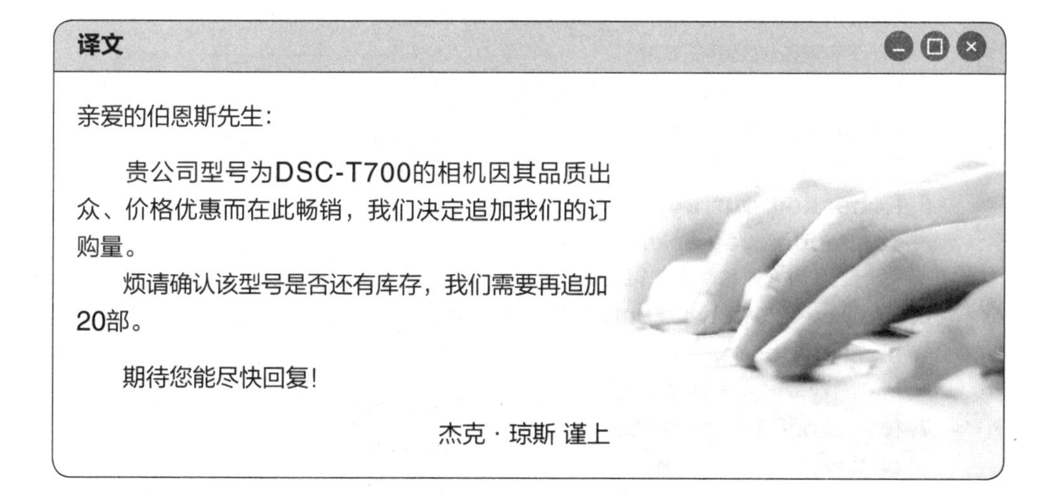
译文

亲爱的伯恩斯先生：

贵公司型号为DSC-T700的相机因其品质出众、价格优惠而在此畅销，我们决定追加我们的订购量。

烦请确认该型号是否还有库存，我们需要再追加20部。

期待您能尽快回复！

杰克・琼斯 谨上

语法重点解析

1 解析重点1 **Because your camera, DSC-T700, sells well here...**

一般我们在与卖家交易的过程中，如果卖家提供的商品销量不错，需要再进货的时候，我们都免不了要夸赞一下对方的商品，这个时候我们可以这样说：Because your camera, DSC-T700, sells well here...（贵公司型号为DSC-T700的照相机销量很好……），这里提到了照相机的型号DSC-T700，很具体（Concreteness），说得清楚明白。

2 解析重点2 **Please check out your inventory to see if you have twenty more for another delivery.**

询问库存情况的时候，我们需要用到一个比较礼貌的句型，那就是 Please check out your inventory to see if...（还请确认一下……是否还有库存），这里的 to see if 后面接从句，具体地提到了需要购买的数量：twenty more（追加二十台）。这样既礼貌（Courtesy），又做到了数目清晰、具体（Concreteness），毫不含糊。

1. The ***current***[5] inventory of the product can't meet the need.
   目前产品库存无法满足订单需求。
2. Would you do an inventory check for us?
   能否麻烦您为我们查看一下库存情况?
3. What type of model do you have in ***stock***[6]?
   你们库存里有什么型号的商品?
4. I just got an ***answer***[7] about the stock we have on hand.
   我刚刚获知我们现有的库存量。
5. I checked our ***supply***[8] of the commodity you asked for.
   我查过了你要的那种商品的库存。
6. At present, we have only a limited stock of goods.
   目前，我们的货物库存有限。
7. I regret not ***receiving***[9] the inventory on time.
   我很遗憾没有及时收到存货清单。

必背关键单词

1. ***camera*** [ˈkæmərə] n. 照相机；摄影机
2. ***purchase*** [ˈpɜːtʃəs] v. 购买
3. ***check*** [tʃek] v. 检查；核对
4. ***inventory*** [ˈɪnvəntrɪ] n. 库存；存货清单；详细目录
5. ***current*** [ˈkʌrənt] adj. 现在的；现行的
6. ***stock*** [stɒk] n. 库存；股票
7. ***answer*** [ˈɑːnsə(r)] n. 回答；回复
8. ***supply*** [səˈplaɪ] n. 供给物；储备物资
9. ***receive*** [rɪˈsiːv] v. 收到；接到

05 咨询未到货商品

这样写就对了

From "Jim Landy" (jlandy@hie.com)　Date Wed., February 25, 2009
To "Ortis Branden" (obranden@akl.com)
Subject Inquiry About Undelivered Goods

Dear Mr. Branden,

We have ***ordered***[1] five computers (No. 4879) on February 20, but we haven't received them yet. Would you tell us **when you will be *delivering*[2] these computers which should have *arrived*[3] a week ago**?
We ***desperately***[4] need them for our new employees.

Please ***respond***[5] without ***delay***[6]!

Yours sincerely,
Jim Landy

译文 

亲爱的布兰登先生：

我们于2月20号在贵公司订购了5台电脑（编号为4879），但至今尚未收到货物。贵公司能否告知这些原本一周前就应该到货的电脑将于何时出货呢?
我们急需这些电脑供新员工使用。

请立即回复!

吉姆·兰迪 谨上

语法重点解析

1 解析重点1 **When will you be delivering these computers?**

询问对方本应送达的商品为何推迟时，不要太过于着急或是责问对方，也许对方遇到了运送上的困难。我们一般会这样询问：When will you...?（你们将何时……？）而不会说：When will we...?（我们将何时……？）很明显，后面的问法会让人听起来很有压力，而前面的问法则设身处地地为对方着想，体现了体贴（Consideration）原则。同时，will be doing sth 表达的是将来某一时间正在进行的动作，常用来表示礼貌（Courtesy）的询问及请求等。

2 解析重点2 **...which should have arrived a week ago**

本来应该到货的商品没有到，我们也许会说：The computers haven't reached us yet. 当然，这样的表达在意思上是没有问题的。但是，如果用 should have done 或是 should have been 来表达，更能显示出本来应该发生的事情却没有发生的意味。which should have arrived a week ago（一周前就应该送达的电脑却没有到货），这样的表达既贴切，又很简洁（Conciseness）。在今后类似的情况中，我们可以多运用虚拟语气。

高频例句

1. **They were *supposed*[7] to arrive three days ago.**
   它们早在三天前就应该送达。
2. **We were informed that we would get the goods *within*[8] one week.**
   你们之前告知我们一周内到货。
3. **I was just informed that the new product we ordered on September 6 hasn't arrived yet.**
   我刚才才知道我公司于9月6日订购的新产品尚未到货。
4. **Follow up the order to ensure the *punctual*[9] arrival of goods.**
   请追踪订单以确保物品准时到货。
5. **We greatly *regret*[10] to say that the goods we ordered haven't reached us yet.**
   非常遗憾，我们至今还未收到订购的产品。

必背关键单词 

1. ***order*** [ˈɔ:də(r)] v. 订购；订货
2. ***deliver*** [dɪˈlɪvə(r)] v. 递送；交付
3. ***arrive*** [əˈraɪv] v. 到达；到来
4. ***desperately*** [ˈdespərətlɪ] adv. 绝望地；不顾一切地
5. ***respond*** [rɪˈspɒnd] v. 回答；回报
6. ***delay*** [dɪˈleɪ] n. 耽搁；延迟
7. ***suppose*** [səˈpəʊz] v. 料想；猜想
8. ***within*** [wɪˈðɪn] prep. （表示时间）不超过
9. ***punctual*** [ˈpʌŋktʃʊəl] adj. 准时的
10. ***regret*** [rɪˈgret] v. 遗憾

06 咨询价格及费用

这样写就对了

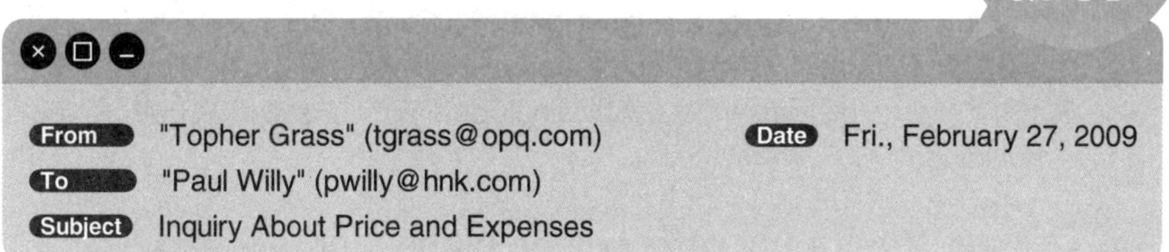

From "Topher Grass" (tgrass@opq.com)　Date Fri., February 27, 2009
To "Paul Willy" (pwilly@hnk.com)
Subject Inquiry About Price and Expenses

Dear Mr. Willy,

I would like to know the price with ***freight***[1] and ***handling***[2] included for the ***laser***[3] ***printer***[4] your company provides lately.
Please let us know the terms on which you can give us some ***discount***[5].

Your early ***offer***[6] will be highly appreciated.

Yours sincerely,
Topher Grass

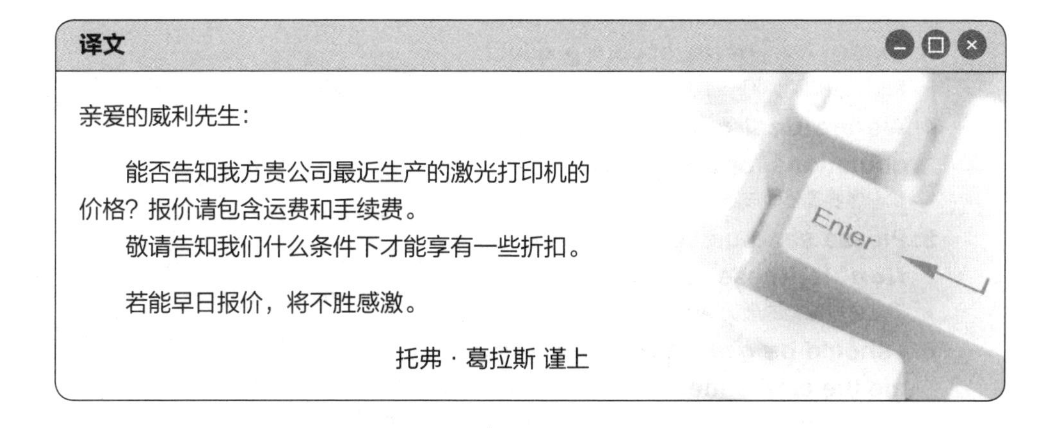

译文

亲爱的威利先生:

能否告知我方贵公司最近生产的激光打印机的价格? 报价请包含运费和手续费。
敬请告知我们什么条件下才能享有一些折扣。

若能早日报价，将不胜感激。

托弗 · 葛拉斯 谨上

语法重点解析

1 解析重点1 **I would like to know the price with freight and handling included for the laser printer.**

在咨询价格及费用的时候，一般会很笼统地问：I would like to know the price for the laser printer（能否告知激光打印机的价格），但是，如果我们更具体一点（Concreteness），把运费、手续费等额外费用都罗列出来，则使对方的报价更明确。with...included 表示“把……包括在内”，与“including＋名词”意思相同。

2 解析重点2 **Your early offer will be highly appreciated.**

在邮件中，我们往往追求简洁明了（Conciseness），所以经常使用一些被动语态。例如：Your early offer will be highly appreciated（若能早日报价，将不胜感激），使用被动语态之后，your early offer 这个句子的重心就突显出来了，比使用 We will highly appreciate it if you send us your early offer. 更加清晰（Clearness）、一目了然。

1. **Will you send us a copy of your catalogue, with details of the prices and terms of payment?**
   可以请您给我方寄一份贵公司的产品目录，并注明价格和付款条件吗？
2. **We would like to make an inquiry about this product.**
   我们想要对该产品进行询价。
3. **We would like to know the price *exclusive*[7] of tax of your product.**
   我们想要知道你们产品不含税的价格。
4. **We are desirous of your lowest quotations for the printer.**
   我们想要贵公司该类打印机的最低报价。
5. **Please send us your best *quotation*[8] for these computers.**
   请把这些电脑的最优惠价格报给我们。
6. **I should be *grateful*[9] if you send me the catalogue.**
   若您能寄给我方一份目录，将不胜感激。
7. **Kindly quote us your lowest prices for the furniture.**
   请把这批家具的最低价格报给我们。

必背关键单词

1. ***freight*** [freɪt] n. 运费
2. ***handling*** [ˈhændlɪŋ] n. 处理；手续费
3. ***laser*** [ˈleɪzə(r)] n. 激光
4. ***printer*** [ˈprɪntə(r)] n. 打印机
5. ***discount*** [ˈdɪskaʊnt] n. 折扣
6. ***offer*** [ˈɒfə(r)] n. 提供；报价
7. ***exclusive*** [ɪkˈskluːsɪv] adj. 排外的
8. ***quotation*** [kwəʊˈteɪʃn] n. 报价
9. ***grateful*** [ˈɡreɪtfl] adj. 感激的

07 咨询公司信息

这样写就对了

From "Willy Miller" (wmiller@hof.com)　Date Sat., February 28, 2009
To "Ryan Collins" (rcollins@jkr.com)
Subject Asking for Information About the Company

Dear Mr. Collins,

I would like to ***request***[1] more information about your company.
Compared with other *similar*[2] products, I am planning to ***invest***[3] in your hearing aids because of their better ***performances***[4]. **I would appreciate any *brochures*[5] or marketing materials with which you could provide me**.

Thank you in ***advance***[6]. I am looking forward to your reply.

Yours sincerely,
Willy Miller

译文

亲爱的柯林斯先生：

我想咨询有关贵公司的更多信息。

与其他同类产品比较之后，发现你们的助听器性能更佳，所以我打算投资贵公司的助听器。如果您能将一些介绍手册或市场销售资料提供给我，将不胜感激。

谨此先向您致谢，并期待您的来信！

威利 · 米勒 谨上

语法重点解析

1 解析重点1 Compared with other similar products,...

在打算投资某公司产品时，肯定要做一番产品之间的比较。compared with other similar products（与其他同类产品作比较）中的 similar 这个单词指出是在同类产品间做了比较，指明调查很科学。措辞严谨这一点与日期、资料准确比起来，更容易被人们忽略。因此，在写作的时候要多加注意。

2 解析重点2 I would appreciate any brochures or marketing materials with which you could provide me.

当我们在询问对方公司的信息时，一般会直接说：I would like to request more information about your company（我想咨询有关贵公司更多的信息），这个句子并没有错误，也是经常使用的句子。但如果能够加上一些更为具体的内容，例如：I would appreciate any brochures or marketing materials with which you could provide me（如果您能将一些介绍手册或市场销售资料提供给我，将不胜感激），会更具体（Concreteness）、清楚（Clearness）。

高频例句

1. I would like to request a copy of your company brochure.
我想要一份贵公司的简介。

2. I am very interested in knowing more about your company.
对于进一步了解贵公司，我非常感兴趣。

3. Thank you in advance for your kind attention.
在此先谢谢您的关照。

4. I am writing to request some information about your company.
我写这封信是想要咨询一些贵公司的相关信息。

5. Is it possible to have a copy of your *annual*[7] report?
您可不可以给我一份贵公司的年度报告？

6. Thank you for your *continued*[8] *support*[9].
谢谢您一如既往的支持。

必背关键单词

1. ***request*** [rɪˈkwest] v. 请求；要求
2. ***similar*** [ˈsɪmələ(r)] adj. 类似的；同类的
3. ***invest*** [ɪnˈvest] v. 投资；抽入（时间、精力等）
4. ***performance*** [pəˈfɔːməns] n. 性能；表现
5. ***brochure*** [ˈbrəʊʃə(r)] n. 介绍手册；说明书
6. ***advance*** [ədˈvɑːns] adj. 事先的；预先的；提前的
7. ***annual*** [ˈænjʊəl] adj. 年度的
8. ***continued*** [kənˈtɪnjuːd] adj. 继续的
9. ***support*** [səˈpɔːt] n. 支持

08 咨询银行业务

这样写就对了

From "Kenny Relly" (krelly@wty.com) Date Sun., March 1, 2009
To "John Doorman" (jdoorman@nhk.com)
Subject Inquiry About Bank Business

Dear Mr. Doorman,

Our company is looking for a new ***bank***[1] which will provide us with good service at a ***reasonable***[2] ***cost***[3].
Please send us your brochure and ***fee***[4] schedule on business service. After we ***review***[5] all the materials from different banks, we will inform you whether we will open our ***account***[6] in your bank.

Looking forward to your reply!

Yours sincerely,
Kenny Relly

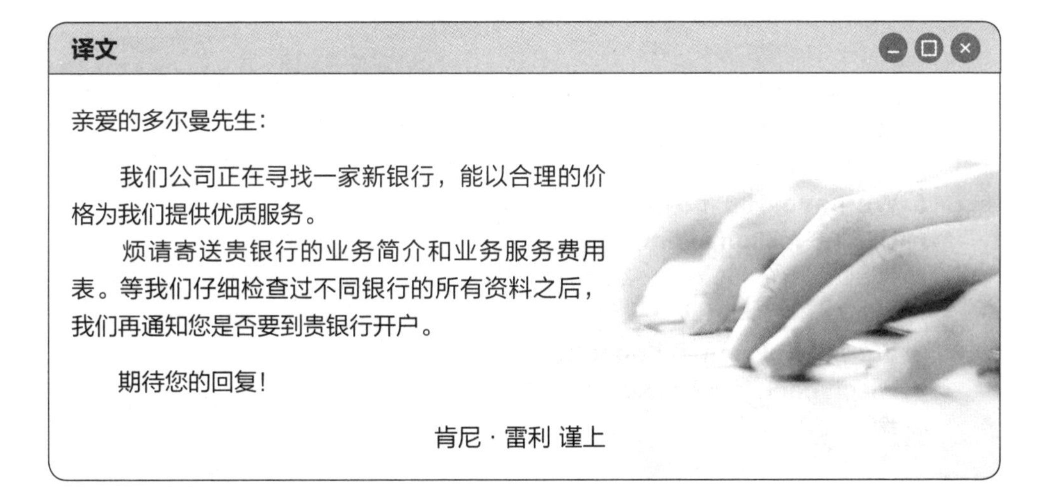

译文

亲爱的多尔曼先生：

我们公司正在寻找一家新银行，能以合理的价格为我们提供优质服务。
烦请寄送贵银行的业务简介和业务服务费用表。等我们仔细检查过不同银行的所有资料之后，我们再通知您是否要到贵银行开户。

期待您的回复！

肯尼 · 雷利 谨上

语法重点解析

1 解析重点1 ...which will provide us with good service at a reasonable cost.

在寻求商业服务的时候，肯定是希望对方收费合理，最常用的说法就是 favorable price（优惠的价格）。但是这个说法，一般用于跟他人或是其他公司进行贸易、讨价还价的情况当中。而且，它还含有可以商榷的意味。银行业务与贸易交往还是有所区别的，银行提供的服务原则性更强，价格浮动比较小。因此，我们使用 at a reasonable cost（以合理的价格）更为妥当。这体现了写作原则中的清晰（Clearness）原则，措辞准确。

2 解析重点2 Please send us your brochure and fee schedule on business service.

在写作的时候，英文句子往往都会涉及到某个主题或特定内容。例如：Please send us your brochure and fee schedule on business service（烦请寄送贵银行的业务简介和业务服务费用表），其中 business service（业务服务）就是前面 brochure and fee schedule 两个名词针对的内容。一般我们会选择用 about 而不是 on，但是相对于 about 来说，on 更为正式，针对性也更强。我们用 on 来表达，意思也会更准确（Correctness）、更明白。

1. **We are writing to several banks to get more information.**
   我们给多个银行写信，希望得到更多的信息。
2. **What services can you offer for our company?**
   你们能为我们公司提供何种服务?
3. **How much does such an account cost?**
   这样一个账户要花费多少钱?
4. **Please tell me the *procedure*[7] for opening an account.**
   请告诉我开立账户的程序。
5. **We have 100 *employees*[8] and US$1.6 million in annual *sales*[9].**
   我们公司拥有100名员工，年销售额达160万美元。
6. **We will hold a meeting to discuss which bank is most suitable for us.**
   我们将开会讨论哪个银行最适合我们。

必背关键单词

1. ***bank*** [bæŋk] n. 银行
2. ***reasonable*** [ˈri:znəbl] adj. 合理的
3. ***cost*** [kɒst] n. 代价；费用
4. ***fee*** [fi:] n. 费用；酬金
5. ***review*** [rɪˈvju:] v. 回顾；仔细检查
6. ***account*** [əˈkaʊnt] n. 账目；账户
7. ***procedure*** [prəˈsi:dʒə(r)] n. 手续；程序
8. ***employee*** [ɪmˈplɔɪi:] n. 雇员；职员
9. ***sales*** [seɪlz] n. 销售额

09 咨询仓库租赁

这样写就对了

From "Ronan Kendy" (rkendy@wit.com)　Date Mon., March 2, 2009
To "Allen Carter" (acarter@got.com)
Subject Inquiry About Leasing a Big Storehouse

Dear Mr. Carter,

Our company is interested in ***leasing***[1] a big ***storehouse***[2] rather than ***purchasing***[3] one. It will mainly be used to store our products shipped from ***foreign***[4] companies. There is a large ***quantity***[5] of them.
Please inform us of the detailed information about the lease, including the total ***area***[6] of the storehouse and its payment requirements.

Looking forward to your reply!

Yours sincerely,
Ronan Kendy

译文

亲爱的卡特先生：

我们公司不想购买仓库，只想租赁一个大型仓库，主要用来存放我们从国外公司运来的产品，这些产品数量很大。
烦请告知我们有关租赁的详细信息，包括仓库总面积以及付款要求。

期待您的回复！

罗南 · 坎迪 谨上

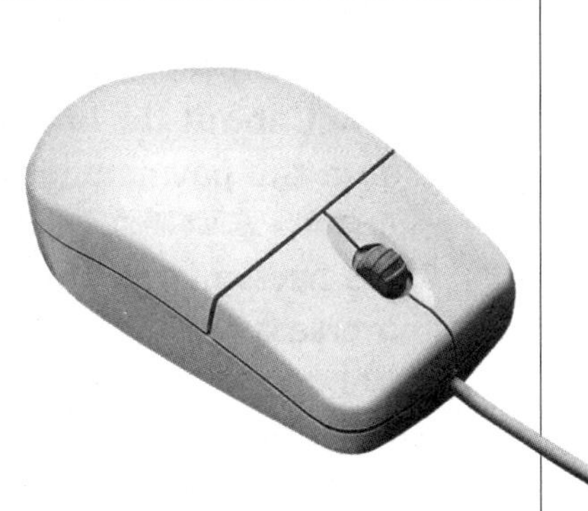

语法重点解析

1 解析重点1 Our company is interested in leasing a big storehouse rather than purchasing one.

有时候，我们很可能是因为某个特定的原因而去寻求商业服务。例如在邮件内文中，寻求租赁仓库方面的服务时，可以说：Our company doesn't want to buy a storehouse. But we want to rent one.（我们公司不想购买一个大仓库。只想租一个。）利用 rather than 或 instead of 把两句话变成一句话，会显得更简洁（Conciseness）。另外，需要注意的是，lease 和 rent 都有“租赁”的意思，而 lease 更强调是带租约的租赁。

2 解析重点2 It will mainly be used to store our products shipped from foreign companies.

寻求商业服务的时候，我们可以告知对方用途或目的。例如上述邮件中是租赁仓库，于是在说明写信目的之后，就补充了这一句：It will mainly be used to store our products shipped from foreign companies（它将主要用来存放我们从国外公司运来的产品），在这里，It will mainly be used to... 就指明了租赁的主要用途。我们还可以说，It will primarily be used for...（它主要用来……），以上两种说法意思相同，只是后面所接内容略有不同。

高频例句

1. **The storehouse will be used to keep our goods.**
   仓库将被用来存放我们的货物。
2. **It must be *tidy*[7] and have a large area.**
   它必须干净，并且面积大。
3. **Please tell us the related information about the storehouse.**
   请告知我们仓库的相关信息。
4. **What about its *location*[8], total area and payment terms?**
   它的地点、总面积及付款方式是什么？
5. **We have a large amount of goods overseas.**
   我们在海外有大量的货物。
6. **A spacious storehouse is a *necessity*[9] for these import cargoes.**
   我们需要一个宽敞的仓库来存放这些进口商品。

必背关键单词

1. ***lease*** [li:s] v. 出租；租借
2. ***storehouse*** [ˈstɔ:haʊs] n. 仓库
3. ***purchase*** [ˈpɜ:tʃəs] v. 购买
4. ***foreign*** [ˈfɒrən] adj. 国外的
5. ***quantity*** [ˈkwɒntətɪ] n. 数目；数量
6. ***area*** [ˈeərɪə] n. 地区；领域；面积；方面
7. ***tidy*** [ˈtaɪdɪ] adj. 整洁的；整齐的
8. ***location*** [ləʊˈkeɪʃn] n. 位置
9. ***necessity*** [nəˈsesətɪ] n. 必需品

10 咨询酒店预订

这样写就对了

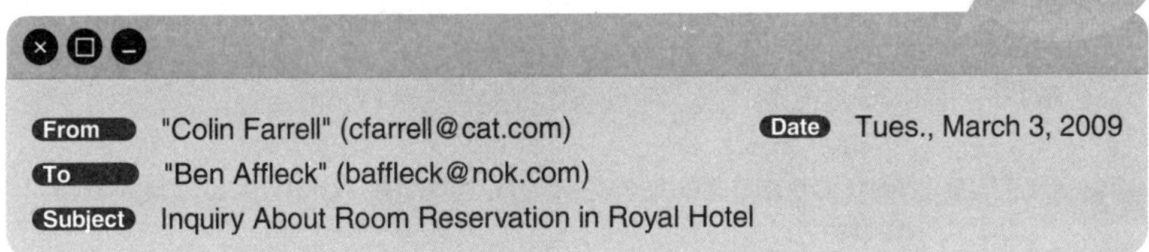
From "Colin Farrell" (cfarrell@cat.com)
Date Tues., March 3, 2009
To "Ben Affleck" (baffleck@nok.com)
Subject Inquiry About Room Reservation in Royal Hotel

Dear Mr. Affleck,

I will be going on a business ***trip***[1] in your city with one ***colleague***[2] from March 6, 2009 to March 12, 2009. As far as I know, your business is ***brisk***[3] all the time, but I would appreciate it if you could ***reserve***[4] two ***single***[5] rooms under my name for us. Thank you in advance.

Looking forward to hearing from you soon!

Yours sincerely,
Colin Farrell

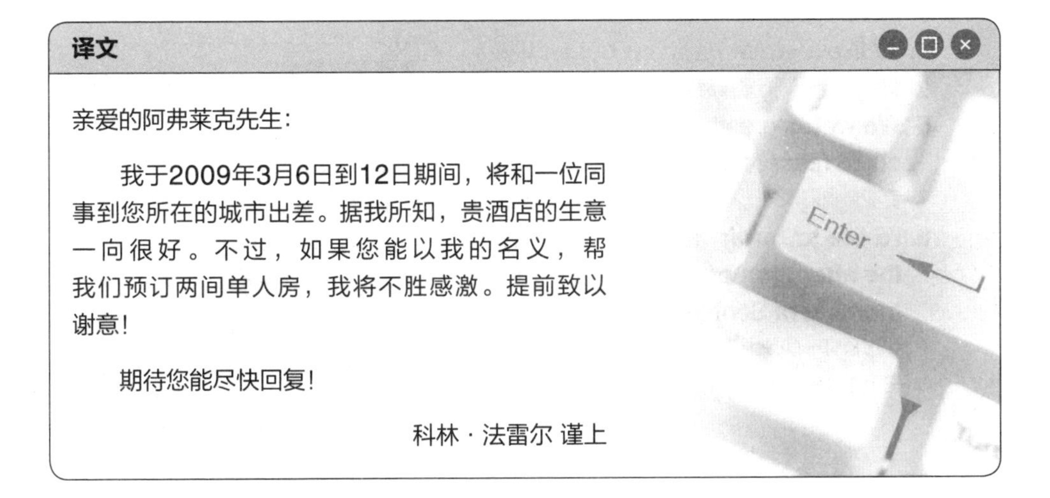
译文

亲爱的阿弗莱克先生：

我于2009年3月6日到12日期间，将和一位同事到您所在的城市出差。据我所知，贵酒店的生意一向很好。不过，如果您能以我的名义，帮我们预订两间单人房，我将不胜感激。提前致以谢意！

期待您能尽快回复！

科林 · 法雷尔 谨上

语法重点解析

1

解析重点1 **I will be going on a business trip in your city with one colleague from March 6, 2009 to March 12, 2009.**

向酒店预订房间的时候，可以顺便提一下预订房间的原因，但也可以省略。其中，需要我们特别说明的则是具体的居住时间。正如上述邮件内文中的句子提到了以上两点，一是说明因为出差需要订房，另一个则说明需要从2009年3月6日开始，一直住到12日。居住时间说的相当具体（Concreteness）。

2

解析重点2 **You could reserve two single rooms under my name.**

预订房间的时候，一般要说明房型要求，比如双人房（double room）或单人房（single room）。而预订两个单人房的说法就是 reserve two single rooms。同时，我们还要注意是以谁的名义来订房。上述的邮件是以寄件人的名义来订房：under my name（以我的名义）。如果是帮别人订房，最好是以别人的名义来订房，以方便酒店查询。

1. **I am planning a trip to New York for two days.**
   我计划到纽约出差两天。
2. **I urgently need a room for tomorrow night.**
   我明晚急需订一个房间。
3. **Do you *serve*[6] breakfast for *deluxe*[7] rooms?**
   你们为豪华客房提供早餐服务吗？
4. **I'd like a *suite*[8] with an ocean view.**
   我想预订一间海景套房。
5. **We would greatly appreciate it if you could reserve a double room for us.**
   若能帮我们预留一间双人房，我们将不胜感激。
6. **I'd like to *book*[9] a single room from the afternoon of September 4 to the morning of September 10.**
   我想预订一间单人房，9月4日下午入住，9月10日上午退房。

必背关键单词 

1. ***trip*** [trɪp] n. 旅行；出行
2. ***colleague*** [ˈkɒliːg] n. 同僚；同事
3. ***brisk*** [brɪsk] adj. 兴旺的；繁荣的
4. ***reserve*** [rɪˈzɜːv] v. 保留；预留
5. ***single*** [ˈsɪŋgl] adj. 单一的
6. ***serve*** [sɜːv] v. 服务；招待
7. ***deluxe*** [dəˈlʌks] adj. 豪华的；高级的；奢华的
8. ***suite*** [swiːt] n. 套房
9. ***book*** [bʊk] v. 登记；预订

Unit 9 请求 Request

01 请求付款

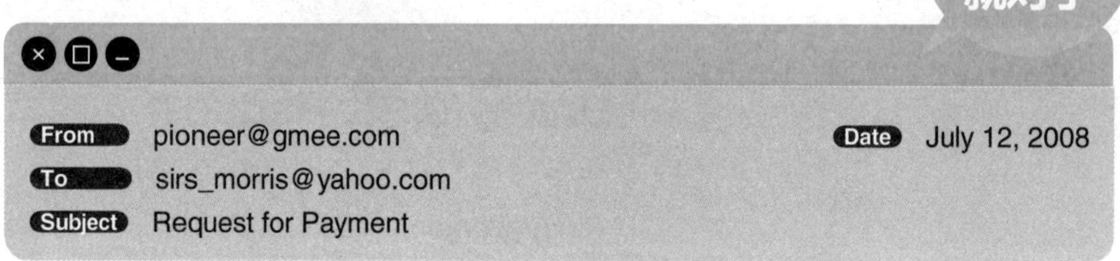

From pioneer@gmee.com
Date July 12, 2008
To sirs_morris@yahoo.com
Subject Request for Payment

Dear Mr. Morris,

Thank you very much for your ***purchase***[1] of our ***products***[2].

For your ***reference***[3], we have attached our ***invoice***[4] number 79023108 for 2,000 US dollars.
Please ***pay***[5] the ***bill***[6] ***within***[7] 15 days upon receipt of the request letter. Please notify us when you make the remittance.

Yours faithfully,
Pioneer Electric Appliance

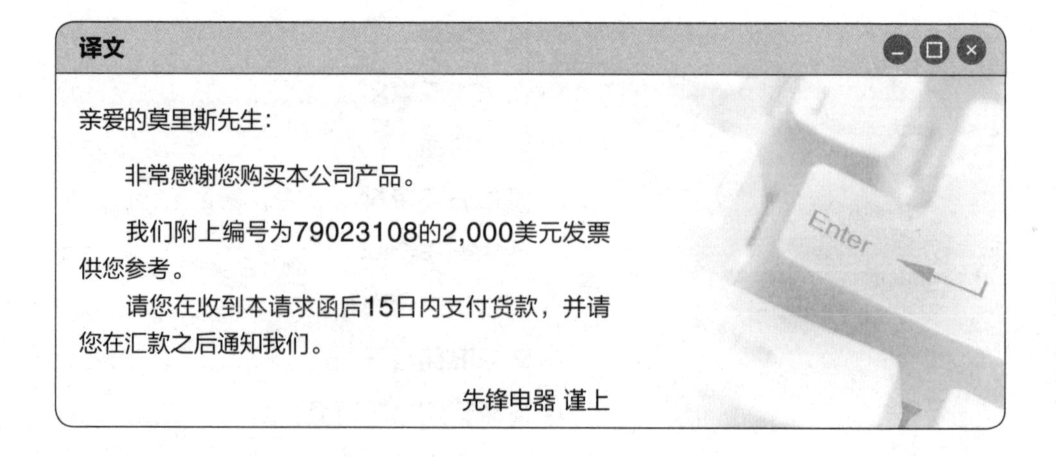
译文

亲爱的莫里斯先生:

非常感谢您购买本公司产品。

我们附上编号为79023108的2,000美元发票供您参考。

请您在收到本请求函后15日内支付货款，并请您在汇款之后通知我们。

先锋电器 谨上

语法重点解析

1 解析重点1 Thank you very much for your purchase of our products.

这句话的意思是“非常感谢您购买本公司产品”。类似的表达方式还有：Thank you for using / purchasing our products（感谢您使用 / 购买我们的产品）；Thank you for choosing our commodities（感谢您选择我们的商品）；Thank you for shopping in our supermarket（感谢您光临本超市）等。

2 解析重点2 make the remittance

短语 make a remittance 的意思是“汇款”，相当于 remit money “付款；划拨款项”。请看以下例句：

Please make a remittance of US$150 for the books you have ordered.（请为您订购的书汇款150美元。）

I remit money to my family through a bank every month.（我每月通过银行给家里汇款。）

1. **We promise to make a *remittance*[8] within a week.**
   我们答应在一星期内汇款。
2. **We will make a remittance within a week in full settlement of our purchase of the goods contracted.**
   我方合约购买货物的货款，将于一周内汇款结清。
3. **Would you please fax the remittance certificate to us?**
   您能将汇款凭证传真给我们吗?
4. **Please pay the bill within seven days upon receipt of the request letter.**
   请在收到请求函后的7日内付款。
5. **Thank you for choosing commodities in our mall.**
   非常感谢您在本商场选购商品。
6. **Please inform us after you make the remittance.**
   请您在汇款之后通知我们。

必背关键单词 

1. ***purchase*** [ˈpɜːtʃəs] n. 购买
2. ***product*** [ˈprɒdʌkt] n. 产品
3. ***reference*** [ˈrefrəns] n. 参考；参照
4. ***invoice*** [ˈɪnvɔɪs] n. 发票；发货单
5. ***pay*** [peɪ] v. 付款
6. ***bill*** [bɪl] n. 账单
7. ***within*** [wɪˈðɪn] prep. 在……之内
8. ***remittance*** [rɪˈmɪtns] n. 汇款

02 请求退款

From emmablake@nec.com
To servicedep@taob.com
Subject Request for Refund
Date August 9, 2008

Dear Service Department,

I am afraid I would like a ***refund***[1] for the ***skirt***[2] I ***returned***[3] earlier. Because of the wrong ***size***[4], I sent the skirt back to you on August 7, the day after it arrived. However, the skirt was ***included***[5] in the bill **charged to my account**. Could you refund me the money of the skirt?

Please ***confirm***[6] receipt of this e-mail by phone.

Yours faithfully,
Emma Blake

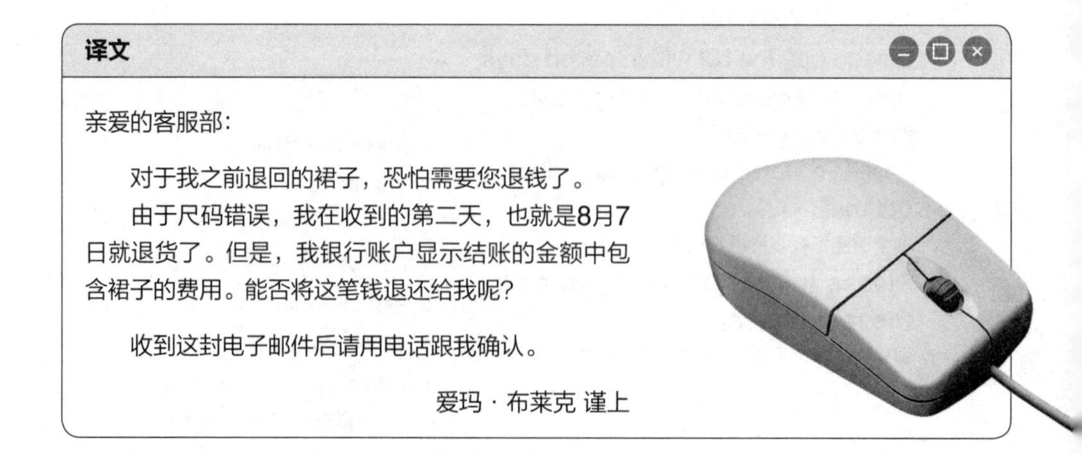

译文

亲爱的客服部：

对于我之前退回的裙子，恐怕需要您退钱了。

由于尺码错误，我在收到的第二天，也就是8月7日就退货了。但是，我银行账户显示结账的金额中包含裙子的费用。能否将这笔钱退还给我呢？

收到这封电子邮件后请用电话跟我确认。

爱玛 · 布莱克 谨上

语法重点解析

1 解析重点1 refund

refund 表示“退还；偿还”，本身有名词和动词两种词性，在此语境中做名词，表示“退款”。表达此意的词还有repay, pay back 等。请看以下例句:

Can you refund the cost of postage in a case like this?（若发生这种情况你能退还邮资吗？）

I lent him £5 on the condition that he must pay me back today.（我借给他5英镑，条件是他今天必须还给我。）

He will pay back the money in monthly installments.（他将按月以分期付款的方式偿还这笔钱。）

2 解析重点2 charge to my account

短语 charge one's account 的意思是“从某人的账户中扣除”。现在利用银行账户自动扣款的支付方式已经非常普遍。请看以下例句:

Freight for the shipment from Shanghai to Beijing is to be charged to your account.（从上海到北京的运费由贵方承担。）

1. **She took the faulty radio back to the shop and demanded a refund.**
   她将有瑕疵的收音机拿回商店，要求退款。
2. **Needless to say, we shall refund any expenses.**
   不用说，我们将退还您所有的费用。
3. **Please *charge*[7] the bill to my *account*[8].**
   请将账单记入我的账上。
4. **Because it was the wrong color, I want to return it.**
   因为颜色错误，所以我想退货。
5. **My father bought a new suit on his charge account.**
   我父亲以赊账的方式买了一套新西装。

必背关键单词

1. ***refund*** [ˈriːfʌnd] v. 偿还；退还
2. ***skirt*** [skɜːt] n. 裙子
3. ***return*** [rɪˈtɜːn] v. 归还；返还
4. ***size*** [saɪz] n. 尺码
5. ***include*** [ɪnˈkluːd] v. 包含；包括
6. ***confirm*** [kənˈfɜːm] v. 确认；证实
7. ***charge*** [tʃɑːdʒ] v. 索价；收费
8. ***account*** [əˈkaʊnt] n. 账户

03 请求寄送价目表

这样写就对了

Dear Customer Service,

I am looking for companies ***selling***[1] ***camera***[2] ***equipment***[3].
I learnt from the Yellow Pages that your corporation ***engages***[4] in ***photographic***[5] equipments, so you have the ***exact***[6] items I need. Could you please send me your price list of cameras with ***diversiform***[7] ***Model***[8] Numbers to us? Thank you very much!

An early reply will be greatly obliged.

Yours faithfully,
Media Adv. Co.

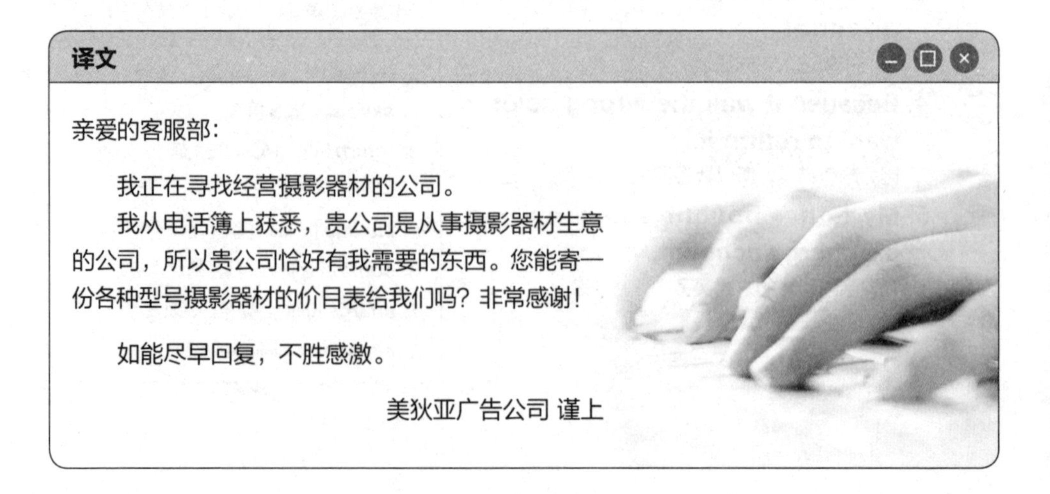

译文

亲爱的客服部：

我正在寻找经营摄影器材的公司。

我从电话簿上获悉，贵公司是从事摄影器材生意的公司，所以贵公司恰好有我需要的东西。您能寄一份各种型号摄影器材的价目表给我们吗？非常感谢！

如能尽早回复，不胜感激。

美狄亚广告公司 谨上

语法重点解析

解析重点1 Yellow Pages

Yellow Pages（黄页）是指刊载公司、厂商电话的电话号码薄。美国以黄色的纸张按企业性质和产品类别编排的工商电话薄为Yellow Pages类；White Pages（白页），则是指用白色纸张刊载私人电话，按个人分类。请看下面的句子：

Let us find their telephone number from the Yellow Pages and make a call.（我们从黄页上查询电话号码，并打电话给他们。）

解析重点2 price list

price list 是“价目表”的意思，可以表达相同意思的词组有 price schedule, rate scale 等。有关“价格”的表达方式还有：price list for...（……的价目表）；fee policy / pricing policy（价格政策）。请看下面的句子：

Can you give me a price list with specification?（你能否给我一份有规格说明的价目单？）

高频例句

1. **He paid only a *quarter*[9] of the list price.**
   他只付了标价的四分之一。
2. **Please enclose your price list and all necessary illustrations.**
   请随函附上贵方的价目表和一切必要的说明。
3. **I am looking for a translation company.**
   我正在寻找一家翻译公司。
4. **I saw your advertisement in the Yellow Pages.**
   我在电话簿上看到了贵公司的广告。
5. **I expect your prompt reply.**
   我期待您尽快答复。
6. **Could you please send me your price list?**
   您能将您的价目表寄给我吗？
7. **Thank you very much for your swift action!**
   非常感谢您能迅速采取行动。

必背关键单词

1. ***sell*** [sel] v. 贩卖；出售
2. ***camera*** [ˈkæmərə] n. 照相机
3. ***equipment*** [ɪˈkwɪpmənt] n. 设备
4. ***engage*** [ɪnˈgeɪdʒ] v. 从事
5. ***photographic*** [ˌfəʊtəˈgræfɪk] adj. 摄影的
6. ***exact*** [ɪgˈzækt] adj. 精确的
7. ***diversiform*** [daɪˈvɜːsɪfɔːm] adj. 多样的；各色各样的
8. ***model*** [ˈmɒdl] n. 模型；型号
9. ***quarter*** [ˈkwɔːtə(r)] n. 四分之一

04 请求送货上门

这样写就对了

From bettyjones@home.com
To custservice@mall.com
Subject Request for Home-Delivery Service
Date May 3, 2008

Dear Service Department,

I have ordered three ***mattresses***[1] on the ***Internet***[2] from your company. And I want to know if you ***provide***[3] home-delivery service. If you can deliver goods to the customers, may I pay ***cash***[4] on delivery or by ***bank***[5] transfer?

Please ***notify***[6] me at your ***earliest***[7] ***convenience***[8]. Thanks a lot!

Yours faithfully,
Betty Jones

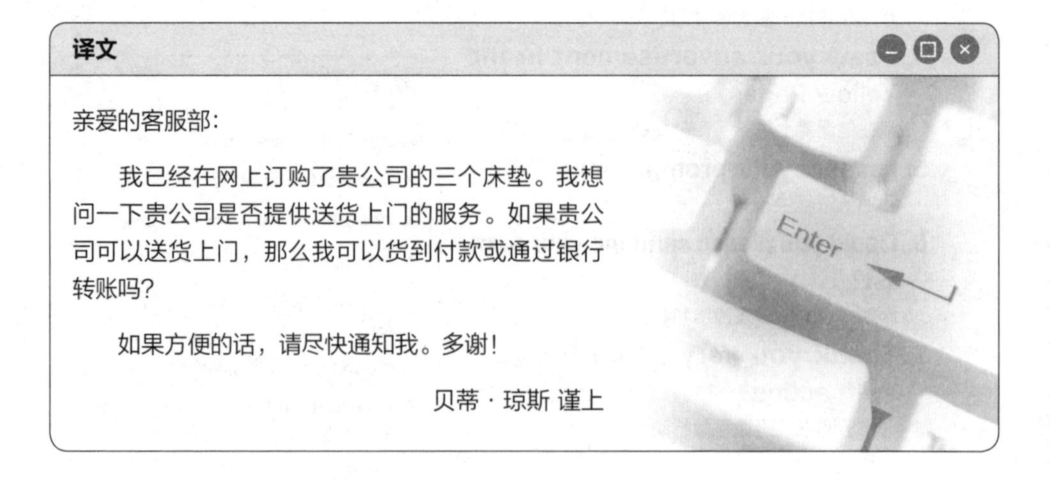

译文

亲爱的客服部:

我已经在网上订购了贵公司的三个床垫。我想问一下贵公司是否提供送货上门的服务。如果贵公司可以送货上门，那么我可以货到付款或通过银行转账吗?

如果方便的话，请尽快通知我。多谢!

贝蒂 · 琼斯 谨上

语法重点解析

1 解析重点1 home-delivery service

短语 home-delivery service 的意思是“送货上门服务”，相同意思还可以说 deliver goods to the customers, provide home-delivery service 等。“上门服务”可以说 door-to-door service。请看下面的句子：

To save customers' time, they started a home-delivery service.（为了节省顾客时间，他们开始提供送货上门服务。）

Our neighborhood service center offers a door-to-door service for those who have special difficulties.（我们社区服务中心针对有特殊困难的客户提供送货上门服务。）

2 解析重点2 bank transfer

bank transfer 是“银行转账”的意思，是当今盛行的网络购物的一种付款方式。网络购物的其他付款方式还有 cash on delivery（缩写为 COD）（货到付款），payment by post（邮寄付款）等。请看下面的句子：

The common international import transaction is via bank transfer.（常见的国际进口交易是通过银行转账进行的。）

1. **Please deliver the goods at your earliest convenience.**
   请尽早送货。
2. **Many stores now deliver goods to your door free of charge.**
   现在很多商店免费送货上门。
3. **Could we pay cash on delivery?**
   我们能货到付款吗?
4. **Door-to-door service is the *consistent*[9] style of our shop.**
   上门服务是本店一贯的做事风格。
5. **Within the province it provides express door-to-door service.**
   在省内提供快递上门服务。
6. **Could you please provide home-delivery service?**
   您能提供送货上门的服务吗?

必背关键单词

1. ***mattress*** [ˈmætrəs] n. 床垫
2. ***Internet*** [ˈɪntənet] n. 互联网
3. ***provide*** [prəˈvaɪd] v. 提供
4. ***cash*** [kæʃ] n. 现金
5. ***bank*** [bæŋk] n. 银行
6. ***notify*** [ˈnəʊtɪfaɪ] v. 通知
7. ***early*** [ˈɜːlɪ] adj. 早的；初期的
8. ***convenience*** [kənˈviːnɪəns] n. 方便
9. ***consistent*** [kənˈsɪstənt] adj. 一贯的

05 请求公司资料

From monicaali@bisu.com

To custservice@if.com

Subject Request for Company Information

Date June 25, 2008

Dear Sir or Madam,

I would like to request ***information***[1] about your company to ***facilitate***[2] my ***research***[3] for a school ***project***[4].
The project ***focuses***[5] on start-up food ***processing***[6] companies and their ***impact***[7] within the industry. I would appreciate any corporate brochures or marketing ***materials***[8] with which you could provide me.

Thank you very much for your help!

Yours sincerely,
Monica Ali

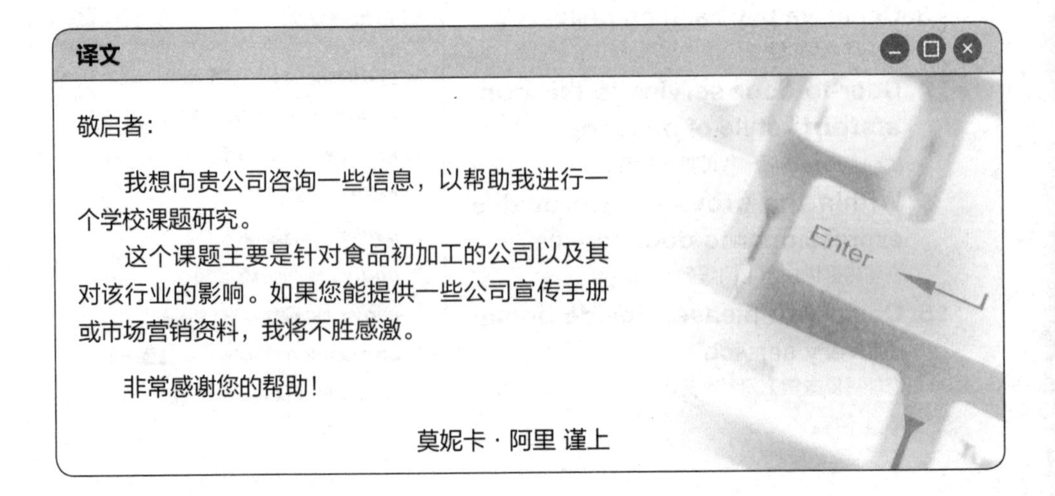

译文

敬启者：

我想向贵公司咨询一些信息，以帮助我进行一个学校课题研究。

这个课题主要是针对食品初加工的公司以及其对该行业的影响。如果您能提供一些公司宣传手册或市场营销资料，我将不胜感激。

非常感谢您的帮助！

莫妮卡·阿里 谨上

语法重点解析

1 解析重点1 facilitate

facilitate 的意思是“促进；使便利；帮助进步”。文中的 facilitate 还可以替换成 help, assist, avail 等。请看下面的句子：

Zip codes are used to facilitate mail service.（邮政编码方便了邮递服务。）

Could you help me take this suitcase upstairs?（你能帮我把这个箱子搬到楼上吗？）

He asked us to assist him in carrying out his plan.（他要求我们帮他执行他的计划。）

2 解析重点2 corporate brochures

短语 corporate brochures 的意思是“公司宣传手册”。corporate，即形容词，表示“公司的；法人的”；brochure 则为名词，表示“小册子；宣传手册”。每个公司的基本资料通常都会印成小册子作为宣传材料，即“公司宣传手册”。请看下面的句子：

Could you give me some brochures for your clothing?（您能给我一些你们服饰的宣传手册吗？）

1. **I would like to request a copy of your company *brochure*[9].**
   我想要一份贵公司的宣传手册。
2. **I wonder if you could give me some information about your comany.**
   请问你可否提供给我一些贵公司的相关资料。
3. **I am very interested in learning more about your company.**
   我对进一步了解贵公司很感兴趣。
4. **Is it possible to obtain a copy of your price list?**
   您可不可以给我一份贵公司的价目表？
5. **Thank you in advance for your kind attention.**
   提前感谢您的关照。
6. **I will be grateful if you will give me a brochure of your company.**
   若您能给我一份贵公司的宣传手册，将不胜感激。

必背关键单词

1. ***information*** [ˌɪnfəˈmeɪʃn] n. 信息
2. ***facilitate*** [fəˈsɪlɪteɪt] v. 促进；使便利
3. ***research*** [rɪˈsɜːtʃ] n. 研究
4. ***project*** [ˈprɒdʒekt] n. 项目；工程
5. ***focus*** [ˈfəʊkəs] v. 聚集；集中
6. ***process*** [ˈprəʊses] v. 加工；处理
7. ***impact*** [ˈɪmpækt] n. 影响
8. ***material*** [məˈtɪərɪəl] n. 资料；材料
9. ***brochure*** [ˈbrəʊʃə(r)] n. 小册子；宣传手册

06 请求开立发票

这样写就对了

From roselear@yahoo.com

To ianui@21c.com

Subject Request for Opening an Invoice

Date June 3, 2008

Dear Mr. Ui,

I need you to ***settle***[1] **the invoice** for my two pairs of high-***heel***[2] shoes bought from your store on June 2. I guess you might have forgotten to ***issue***[3] the invoice for me because of ***busy***[4] work.
Please give your ***attention***[5] to this ***problem***[6] at once, as I would not like to see something like this have ***effect***[7] on your ***credit***[8] **standing**.

Yours faithfully,
Rose Lear

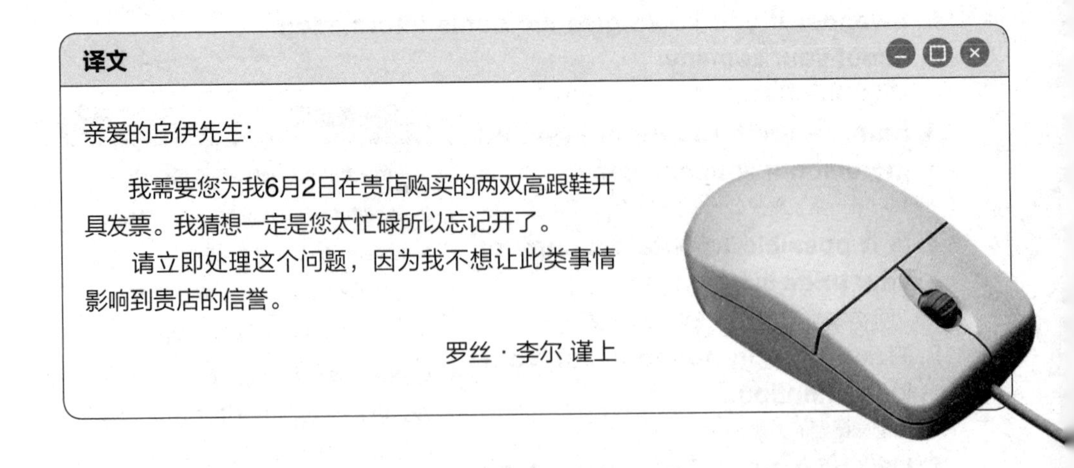

译文

亲爱的乌伊先生:

我需要您为我6月2日在贵店购买的两双高跟鞋开具发票。我猜想一定是您太忙碌所以忘记开了。

请立即处理这个问题，因为我不想让此类事情影响到贵店的信誉。

罗丝 · 李尔 谨上

语法重点解析

1 解析重点1 settle the invoice

settle an invoice 的意思是“开发票”，invoice 意为“发票”。发票是指在购销商品、提供或者接受服务以及从事其他经营活动中，开具、收取的收付款凭证。它是消费者的购物凭证，是纳税人进行经济活动的重要凭证，也是财政、税收、审计等部门进行财务税收检查的重要依据。表达“开发票”还可以用 issue an invoice, make out an invoice, write a receipt 等。

2 解析重点2 credit standing

短语 credit standing 的意思是“信誉”。类似的表达方法还有：credit worthiness（信誉；信用价值）；goodwill（信誉；声誉）；credibility（可信性；信誉）；prestige（名望；声望；威望）；reputation（名誉；信誉）等。请看下面的句子：

His credit standing is highest in this area.（在这个地区他的信用等级是最高的。）

Credibility is hard to get but easy to lose.（信用不易建立，却极易失去。）

1. **I hope you could give me an invoice for my payment at once.**
   我希望您立即为我的付款开具发票。
2. **I have called you several times to request the invoice.**
   我已经给你打了好几次电话索要发票。

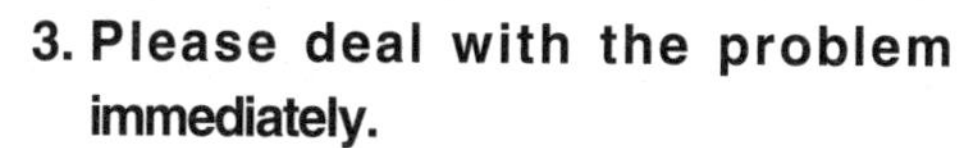

3. **Please deal with the problem immediately.**
   请立即处理这个问题。
4. **We will make out an invoice for selling *immovable*[9] properties for you.**
   我们将为您开具不动产销售发票。
5. **Wait a minute, please. I'll make out an invoice for you.**
   请稍等，我会为您开张发票。
6. **Thank you for turning your attention to the problem.**
   感谢您对此事予以关注。

必背关键单词

1. ***settle*** [ˈsetl] v. 解决；安排
2. ***heel*** [hiːl] n. 脚后跟
3. ***issue*** [ˈɪʃuː] v. 发出；开立；签发
4. ***busy*** [ˈbɪzɪ] adj. 忙碌的
5. ***attention*** [əˈtenʃn] n. 注意力，关心
6. ***problem*** [ˈprɒbləm] n. 难题
7. ***effect*** [ɪˈfekt] n. 影响
8. ***credit*** [ˈkredɪt] n. 信用；荣誉
9. ***immovable*** [ɪˈmuːvəbl] adj. 不可移动的；固定的

07 请求追加投资

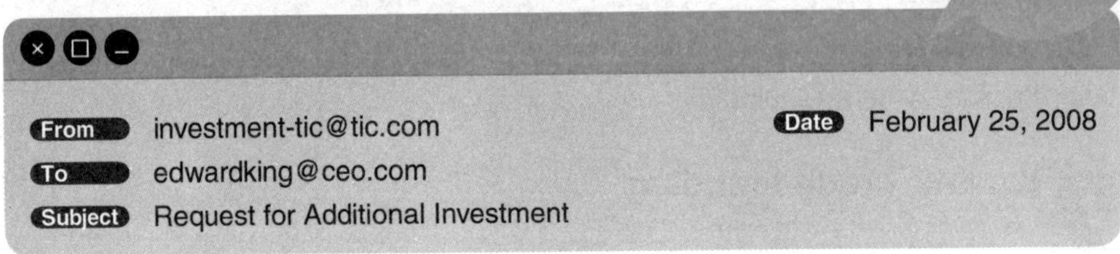

From investment-tic@tic.com
To edwardking@ceo.com
Subject Request for Additional Investment
Date February 25, 2008

Dear Mr. Edward,

I am very glad that ***Talent***[1] Investment Company has been ***operating***[2] quite well since its ***establishment***[3].
However, we are now ***confronting***[4] a ***fiscal***[5] crisis because of the ***global***[6] financial crisis. So I would like to know if it is possible for you to make ***additional***[7] investment to help us bridge over the current difficulty.

Thank you so much for your careful consideration and we hope for your support!

Yours faithfully,
TIC

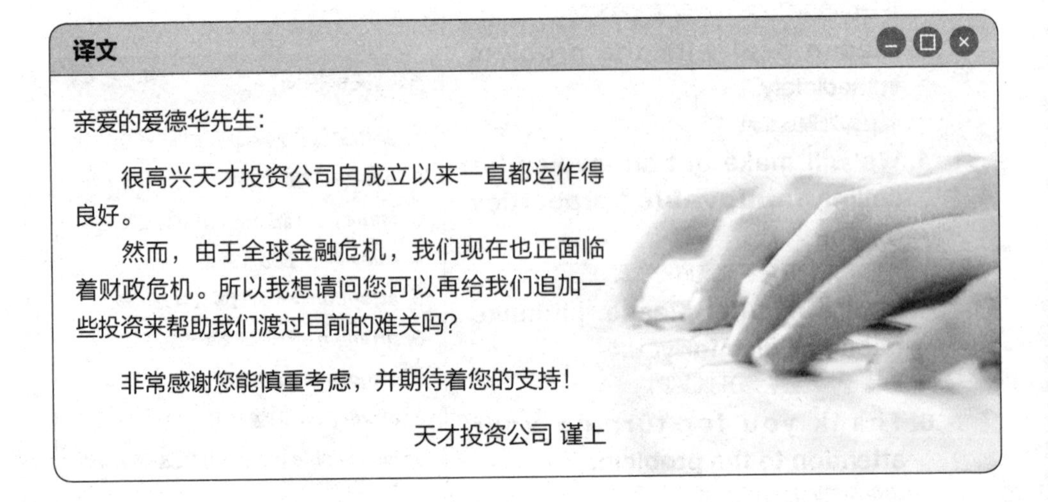

译文

亲爱的爱德华先生：

很高兴天才投资公司自成立以来一直都运作得良好。

然而，由于全球金融危机，我们现在也正面临着财政危机。所以我想请问您可以再给我们追加一些投资来帮助我们渡过目前的难关吗？

非常感谢您能慎重考虑，并期待着您的支持！

天才投资公司 谨上

语法重点解析

1 解析重点1 **We are now confronting a fiscal crisis because of the global financial crisis.**

这句话的意思是“由于全球金融危机，我们现在正面临着财政危机”。其中 confront 为谓语动词，意为“面临；遭遇”，表达“面临”的短语还有 be faced with, be confronted with, be up against 等。请看下面的句子：

The new system will be confronted with great difficulties at the start.（这种新的制度一开始将会面临很大的困难。）

You'll be up against it if you don't pass the test.（如果你考试不及格的话，你将面临困难。）

2 解析重点2 **bridge over**

短语 bridge over 的意思是“渡过（难关）”。表达“渡过（难关）”还可以用 tide over, pull through, weather the storm 等。请看下面的句子：

He did a lot to bridge over his difficulties.（为渡过难关他做了许多努力。）

If he could muster up more strength, he might pull through.（如果他再加把劲，或许就可以渡过难关了。）

The next year or two will be very difficult for our firm, but I think we will weather the storm.（今后一两年我们公司会很困难，但是我认为我们会渡过难关的。）

高频例句

1. **He made a large investment in the business enterprise.**
   他对那个企业投入了大量资金。
2. **We are confronted with great fiscal trouble now.**
   我们正面临着严重的财政危机。
3. **We would like to enlarge the scale of our company.**
   我们想扩大公司规模。
4. **Could you invest more money in our corporation?**
   您能追加对我们公司的投资金额吗？

必背关键单词 

1. ***talent*** ['tælənt] n. 才能；天才；天资
2. ***operate*** ['ɒpəreɪt] v. 运转；操作
3. ***establishment*** [ɪ'stæblɪʃmənt] n. 建立
4. ***confront*** [kən'frʌnt] v. 面临
5. ***fiscal*** ['fɪskl] adj. 财政的
6. ***global*** ['gləʊbl] adj. 全球性的；全局的
7. ***additional*** [ə'dɪʃənl] adj. 额外的；附加的；追加的

08 请求会面

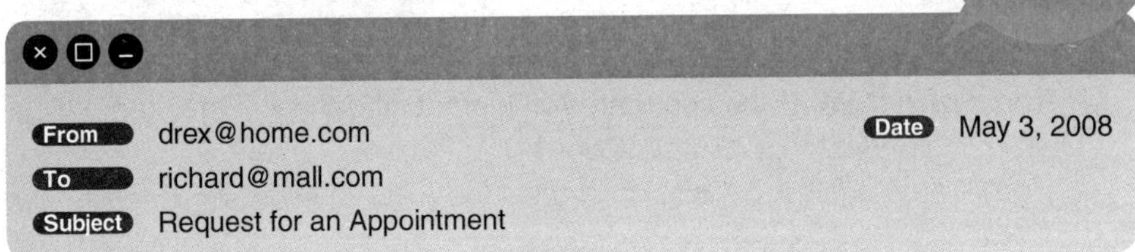

From drex@home.com
To richard@mall.com
Subject Request for an Appointment
Date May 3, 2008

Dear Mr. Richard,

I am going to visit Los Angeles on Saturday, October 4, 2008. By the way, may I have the **opportunity** of paying you a visit with the ***purpose***[1] to discuss the ***current***[2] ***financial***[3] ***crisis***[4]?
May I ***suggest***[5] 4:00 p.m. on Sunday, October 5, as a ***convenient***[6] time for my visit? **If it is not fine with you, perhaps you would be kind enough to let me know a time convenient for you.**

Hoping to meet you soon!

Sincerely yours,
David Rex

译文

亲爱的理查德先生：

我将于2008年10月4日，即星期六前往洛杉矶。我可以有这个机会顺道去拜访您，并讨论一下当今的金融危机吗？

如果您10月5日，即星期日下午4点方便的话，我可以去拜访您吗？如果不行的话，可以恳请您告知我方便的时间吗？

期待着与您会面！

大卫·雷克斯 谨上

语法重点解析

1 解析重点1 opportunity

opportunity 是“机会；时机”的意思，“机会”还可以用 chance 来表达。但是 opportunity 和 chance 是有区别的：opportunity 更侧重指非常好、非常难得的机会，即“良机”，而 chance 仅指一般的“机会”。在此信中，为了请求会面，用 opportunity 则更有诚意。请对照下面的句子：

We have waited for such a good opportunity for so many years!（这样好的机会我们等了好多年了！）

Please give me a chance to explain.（请给我个机会让我解释一下。）

2 解析重点2 If it is not fine with you, perhaps you would be kind enough to let me know a time convenient for you.

这句话的意思是“如果不行的话，可以恳请您告知我方便的时间吗？”写信者在此句话之前就约定了见面的时间，加上这句话的意思是说如果对方在约定的时间可以见面的话就见，如果不方便的话还可以再约时间，表现出了写信者的诚意，这正符合了英文书信写作中的 Consideration（体贴）原则。

1. **I would like to visit your office to discuss our *plan*[7].**
   我想拜访贵公司，讨论我们的计划。
2. **Could you meet with me between April 10 and 14 if possible?**
   如果可以的话，能否在4月10日到14日中的一天会面呢？
3. **Could we make an *appointment*[8] on this Saturday?**
   我们这个星期六会面可以吗？
4. **May I make an appointment?**
   我可以预约？
5. **When is your next appointment?**
   你下一次的预约是什么时候？
6. **Can we reschedule our appointment?**
   我们能不能更改一下会面的时间？

必背关键单词

1. ***purpose*** ['pɜːpəs] n. 目的
2. ***current*** ['kʌrənt] adj. 现在的；当前的
3. ***financial*** [faɪ'nænʃl] adj. 金融的；财政的；账务的
4. ***crisis*** ['kraɪsɪs] n. 危机
5. ***suggest*** [sə'dʒest] v. 建议；提出
6. ***convenient*** [kən'viːnɪənt] adj. 方便的
7. ***plan*** [plæn] n. 计划
8. ***appointment*** [ə'pɒɪntmənt] n. 约会；约定

09 请求延期付款

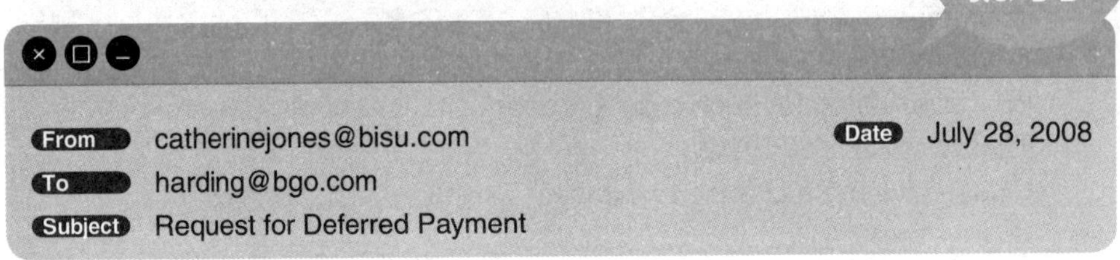
From catherinejones@bisu.com
Date July 28, 2008
To harding@bgo.com
Subject Request for Deferred Payment

Dear Mr. Harding,

I am ***terribly***[1] sorry for the ***overdue***[2] balance on our account. It turns out that the ***last***[3] invoice we received from you was ***somehow***[4] ***misplaced***[5]. Would it be possible to ***grant***[6] us a 5-day payment ***extension***[7]? I can ensure that you will receive the outstanding balance by next Thursday.

I appreciate your understanding very much.

Yours faithfully,
Catherine Jones

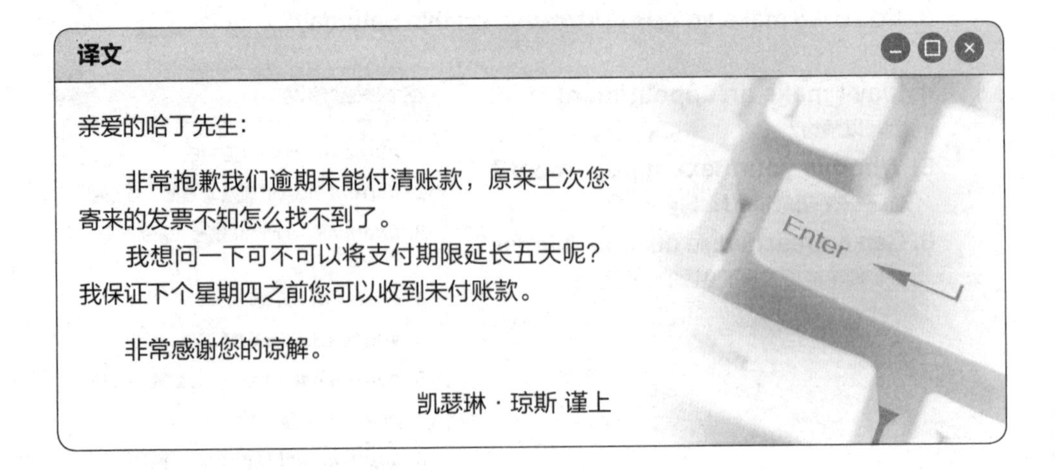
译文

亲爱的哈丁先生：

非常抱歉我们逾期未能付清账款，原来上次您寄来的发票不知怎么找不到了。

我想问一下可不可以将支付期限延长五天呢？我保证下个星期四之前您可以收到未付账款。

非常感谢您的谅解。

凯瑟琳·琼斯 谨上

语法重点解析

1 解析重点1 It turns out that the last invoice we received from you was somehow misplaced.

这句话的意思是“结果是上次您寄来的发票不知怎么找不到了”。这是在解释请求延迟支付的原因。It turns out that...的意思是“结果是……”；somehow 意为“不明原因；不知怎么地”。写信者用 It turns out that... 和 somehow 非常婉转地表达了不能按时支付货款是有客观原因的，因而使对方比较容易接受。

2 解析重点2 outstanding

outstanding 有“杰出的；显著的；未偿清的；未完成的”意思，在此语境中则是“未付款的”意思，相当于 unpaid。请对照下面的句子：

Einstein was an outstanding scientist.（爱因斯坦是位杰出的科学家。）

The outstanding debts must be paid by the end of the month.（未偿债务必须在月底前偿还。）

There is an unpaid bill on my table.（我桌上有张未付的账单。）

高频例句

1. **I understand that the payment for our last order is due next Monday.**
   我知道我们上次订单的款项应于下个星期一支付。
2. **Would it be possible to extend the payment deadline until the end of the month?**
   可不可以将支付期限延长到月底？
3. **We would greatly appreciate it if you could grant us a 7-day payment extension.**
   如果您能将付款期限延迟7天，我们将不胜感激。
4. **I was wondering whether it is possible to extend this period by one week.**
   我想知道能不能延期一周。

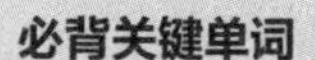

必背关键单词

1. ***terribly*** [ˈterəblɪ] adv. 非常
2. ***overdue*** [ˌəʊvəˈdjuː] adj. 逾期的；到期未付的
3. ***last*** [lɑːst] adj. 最近的；最后的
4. ***somehow*** [ˈsʌmhaʊ] adv. 不知怎地；用某种方法
5. ***misplace*** [ˌmɪsˈpleɪs] v. （因记不起放在哪）丢失；放错地方
6. ***grant*** [grɑːnt] v. 授予；同意
7. ***extension*** [ɪkˈstenʃn] n. 延长；扩充；延期

10 请求推荐客户

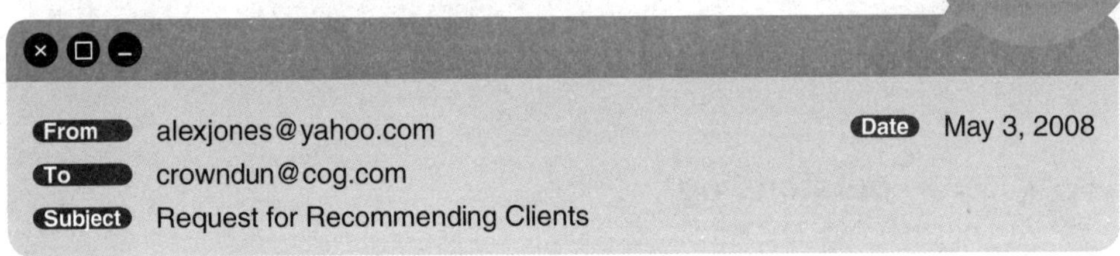
From alexjones@yahoo.com
Date May 3, 2008
To crowndun@cog.com
Subject Request for Recommending Clients

Dear Mr. Crown,

As you know, DongHuang Corporation is ***expanding***[1].
Since you have ***expressed***[2] that you are very ***satisfied***[3] with our service, I ***wonder***[4] if you **could be so kind as to** ***recommend***[5] some ***potential***[6] clients to us for ***activating***[7] **business**.

We deeply appreciate any suggestions you may have to offer.

Sincerely yours,
Alex Jones

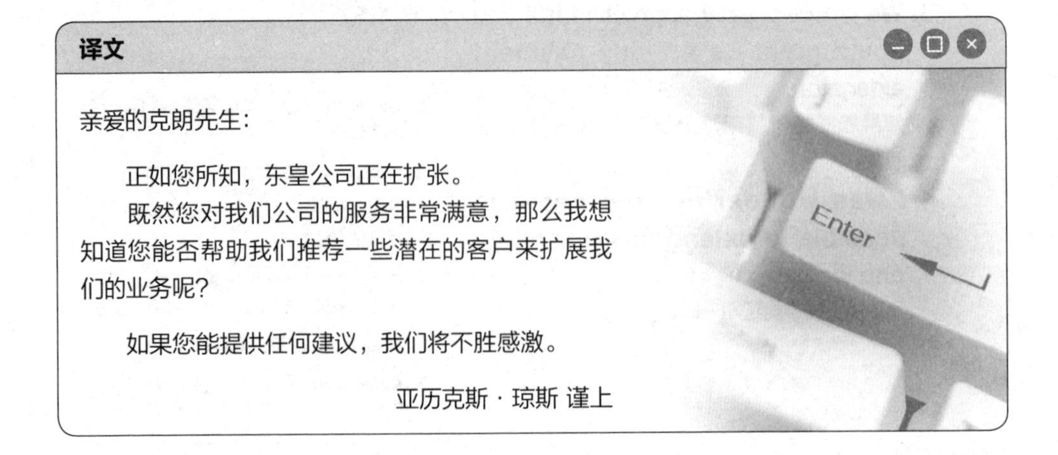
译文

亲爱的克朗先生:

正如您所知，东皇公司正在扩张。
既然您对我们公司的服务非常满意，那么我想知道您能否帮助我们推荐一些潜在的客户来扩展我们的业务呢?

如果您能提供任何建议，我们将不胜感激。

亚历克斯 · 琼斯 谨上

语法重点解析

1 解析重点1 could be so kind as to...

could be so kind as to... 意思是“如果您能……（某人会非常感激）？”是请求别人做某事时常会用到的句型。类似委婉的表达方法还有：Would you please...? / Would you be so kind to...?；Could you...? / Could I ask you to...? 请看下面的句子：

Would you be so kind as to lend me some money?（你能够借我一点儿钱吗？）

Would you please tell us about the payment terms?（能请您告诉我们付款方式吗？）

2 解析重点2 activate business

短语 activate business 的意思是“扩展业务”，类似的表达还有：expand business, develop business, branch out等。请看下面的句子：

I'd like to find a partner in China to expand my business.（我想在中国找一个合作伙伴拓展业务。）

From selling train tickets, the company branched out into package holidays.（公司业务从发售火车票扩展到跟团旅游。）

1. **Could you be kind enough to introduce some clients for us?**
   您能为我们介绍一些客户吗？
2. **The customer was impressed by our machines' performance.**
   客户对我们机器的性能印象深刻。
3. **I wonder if you could be so kind as to recommend some customers for us.**
   我想知道您是否能给我们推荐一些客户。
4. **Could I ask you to offer some favorable suggestions?**
   我能请您提供一些有利的建议吗？
5. **Our company is on its way to prosperity.**
   本公司正走向繁荣之路。

必背关键单词

1. ***expand*** [ɪk'spænd] v. 扩张
2. ***express*** [ɪk'spres] v. 表达
3. ***satisfy*** ['sætɪsfaɪ] v. 满足；使满意
4. ***wonder*** ['wʌndə(r)] v. 惊奇；想知道；怀疑
5. ***recommend*** [ˌrekə'mend] v. 推荐
6. ***potential*** [pə'tenʃl] adj. 潜在的
7. ***activate*** ['æktɪveɪt] v. 激活；使活动

11 请求变更日期

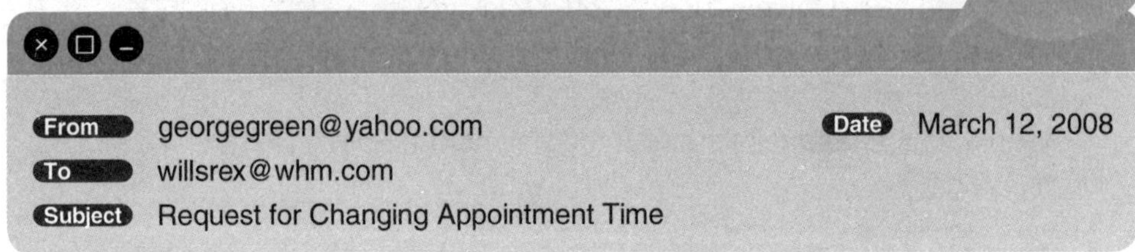

From georgegreen@yahoo.com
Date March 12, 2008
To willsrex@whm.com
Subject Request for Changing Appointment Time

Dear Mr. Wills,

I ***regret***[1] that I won't be able to make it to the meeting we set up at your ***firm***[2] on March 14.
However, I am wondering if I could visit your office on ***either***[3] of the ***following***[4] two days, March 15 or 16, ***since***[5] I would like to meet with you as ***soon***[6] as possible.

Please ***contact***[7] me at your earliest ***convenience***[8].

Yours truly,
George Green

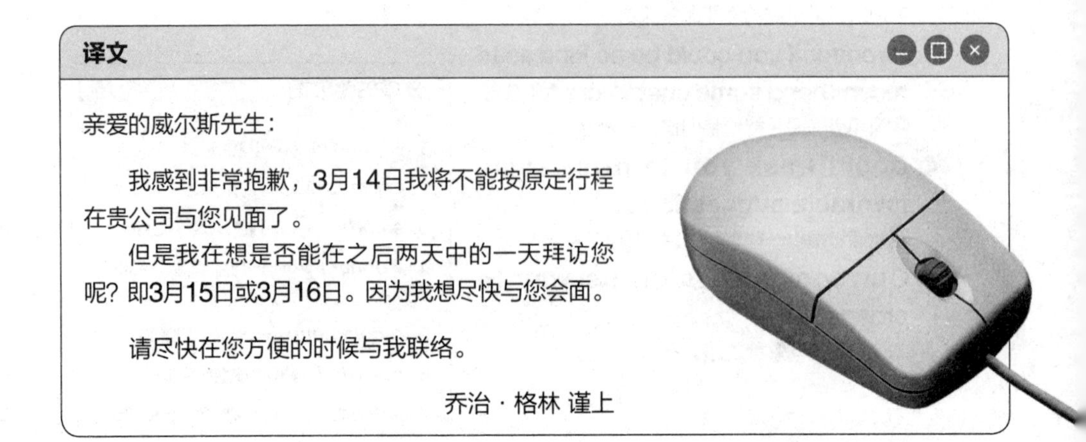

译文

亲爱的威尔斯先生：

我感到非常抱歉，3月14日我将不能按原定行程在贵公司与您见面了。

但是我在想是否能在之后两天中的一天拜访您呢？即3月15日或3月16日。因为我想尽快与您会面。

请尽快在您方便的时候与我联络。

乔治 · 格林 谨上

语法重点解析

1 解析重点1 the meeting we set up

the meeting we set up 的意思是“计划好的会面”。这句话还可以这样表达：the meeting we set, the meeting we arranged, the meeting we scheduled 等。请看下面的句子：

I am afraid I can not meet you at the time we scheduled before.（我恐怕不能在我们之前约定的时间与您见面了。）

2 解析重点2 as soon as possible

短语 as soon as possible 的意思是“尽快；越快越好”，表达同样意思的短语还有 as quickly as possible。请看下面的句子：

I asked her to get in touch with Henry as soon as possible.（我要求她尽快与亨利联系。）

The assistant wrapped up the clothes for her as quickly as possible.（这个店员以最快的速度为她把衣服包好。）

高频例句

1. **Could I schedule an appointment with you on May 10?**
   与您会面的日期订在5月10日可以吗？
2. **I am looking forward to meeting you soon.**
   我期待尽快与您会面。
3. **Could you let me know what time is convenient for you?**
   您什么时候方便能告诉我吗？

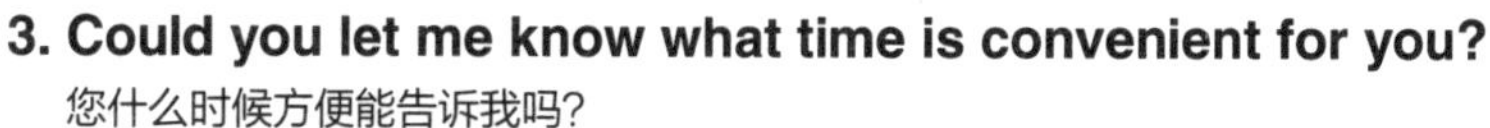

4. **I am very sorry that I have changed the time of the appointment.**
   对于约会时间的改变我感到非常抱歉。
5. **Could we *arrange*[9] another time of the appointment?**
   我们再安排个时间见面好吗？
6. **What about meeting at 4:00 p.m. on September 21?**
   我们在9月21日下午4点见面如何？
7. **Thank you very much for your understanding.**
   非常感谢您的谅解。

必背关键单词

1. ***regret*** [rɪˈgret] v. 后悔；遗憾；抱歉
2. ***firm*** [fɜːm] n. 商行；公司
3. ***either*** [ˈaɪðə(r)] prep. （两者之中）任一的
4. ***following*** [ˈfɒləʊɪŋ] adj. 下列的；后面的
5. ***since*** [sɪns] conj. 因为
6. ***soon*** [suːn] adv. 不久；很快
7. ***contact*** [ˈkɒntækt] v. 接触；联系
8. ***convenience*** [kənˈviːnɪəns] n. 方便
9. ***arrange*** [əˈreɪndʒ] v. 安排

12 请求退货

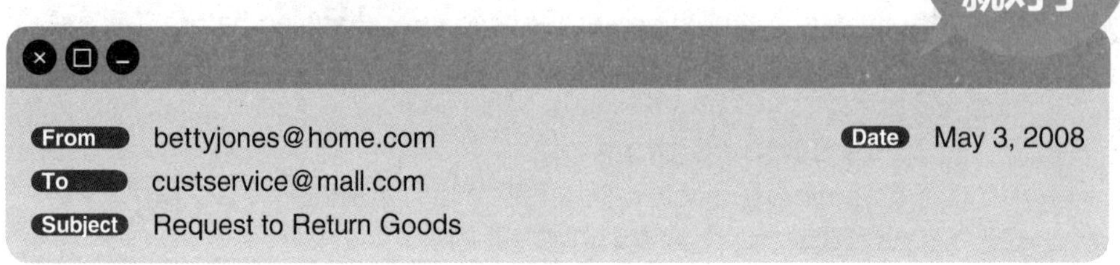

From bettyjones@home.com Date May 3, 2008
To custservice@mall.com
Subject Request to Return Goods

Dear Mr. Peter,

I would like to ***return***[1] the ten window ***curtains***[2] that arrived today.
The ***orange***[3] color is a bit darker than I ***expected***[4]. May I ***exchange***[5] them for ***brighter***[6] ones? I will send them back **if this is all right with you. Please let me know *whether*[7] you could send me other *samples*[8]**.

I am looking forward to hearing from you soon.

Yours faithfully,
Betty Jones

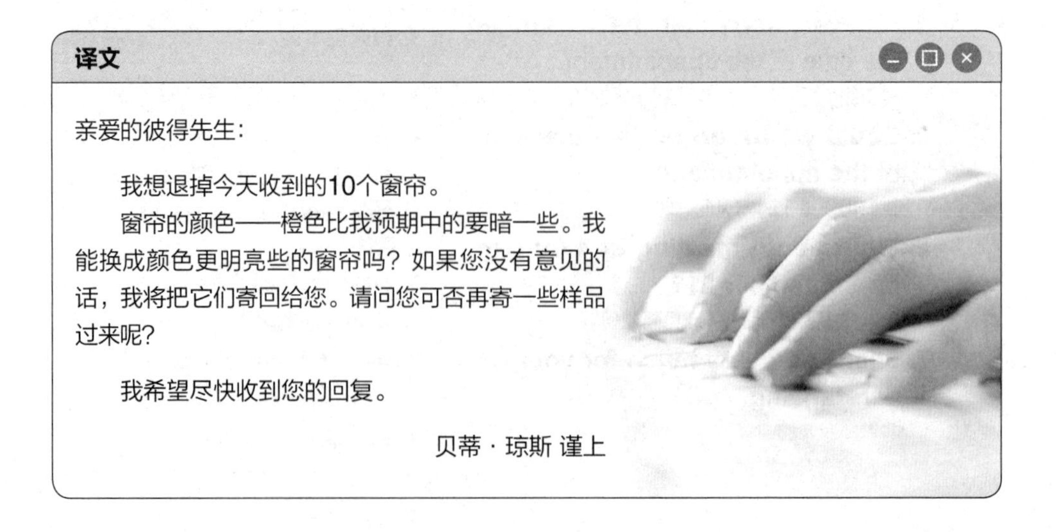

译文

亲爱的彼得先生：

我想退掉今天收到的10个窗帘。

窗帘的颜色——橙色比我预期中的要暗一些。我能换成颜色更明亮些的窗帘吗？如果您没有意见的话，我将把它们寄回给您。请问您可否再寄一些样品过来呢？

我希望尽快收到您的回复。

贝蒂 · 琼斯 谨上

语法重点解析

1 解析重点1 if this is all right with you

if this is all right with you 的意思是“您方便的话”，其中的 all right 可以换成 OK。这是一种很委婉的表达方式。类似的表达方法有：If you don't mind（如果您不介意的话），I would like to...（我想……可以吗？），Could you...（……行吗？）还可以用更委婉的表达方式：If it's not too much trouble, I would like to...（如果不是很麻烦的话，可以……吗？）

2 解析重点2 Please let me know whether you could send me other samples.

这句话的意思为“请问您可否再寄一些样品过来呢”，其中 whether 可以引导名词性从句，意为“是否”。表达“是否”还可以用 if，所以这句话还可以这样说：Please let me know if you could send me other samples. 请看下面的句子：
She was in doubt whether she was right.（她对于自己是否正确感到怀疑。）
I will see if he wants to talk to you.（我去了解一下他是否想和你谈话。）

1. **I would like to return the goods which arrived today.**
   我想把今天到的货退回去。
2. **I am afraid you have *confused*[9] the size I want with the size he wants.**
   我担心你把我和他要的尺寸弄混了。
3. **Could I exchange the goods for different ones?**
   我能把货物换成别的货物吗？
4. **The color is wrong.**
   颜色不对。
5. **Would it be possible to change it to other products?**
   我能换成其他的产品吗？
6. **I will change the time if this is all right with you.**
   如果您没有意见的话，我将改期。

必背关键单词 

1. ***return*** [rɪ'tɜːn] v. 返回；退还
2. ***curtain*** ['kɜːtn] n. 窗帘
3. ***orange*** ['ɒrɪndʒ] adj. 橙色的
4. ***expect*** [ɪk'spekt] v. 预期；期望
5. ***exchange*** [ɪks'tʃeɪndʒ] v. 交换；兑换
6. ***bright*** [braɪt] adj. 明亮的
7. ***whether*** ['weðə(r)] conj. 是否
8. ***sample*** ['sɑːmpl] n. 样品
9. ***confuse*** [kən'fjuːz] v. 使混乱；使困惑

13 请求澄清事实

这样写就对了

From "Tom Reeves" (treeves@ilike.com)
Date Mon., September 1, 2008
To "Martin Law" (mlaw@cheese.com)
Subject Request for Clarifying Matters

Dear Mr. Law,

We have got your e-mail in which you ***complained***[1] about the quality of food in our supermarket, and we have to say that your ***remark***[2] has had a bad ***effect***[3] on our ***daily***[4] business to a certain degree.

As far as we know, the reason that your food ***spoiled***[5] is because you didn't put them immediately in the ***fridge***[6]. We do hope you can eliminate any ***negative***[7] effects caused by your remark. Thank you for you cooperation.

Yours sincerely,
Tom Reeves

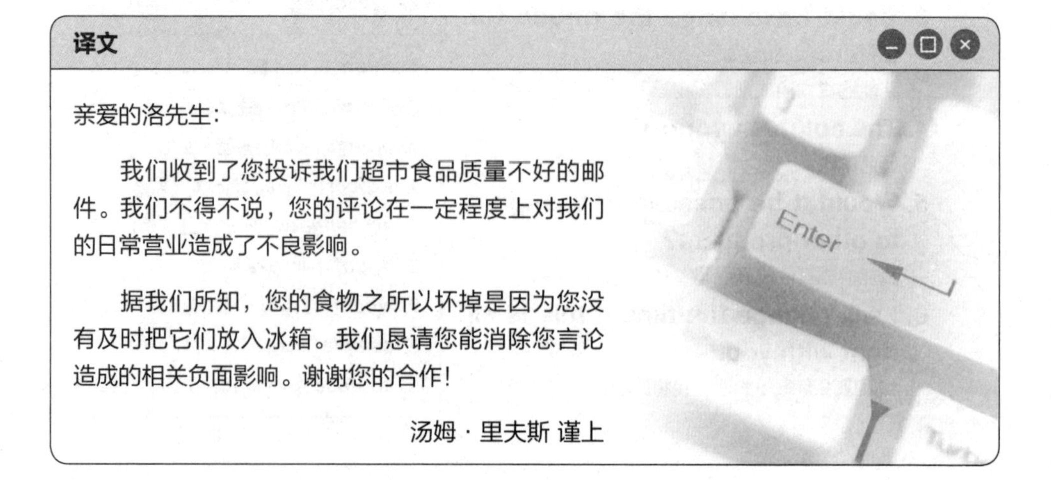

译文

亲爱的洛先生：

我们收到了您投诉我们超市食品质量不好的邮件。我们不得不说，您的评论在一定程度上对我们的日常营业造成了不良影响。

据我们所知，您的食物之所以坏掉是因为您没有及时把它们放入冰箱。我们恳请您能消除您言论造成的相关负面影响。谢谢您的合作！

汤姆 · 里夫斯 谨上

语法重点解析

1 解析重点1 **Your remark has had a bad effect on our daily business to a certain degree.**

有时候顾客的言论可以影响其他购买者的判断能力。如果己方卖出的商品出了质量问题，就要及时解决。但如果不是己方的错误，而是由于顾客自身使用不当或是保存不当出现的状况，我们也要要求其澄清事实，消除负面影响。要注意have a bad effect on...（对……有不好影响）中用的介词是 on。

2 解析重点2 **The reason that your food spoiled is because you didn't put them immediately in the fridge.**

当客户对我们的服务或是产品出现不满情绪的时候，我们一定要耐心地解释清楚。因此在解释的时候，就可以用上这个说法：The reason that...is because...（之所以……，是因为……）。

1. **You should *clarify*[8] the facts related to this problem.**
   你应当澄清与此问题有关的事实。
2. **You must set the facts straight so that my company isn't charged unfairly.**
   你们必须弄清事实，以免我公司受到不公正的指控。
3. **You should publicize my company's products rightly.**
   你应该正确宣传我公司的产品。
4. **You have to actively cooperate with our company.**
   你要积极配合我公司的工作。
5. **You must clarify the truth in order to restore reputation.**
   你必须澄清事实以挽回公司声誉。
7. **We think that your remark isn't *reasonable*[9] enough.**
   我们认为您的评论不够合理。

必背关键单词 

1. ***complain*** [kəmˈpleɪn] v. 抱怨；投诉
2. ***remark*** [rɪˈmɑ:k] n. 话语；评论
3. ***effect*** [ɪˈfekt] n. 影响；效果
4. ***daily*** [ˈdeɪlɪ] adj. 每日的；日常的
5. ***spoil*** [spɔɪl] v. 宠坏；损坏；使（食物）腐坏
6. ***fridge*** [frɪdʒ] n. 冰箱
7. ***negative*** [ˈnegətɪv] adj. 否定的；消极的；沮丧的
8. ***clarify*** [ˈklærəfaɪ] v. 澄清
9. ***reasonable*** [ˈri:znəbl] adj. 合理的

14 请求协助

这样写就对了

From "Terry Affleck" (taffleck@cute.com)　Date Tues., September 2, 2008
To "Davis Farrell" (dfarrell@music.com)
Subject Asking for Help

Dear Mr. Farrell,

We are going to hold a large meeting in our city ***hall***[1] on Tuesday.
If possible, we would like to use a ***projector***[2] to give a ***slide***[3] ***show***[4]. We also need several people to pass out the ***pamphlets***[5] to the members ***present***[6] during the ***meeting***[7].
We would very much appreciate it if you would solve these problems for us.

Looking forward to hearing from you.

Yours sincerely,
Terry Affleck

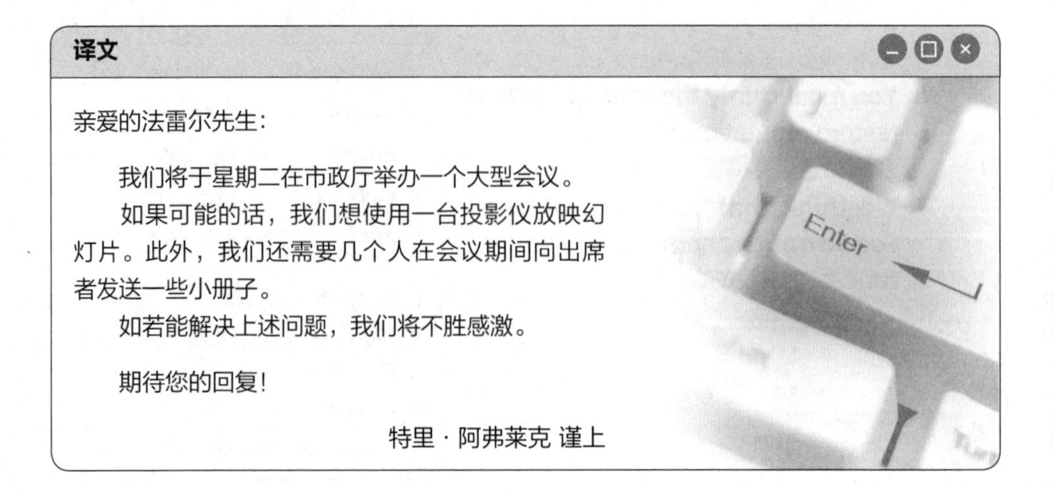

译文

亲爱的法雷尔先生：

我们将于星期二在市政厅举办一个大型会议。

如果可能的话，我们想使用一台投影仪放映幻灯片。此外，我们还需要几个人在会议期间向出席者发送一些小册子。

如若能解决上述问题，我们将不胜感激。

期待您的回复！

特里 · 阿弗莱克 谨上

语法重点解析

1 解析重点1 **We are going to hold a large meeting in our city hall on Tuesday.**

说明事件的时候，我们务必要把时间地点说清楚（Clearness）。We are going to hold a large meeting in our city hall on Tuesday.（我们将于星期二在市政厅举办一个大型会议）在说明时间、地点的同时，也点明了主题。

2 解析重点2 **If possible, we would like to use a projector...**

一般我们说到希望能用上某物，就会想到 hope 和 use 这两个词语。我们会说，we hope to use a projector（我们希望能用投影仪），我们也可以更加委婉地提出这个要求：If possible, we would like to use a projector（如果可能的话，我们想使用投影机），这里的 if possible 和 we would like to... 都是很委婉的说法。

高频例句

1. **We made a request to them for *aid*[8].**
   我们请求他们援助。
2. **We are very grateful for your assistance.**
   我们十分感谢你的协助。
3. **We have to prepare the necessary equipment for the performance.**
   我们不得不为表演准备必需的设备。
4. **Could you provide us with the lighting equipments?**
   您能为我们提供一下照明设备吗?
5. **Who will be responsible for providing us with *facilities*[9]?**
   谁将负责给我们提供设备?
6. **Only a few had been asked to arrange the meeting.**
   只要求了少数人负责筹备会议。
7. **The members were decorating the meeting place.**
   会员们正在布置会场。

必背关键单词

1. ***hall*** [hɔ:l] n. 大厅；过道；走廊
2. ***projector*** [prəˈdʒektə(r)] n. 放映机；投影仪
3. ***slide*** [slaɪd] n. 幻灯片
4. ***show*** [ʃəʊ] n. 显示；展示
5. ***pamphlet*** [ˈpæmflət] n. 小册子
6. ***present*** [ˈpreznt] adj. 出席的；到场的
7. ***meeting*** [ˈmi:tɪŋ] n. 会议
8. ***aid*** [eɪd] n. 援助
9. ***facility*** [fəˈsɪlətɪ] n. 设备；设施

15 请求归还资料

From "Jason Depp" (jdepp@global.com)
Date Wed., September 3, 2008
To "Kenny Rice" (kreeves@nice.com)
Subject Asking for Returning Materials

Dear Mr. Rice,

Could you please ***return***[1] the materials which I ***lent***[2] you last week?
As they are very important documents for my recent business, I really need you to give them ***back***[3] as soon as you can.
I have already *reminded*[4] you about them several times. Please don't let such an *unpleasant*[5] *thing*[6] come between us.

Looking forward to hearing from you soon.

Yours sincerely,
Jason Depp

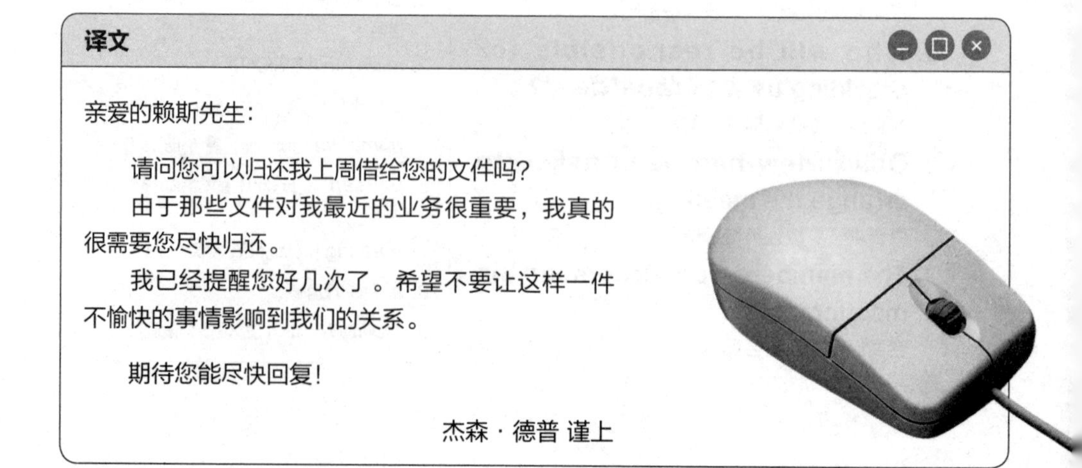

译文

亲爱的赖斯先生：

请问您可以归还我上周借给您的文件吗?

由于那些文件对我最近的业务很重要，我真的很需要您尽快归还。

我已经提醒您好几次了。希望不要让这样一件不愉快的事情影响到我们的关系。

期待您能尽快回复！

杰森 · 德普 谨上

语法重点解析

1 解析重点1 I have already reminded you about them several times.

一般对别人下最后通牒或是发出强烈要求其归还物品前，我们还是要尽量给一些提醒。当提醒不奏效时，我们再义正言辞地要求他们必须归还。remind sb about / of sth意为“提醒某人某事”。

2 解析重点2 Please don't let such an unpleasant thing come between us.

有的时候，对方对你的归还请求不予理睬。这种情况下，你要让他明白事情的严重性，Please don't let such an unpleasant thing come between us（希望不要让这样一件不愉快的事情影响到我们的关系），这句话点明了严重后果，同时也给对方施加了一点压力。

1. **Would you please return my company's important *documents*[7]?**
   能请您归还我公司的重要文件吗？
2. **We hope you will return the company's documents.**
   我们希望您归还公司的文件。

3. **If you can return the document, we would be *grateful*[8].**
   如果您能归还文件，我们将不胜感激。
4. **You are our most trusted partner; I believe you will *immediately*[9] return our documents.**
   您是我们最信赖的合作伙伴，我相信您会立即归还我们的文件。
5. **I don't want something like this to get in the way of our relationship.**
   我不想让这种事情影响我们的关系。
6. **Please tell me why you didn't give them back.**
   请告诉我你没有归还它们的原因。

必背关键单词

1. ***return*** [rɪˈtɜːn] v. 归还；返回
2. ***lend*** [lend] v. 借出
3. ***back*** [bæk] adv. 向后地；往回
4. ***remind*** [rɪˈmaɪnd] v. 提醒
5. ***unpleasant*** [ʌnˈpleznt] adj. 使人不愉快的；不合意的
6. ***thing*** [θɪŋ] n. 东西；物体
7. ***document*** [ˈdɒkjʊmənt] n. 文件
8. ***grateful*** [ˈgreɪtfl] adj. 感激的
9. ***immediately*** [ɪˈmiːdɪətlɪ] adv. 立即地

16 请求制作合同

这样写就对了

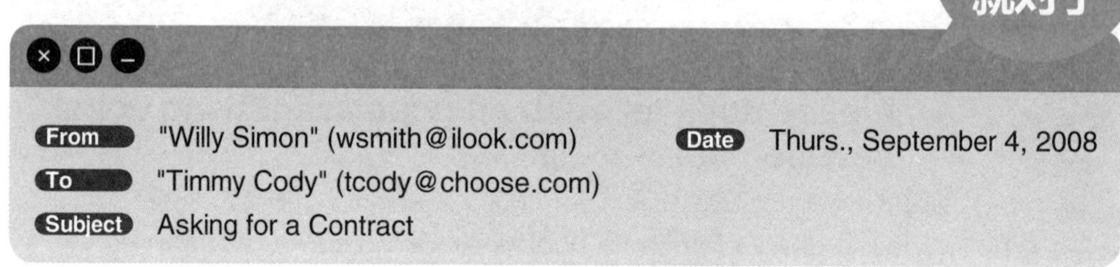
From "Willy Simon" (wsmith@ilook.com)　Date Thurs., September 4, 2008
To "Timmy Cody" (tcody@choose.com)
Subject Asking for a Contract

Dear Mr. Cody,

We both have agreed to all the ***terms***[1], so there shouldn't be any problem with the ***contract***[2]. We are waiting for you to send us the copy of the contract.
If you can send it to us at ***once***[3], we can ***execute***[4] the project next week. **Any *delay*[5] would *mess*[6] up the whole plan.**
Please keep us informed if there is any change in the plan.

Looking forward to hearing from you soon.

Yours sincerely,
Willy Simon

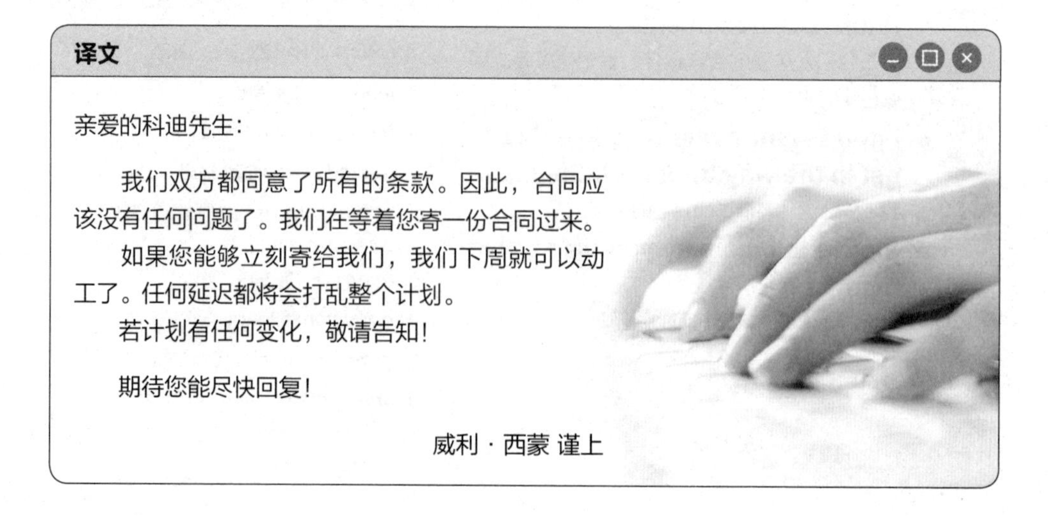
译文

亲爱的科迪先生:

我们双方都同意了所有的条款。因此，合同应该没有任何问题了。我们在等着您寄一份合同过来。

如果您能够立刻寄给我们，我们下周就可以动工了。任何延迟都将会打乱整个计划。

若计划有任何变化，敬请告知!

期待您能尽快回复!

威利 · 西蒙 谨上

语法重点解析

1 解析重点1 **Any delay would mess up the whole plan.**

英文跟中文有一个很大的区别。那就是，在中文里，我们习惯用人作为主语和动作的实施者，而在英文中，以某事或某物作主语是很常见的事情。例如：Any delay would mess up the whole plan（任何延迟都将会打乱整个计划），其意思很清楚明白，也说出了事情的重要性。

2 解析重点2 **Please keep us informed if there is any change in the plan.**

一般我们要求对方能及时告知计划变动时会说：Please inform us if there is any change in the plan.（如若计划有任何变化，请告知！）也可以这样表达：Please keep us informed if there is any change in the plan.（如若计划有任何变化，敬请告知！）

高频例句

1. **We agree with all of your terms. And we will send our man as soon as possible to sign a contract.**
   我们同意您的所有条款，我方将尽快派人与你方签订合同。
2. **If you have no *objection*[7] to the plan, we request you to sign a contract.**
   如果你方对此计划没有异议，我方请求签订合同。
3. **If you sign a contract with us ASAP, we will ship as soon as possible.**
   如果你方尽早与我方签订合同，我方就可以尽早发货。
4. **If you sign a contract with us, we will give you a special discount.**
   如果你方能与我方签订合同，我们将给予你方特别的优惠。
5. **For the terms of the contract we have *agreed*[8] to, please sign a contract as soon as possible, and we can *guarantee*[9] the timely production.**
   对于合约条款我们已经达成共识，请尽早签订合约，如此我方可以保证及时生产。

必背关键单词

1. ***term*** [tɜ:m] n. 条款
2. ***contract*** [ˈkɒntrækt] n. 契约；合同
3. ***once*** [wʌns] n. 一次
4. ***execute*** [ˈeksɪkju:t] v. 实行；执行
5. ***delay*** [dɪˈleɪ] n. 耽搁
6. ***mess*** [mes] v. 弄乱
7. ***objection*** [əbˈdʒekʃn] n. 反对
8. ***agree*** [əˈgri:] v. 同意
9. ***guarantee*** [ˌgærənˈti:] v. 保证

17 请求返还合同

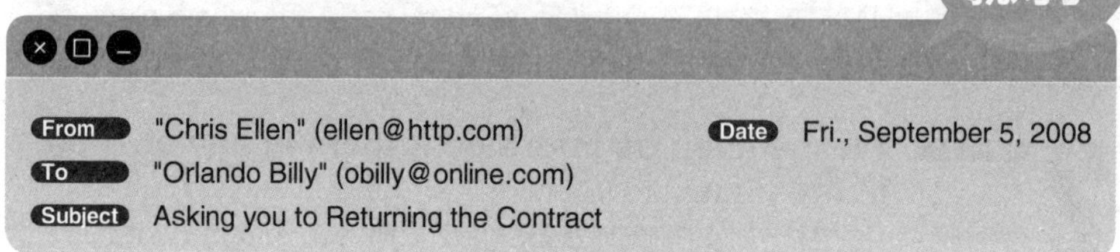

From "Chris Ellen" (ellen@http.com)　Date Fri., September 5, 2008
To "Orlando Billy" (obilly@online.com)
Subject Asking you to Returning the Contract

Dear Mr. Billy,

Have you ***gone***[1] over the contract and ***found***[2] everything in order? We have not contacted you until now, as you ***instructed***[3] us to wait for your ***response***[4].
Could you send us a copy of the ***signed***[5] contract as soon as possible? We are eager to start our work soon.

Looking forward to hearing from you soon.

Yours sincerely,
Chris Ellen

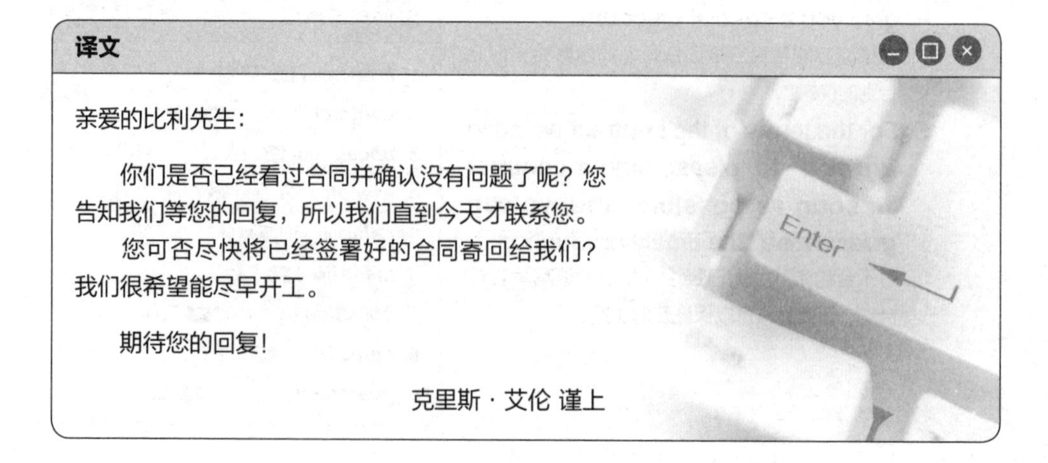

译文

亲爱的比利先生：

你们是否已经看过合同并确认没有问题了呢？您告知我们等您的回复，所以我们直到今天才联系您。

您可否尽快将已经签署好的合同寄回给我们？我们很希望能尽早开工。

期待您的回复！

克里斯 · 艾伦 谨上

语法重点解析

1 解析重点1 **Have you gone over the contract and found everything in order?**

把合同寄给对方之后，我们需要询问对方是否已经看过合同，签订好合同，并要求把签好的合同寄回。这句话中的 go over（仔细检查）还可以用另外一个单词来替换，就是 review（检查；复审）。

2 解析重点2 **We have not contacted you until now, as you instructed us to wait for your response.**

一般贸易往来的时候，对方有时会要求等待回复。这时，我们就要尊重对方的要求。you instructed us to wait for your response（您告知我们等您的回复）中的 instruct 一般是指上级对下属发出命令或指示，用在这里含有尊重对方，遵照对方要求的意思，比较礼貌。

高频例句

1. **Did you have the chance to *review*[6] the contract?**
   不知您有没有机会将合同再看一次？
2. **I *refrained*[7] from contacting you about this matter until today.**
   我直到今天才克制不住就此事联系您。
3. **We plan to begin the *construction*[8] as early as we can.**
   我们计划尽早动工建设。
4. **Please sign and return one copy of this contract.**
   烦请签署好合同并将其中一份寄回。
5. **Please send us the contract at your earliest convenience.**
   请尽早将合同寄给我们。
6. **Please return the contract right away.**
   请尽快把合同寄回。
7. **We hope there is no delay in the project.**
   我们不希望工程发生任何延误。

必背关键单词 

1. ***go*** [gəʊ] v. 去；走
2. ***find*** [faɪnd] v. 找到；发现
3. ***instruct*** [ɪnˈstrʌkt] v. 教导；指令
4. ***response*** [rɪˈspɒns] n. 回应；答复
5. ***sign*** [saɪn] v. 签名；签字
6. ***review*** [rɪˈvjuː] v. 回顾；检查
7. ***refrain*** [rɪˈfreɪn] v. 抑制；克制
8. ***construction*** [kənˈstrʌkʃn] n. 建筑；结构

18 请求商品目录

这样写就对了

From	"Willy Simon" (wsimon@find.com)	Date	Sat., September 6, 2008
To	"Peter Walker" (pwalker@want.com)		
Subject	Request for Catalogue		

Dear Mr. Walker,

I would like to have your latest catalogue so that I can purchase ***mobile***[1] ***phone***[2] ***accessories***[3].
I will contact you regarding any items I may be interested in after receiving the catalogue. Thank you in advance.

Looking forward to hearing from you soon.

Yours sincerely,
Willy Simon

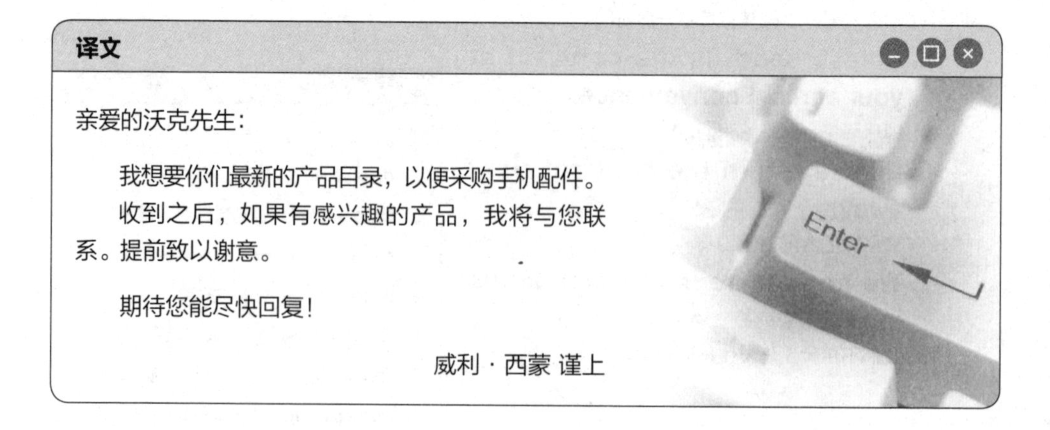

译文

亲爱的沃克先生：

我想要你们最新的产品目录，以便采购手机配件。

收到之后，如果有感兴趣的产品，我将与您联系。提前致以谢意。

期待您能尽快回复！

威利 · 西蒙 谨上

语法重点解析

1 解析重点1 I would like to have your latest catalogue so that...

我们表达“想要做什么”，可以用 want to do sth。如果我们想要更委婉一些，则用 would like to do sth。而若想表达自己做这件事的目的，可以用 to, in order to , so that，for the purpose of。其中，用 to 是最简洁明了的一种方式。需要注意的是，这些单词和短语后面接的动词形态会有所不同。

2 解析重点2 I will contact you regarding any items I may be interested in.

一般我们联系对方，肯定是因为某件事需要跟对方交流。那么，“关于”某件事情应该如何表达呢？一般，我们会想到用 about。其实，还有很多表示“有关……”的单词和短语，例如：regarding, concerning, as to, with regard to。

1. **Could you send us your new *fall*[4] catalogue?**
   能否寄给我们你方新出的秋季商品目录？
2. **Could you provide us with a catalogue or something that tells me about your products?**
   你能否提供产品目录或者介绍贵公司产品的材料？
3. **Your company *reissued*[5] the catalogue with a new price *list*[6].**
   你们公司重新发行了带有新价格表的商品目录。
4. **Please send us a new catalogue of your *merchandise*[7].**
   请寄给我们一份你方新的商品目录。
5. **I shall be glad if you will send me your catalogue together with a quotation.**
   如果您能在寄送目录时附上报价，我将不胜感激。
6. **Can I have a list of your products?**
   能给我一份贵公司的产品目录吗？
7. **Please send me your *current*[8] catalogue ASAP.**
   请将现有目录尽快寄来。

必背关键单词

1. ***mobile*** [ˈməʊbaɪl] adj. 可移动的
2. ***phone*** [fəʊn] n. 电话
3. ***accessory*** [əkˈsesərɪ] n. 附件；零件
4. ***fall*** [fɔːl] n. 秋天
5. ***reissue*** [ˌriːˈɪʃuː] v. 再版；再印
6. ***list*** [lɪst] n. 清单；目录；列表
7. ***merchandise*** [ˈmɜːtʃəndaɪs] n. 商品
8. ***current*** [ˈkʌrənt] adj. 流通的；目前的

19 请求订购办公用品

这样写就对了

From perry0411@apl.com
To wright2008@tom.com
Subject Order of Office Supplies
Date Sun., September 7, 2008

Dear Mr. Wright,

I learned from your advertisement on the Internet that your company ***supplies***[1] ***digital***[2] projectors.
In order to ***facilitate***[3] our sales team in delivering presentations to current and ***prospective***[4] clients, I want to buy a digital projector from your company. The attached application which details the projector's intended use and lists product information, including model, price and technical ***specifications***[5]. Could you deliver what I want to our company before February 28?

Thank you very much for your ***consideration***[6].

Yours sincerely,
Donald Perry

译文

亲爱的赖特先生：

在网络上看到贵公司的广告，得知贵公司提供数码投影仪。

为了便于我们的销售团队向客户（包括现有客户和潜在客户）演示产品，想向贵公司订购一部数码投影仪。附件里有一份申请书，详细写明了投影仪的具体用途和产品的具体信息，包括：型号、价格和技术规格。您能在2月28日前将我需要的产品送至我公司吗?

非常感谢您的关照。

唐纳德·佩里 谨上

语法重点解析

1 解析重点1 I want to buy a digital projector from your company.

I would like to apply for the purchase of a digital projector from your company.（我想从贵公司订购一台数字投影仪。）这句话本身并无错误，但是根据英文书信的7C原则中的 Conciseness（简洁），此句就显得有些啰嗦。所以上面的句子我们可以说：I want to buy a digital projector from your company.（我想从贵公司买一台数字投影仪。）

2 解析重点2 Could you deliver what I want to our company before February 28?

Could you deliver what I want to our company before the 28th?（您能否于28号之前将我要的产品运至我公司呢？）此句本身并无语法或拼写错误，但是在商务书信中要遵循英文书信的7C原则中的 Concreteness（具体），此句只说 before the 28th，那么到底是本月28号还是其他月份的28号呢？应该将具体日期明确。所以上面的句子我们可以说：Could you deliver what I want to our company before February 28?（您能否于2月28日之前将我要的产品运至我公司呢？）

高频例句

1. **I learned from the commercial on TV that your company offers delivery service.**
   我在电视广告上获悉贵公司提供送货服务。
2. **I would like to buy a new photocopier for our department.**
   我想为本部门购买一台新复印机。
3. **I shall be grateful if you will favor me with an *early*[7] reply.**
   若能早日回复，我将不胜感激。
4. **Could you provide home delivery service?**
   您能提供送货上门服务吗？
5. **I heard that your company offers digital products.**
   我听说贵公司提供数码产品。

必背关键单词

1. ***supply*** [səˈplaɪ] v. 提供
2. ***digital*** [ˈdɪdʒɪtl] adj. 数字的；数码的
3. ***facilitate*** [fəˈsɪlɪteɪt] v. 使便利
4. ***prospective*** [prəˈspektɪv] adj. 预期的；未来的；可能的
5. ***specification*** [ˌspesɪfɪˈkeɪʃn] n. 规格；详述；说明书
6. ***consideration*** [kənˌsɪdəˈreɪʃn] n. 体贴；关心；考虑
7. ***early*** [ˈɜːlɪ] adj. 早的；早期的

20 请求客户反馈

这样写就对了

From "Adam Brody" (abrody@facebook.com) Date Mon., September 8, 2008
To "Wentworth Miller" (wmiller@international.com)
Subject Request for Information Feedback

Dear Mr. Miller,

Thank you for purchasing our products.
To help ***improve*** [1] our services, **would you be so *kind*** [2] **as to take a few *minutes*** [3] **to *answer*** [4] **the following *questions*** [5]? **The more *feedback*** [6] **we get from you, the better for all customers.**
Thank you for your assistance.

Looking forward to hearing from you soon.

Yours sincerely,
Adam Brody

译文

亲爱的米勒先生：

感谢您购买我们的产品。
为了帮助我们改善服务，您可否花几分钟时间回答下面几个问题呢？从您这得到的反馈越多，对所有顾客而言就越有利。
感谢您的帮助。

期待您能尽快回复！

亚当 · 布罗迪 谨上

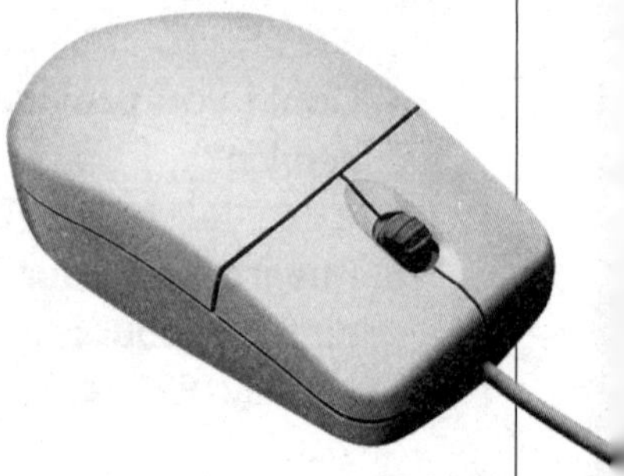

语法重点解析

1 解析重点1 **Would you be so kind as to take a few minutes to answer the following questions?**

请求顾客做某事时，我们可以像上述邮件那样，使用 Would you be so kind as to...（您能不能……？）这种问法虽然显得有点啰嗦，但是很礼貌。“花费时间做某事”，则可以用 take...to do sth 或者 spend...doing sth 来表达。

2 解析重点2 **The more feedback we get from you, the better for all customers.**

The more..., the better... 这个句型的使用，使整个句子结构清晰，同时意思简洁（Clearness）明了。例如：
The more you practice, the better you can speak English.（你练习越多，英语讲得越好。）

1. **We welcome feedback from people who use the goods we produce.**
   我们公司欢迎用户反馈产品信息。
2. **We need more feedback from the consumer in order to improve our goods.**

   我们需要从消费者那里得到更多反馈以提高产品品质。
3. **There was much feedback from our questionnaire.**

   我们的问卷调查有很多反馈信息。
4. **We need to collect and file the feedback from customers.**
   我们需要收集整理来自客户的反馈。
5. **We will collect feedback from our users, subsequently *organize*[7] and *analyze*[8] the collected data.**
   我们将收集用户的反馈，随后将组织和分析这些资料。
6. **Our company *encourages*[9] customer feedback in regard to products and services in the hotel.**
   我们公司鼓励客户对酒店的产品和服务提供相关的反馈。

必背关键单词  

1. ***improve*** [ɪmˈpruːv] v. 改善
2. ***kind*** [kaɪnd] adj. 仁慈的；友好的
3. ***minute*** [ˈmɪnɪt] n. 分；片刻
4. ***answer*** [ˈɑːnsə(r)] v. 回答
5. ***question*** [ˈkwestʃən] n. 疑问；问题
6. ***feedback*** [ˈfiːdbæk] n. 反馈
7. ***organize*** [ˈɔːɡənaɪz] v. 组织；安排
8. ***analyze*** [ˈænəlaɪz] v. 分析；细察
9. ***encourage*** [ɪnˈkʌrɪdʒ] v. 鼓励

Unit 10 催促 Urging

01 催促寄送样品

From abco@yahoo.com
To smpfrom@tom.com
Subject Please Send the Samples Soon
Date March 21, 2008

Dear Mr. Brown,

Please inform us if the ***samples***[1] we requested have ***already***[2] been sent.
It has been two weeks since we received your letter ***confirming***[3] that the samples will be ***delivered***[4] to us, but they have not arrived yet.
I would like to ***remind***[5] you that we are holding a meeting next week to select the ***items***[6] we will buy in the next year, and your products will be ***excluded***[7] from our purchasing list if the samples can't arrive in time. Please be quick!

Sincerely yours,
ABC Company

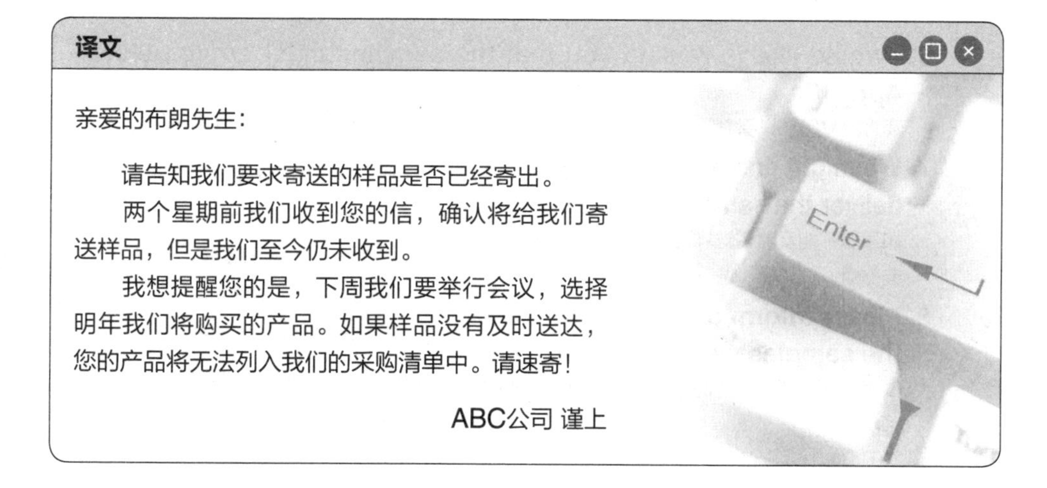

译文

亲爱的布朗先生:

请告知我们要求寄送的样品是否已经寄出。

两个星期前我们收到您的信，确认将给我们寄送样品，但是我们至今仍未收到。

我想提醒您的是，下周我们要举行会议，选择明年我们将购买的产品。如果样品没有及时送达，您的产品将无法列入我们的采购清单中。请速寄!

ABC公司 谨上

语法重点解析

1 解析重点1 I would like to remind you that...

此句型是提醒对方某事时经常用到的句型。remind 的意思是“提醒；使想起”。请看以下例句：

I would like to remind you that our office is in want of a computer.（我想提醒您一下，我们办公室急需一台电脑。）

I'd like to remind you that the lecture is at 6 o'clock.（我想提醒你讲座6点钟开始。）

2 解析重点2 hold a meeting

hold a meeting 的意思是“召开会议；开会”。hold 在此意为“举行”。表达“举行”除了可以用 hold外，还可以用 give, have, throw 等动词，但是 have, throw 不正式，应尽量少用于商务书信中。

高频例句

1. I would like you to confirm whether the sample has been sent.
我想向您确认一下样品是否已寄出。

2. Please advise us if the samples we ordered have been shipped.
请告知我们订购的样品是否已经装运。

3. It has been one month since we got the letter from you.
自从上次收到您的信，已经过了1个月了。

4. Let me know if you have sent the samples to us.
如果您已经寄送了样品，请告知我们。

5. I hope we can get the samples as soon as possible.
我希望我们能尽快收到样品。

6. I would like to remind you that the samples must arrive before April 10.
我想提醒您，样品必须在4月10日之前送达。

7. We might end the deal if you can't deliver the samples to us in time.
如果您不能及时寄送样品的话，我们可能会结束交易。

8. Please inform us if you have sent the samples we requested.
请通知我们您是否已经寄送了我们需要的样品。

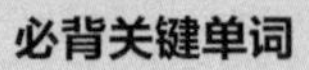

1. ***sample*** [ˈsɑːmpl] n. 样品
2. ***already*** [ɔːlˈredɪ] adv. 已经
3. ***confirm*** [kənˈfɜːm] v. 确认；证实
4. ***deliver*** [dɪˈlɪvə(r)] v. 寄送
5. ***remind*** [rɪˈmaɪnd] v. 提醒
6. ***item*** [ˈaɪtəm] n. 项目；物品
7. ***exclude*** [ɪkˈskluːd] v. 不包括；排除；除……外

02 催促返还所借资料

这样写就对了

From fiona0309@yahoo.com　Date August 22, 2008
To bobgreen@tom.com
Subject Please Return the Documents Soon

Dear Bob,

I want to ask you if you could please ***return***[1] the ***documents***[2] I lent you last month.
As I ***explained***[3] to you earlier, they are very important to me. I really need you to return them as soon as possible.
I have already ***reminded***[4] you of this matter for ***several***[5] times. Please tell me ***frankly***[6], what **on earth** is the ***reason***[7] that you can't return them? In fact, I really don't want something like this to **get in the way of** our relationship.

Yours truly,
Fiona

译文

亲爱的鲍勃：

我想请问您，能否把我上个月借给您的文件还给我。
正如我之前向您解释的，那些文件对我非常重要。我真的需要您尽快归还它们。
关于这件事，我已经提醒您好几次了，能坦白地告诉我到底是什么原因使您不能归还吗？事实上，我真的不想因为这件事影响到我们之间的关系。

菲奥娜 谨上

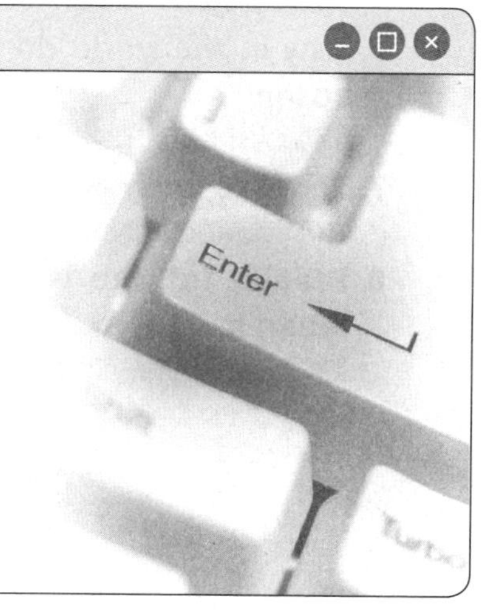

语法重点解析

1 解析重点1 on earth

on earth 在此并不是指“世界上；人世间”，而是指“究竟；到底”，置于疑问词后，用来加强语气。in the word 也有相同的用法，请看以下例句：

What on earth do you mean?（你到底是什么意思？）

What in the world are you doing here?（你到底在这里做什么啊？）

2 解析重点2 get in the way of

get in the way of 是“阻止；妨碍”的意思，等同于 in the way of, in one's way。请看以下例句：

Her social life gets in the way of her study.（她的社交生活妨碍了学业。）

Nothing can stand in the way of love.（什么也阻挡不了爱情。）

He is always in my way.（他总是碍我的事。）

1. Could you give the documents back to me?
您能把文件还给我吗？

2. You have kept my document for two months.
我的文件已经在你那里放了两个月了。

3. Can you tell me why you can't return it?
您可以告诉我为什么不能归还它吗？

4. Let me know when you are planning to give it back.
请告诉我您打算什么时候将它归还。

5. I hope you could return them before this weekend.
我希望您能在这个周末之前将它们归还。

6. Why in the world do you keep it so long?
您到底为什么用了它这么久？

7. They are really important to me.
它们对我真的很重要。

8. Please tell me when you can return it.
请告诉我您什么时候可以归还。

必背关键单词 

1. ***return*** [rɪˈtɜ:n] v. 归还；返还
2. ***document*** [ˈdɒkjʊmənt] n. 文件
3. ***explain*** [ɪkˈspleɪn] v. 解释
4. ***remind*** [rɪˈmaɪnd] v. 提醒；使想起
5. ***several*** [ˈsevrəl] adj. 几个的；各自的
6. ***frankly*** [ˈfræŋklɪ] adv. 坦白地
7. ***reason*** [ˈri:zn] n. 原因

03 催促寄送商品目录

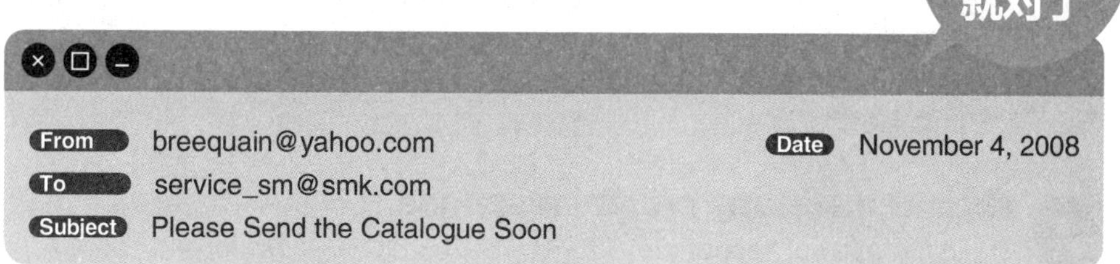

From breequain@yahoo.com Date November 4, 2008
To service_sm@smk.com
Subject Please Send the Catalogue Soon

Dear Service Department,

I would like to inquire if you have sent us the ***Supermarket***[1] ***Catalogue***[2]. I ***requested***[3] it three weeks ago. I haven't received it ***yet***[4]. Could you tell me what the problem is? If there is no problem at all, please send it to me ASAP. I am afraid that **I will have to *consider***[5] other ***products***[6] from other supermarkets if you do not **make any prompt response**.

Yours sincerely,
Bree Quain

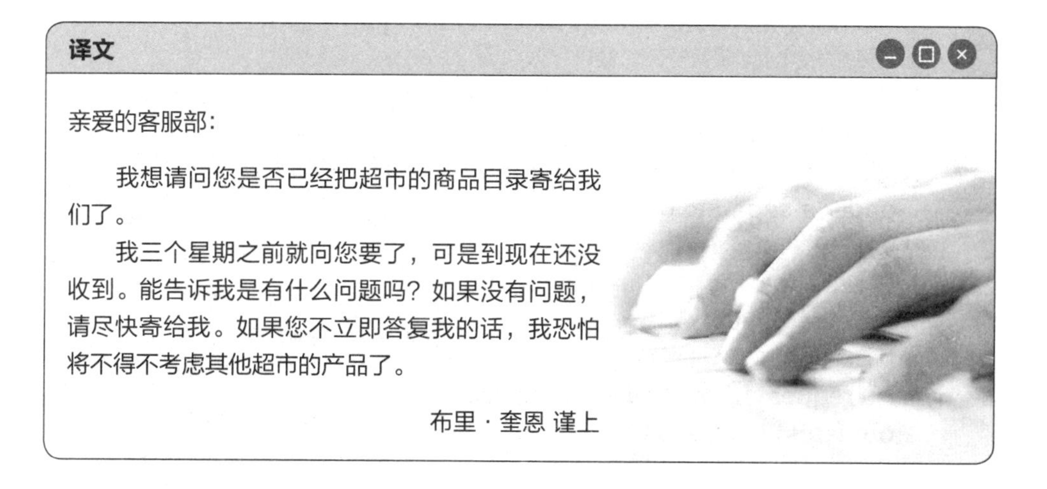

译文

亲爱的客服部:

我想请问您是否已经把超市的商品目录寄给我们了。

我三个星期之前就向您要了，可是到现在还没收到。能告诉我是有什么问题吗？如果没有问题，请尽快寄给我。如果您不立即答复我的话，我恐怕将不得不考虑其他超市的产品了。

布里 · 奎恩 谨上

1 解析重点1 I will have to...

I will have to... 的意思是“我将不得不；我将必须”，是非常直接的表达方式，有给对方施压的感觉，有时也有不情愿的意味。前面的 I am afraid that...（我恐怕……）则显得语气委婉、柔和。请看以下例句：

I will have to be thrifty if I am going to get through school.（如果我想读到毕业，就非得节俭不可。）

2 解析重点2 make any prompt response

make prompt response 的意思是“立即做出回应”。prompt 做形容词是“立刻的；行动迅速的”，response 则做名词，表示“回答；答复”，make no response 则为“不回复”的意思。if you do not make any prompt response 等同于 if you do not reply immediately（如果你不立刻答复）。

1. **The store sent us a new catalogue of its *merchandise*[7].**
   商店寄给了我们一份新的商品目录。
2. **I shall be glad if you will send me your catalogue.**
   如果您能寄送商品目录给我，我会非常高兴。
3. **We have *forwarded*[8] you our new catalogue today.**
   我们今天已经把新的商品目录寄给您了。
4. **We will mail you our most recent catalogue.**
   我们将寄给您我们最新的商品目录。
5. **You can know more about our items through the catalogue.**
   你可以通过商品目录了解更多我们的商品。
6. **We are sending you a catalogue under a separate cover.**
   商品目录将单独邮寄给您。
7. **Please fax me the layout for the new catalogue.**
   请将新目录的版式传真给我。
8. **As requested, we are sending you our latest catalogue.**
   按照您的要求，现寄上最新的商品目录。

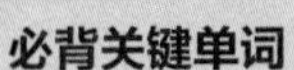

1. ***supermarket*** [ˈsu:pəmɑ:kɪt] n. 超市
2. ***catalogue*** [ˈkætəlɒg] n. 商品目录
3. ***request*** [rɪˈkwest] v. 要求；请求
4. ***yet*** [jet] adv. 还；仍然
5. ***consider*** [kənˈsɪdə(r)] v. 考虑；深思
6. ***product*** [ˈprɒdʌkt] n. 产品
7. ***merchandise*** [ˈmɜ:tʃəndaɪs] n. 商品
8. ***forward*** [ˈfɔ:wəd] v. 发送；递送

04 催促出货

From tonysmith@yahoo.com
To suitsco@21n.com
Subject Please Hurry with the Shipment
Date October 12, 2008

Dear Mr. Thomas,

Regarding the sales contract No.3624, covering 400 ***dozen***[1] sport suits, we wish to remind you that we have had no news from you about ***shipment***[2] of the goods.
As we ***mentioned***[3] in our last letter, we are in ***urgent***[4] need of the goods. And if you are not able to supply them in time, we may be ***compelled***[5] to seek an ***alternative***[6] ***source***[7] of supply.

We look forward to receiving your shipping notice, by fax, within the next seven days.

Yours faithfully,
Tony Smith

译文

亲爱的托马斯先生：

关于销售合同号3624订购的400打运动服，我们想提醒您我们还没有收到贵公司的发货通知。

正如我们上一封电子邮件所提到的，我们急需此批货物。如贵公司未能及时供货，本公司可能被迫寻求其他替代货源。

我们希望您能在7天内将出货通知传真给我方。

托尼 · 史密斯 谨上

语法重点解析

1 解析重点1 **We are in urgent need of the goods.**

这句话的意思是“我们急需这批货物”。in urgent need of 是“急需”的意思，表达相同意思的短语还有 in dire need of。请看以下例句：

We are in urgent need of these two grades of goods.（我们急需这两种等级的货物。）

My brother is in dire need of a good lawyer.（我弟弟急需一位好律师。）

2 解析重点2 **be compelled to**

短语 be compelled to do sth 是指“被迫做某事”，与其意思相近的短语还有 be forced to do sth，既表明了催促对方是不得已而为之，也表明了事情的紧迫性，又不显得语气过于生硬。请看以下例句：

He was compelled to bring this action.（他是不得已才起诉的。）

Sometimes we are forced to tell a white lie.（有时我们被迫讲些善意的谎言。）

1. **In *respect*[8] to our contract No. 246, we wish to bring the fact to your attention.**
   关于双方合同第256条，我们希望您注意到这一事实。
2. **We still have received no news from you about the shipment of the goods.**
   我们仍然没有收到关于这批货物的装运消息。
3. **The goods are being demanded by our customers.**
   我们的客户一再催促这批货物。
4. **The goods should have arrived here three months ago.**
   这批货物本该在3个月前抵达的。
5. **Please understand and resolve this serious and urgent matter ASAP.**
   望贵公司体谅并尽快解决此迫切而严重的问题。
6. **We are in dire need of the goods.**
   我们急需这批货物。
7. **Your prompt attention and the earliest possible shipment are greatly desired.**
   非常希望您能关注这一情况并尽早发货。

必背关键单词

1. ***dozen*** [ˈdʌzn] n. 一打；十二个
2. ***shipment*** [ˈʃɪpmənt] n. 装运
3. ***mention*** [ˈmenʃn] v. 提及
4. ***urgent*** [ˈɜːdʒənt] adj. 急迫的；紧急的
5. ***compel*** [kəmˈpel] v. 强迫；迫使
6. ***alternative*** [ɔːlˈtɜːnətɪv] adj. 两者择一的；替代的
7. ***source*** [sɔːs] n. 来源
8. ***in respect to*** [ɪn rɪˈspekt tu] phr. 关于

05 催促寄送货品

From michaelcole@yahoo.com
To young_abc@21n.com
Subject Please Hurry with the Delivery
Date March 3, 2009

Dear Mr. Young,

We have been waiting for the goods we ***ordered***[1] on February 26, 2009, shipment ***notification***[2] No. 5216-03A. You said they would be delivered by the ***beginning***[3] of this week, but they still have not arrived yet.
Our business conditions ***specify***[4] a ***delivery***[5] date of Mar 2. If the delivery is delayed any further, we will have to ***reconsider***[6] our plans to **deal with** you in the future.

Please ***track***[7] the order immediately. **Thanks very much for your *earnest*[8] efforts in taking care of this matter.**

Yours truly,
Michael Cole

译文

亲爱的杨先生：

我们一直在等2009年2月26日订购的货物，装箱通知单的编号为5216-03A。您说本周初就会发货，但是到现在仍没有到货。

我们的贸易条件清楚地说明发货日期是3月2日。如果发货再有延误，我们将不得不重新考虑以后和您的交易计划。

请立即追踪订单。非常感谢您能认真处理此事。

迈克尔 · 科尔 谨上

语法重点解析

1 解析重点1 **deal with**

短语 deal with 有“应付；处理；与……交易；和……做买卖”的意思，这里是指“与……交易；和……做买卖”，相当于 do business with, trade with。请看以下例句：

We have dealt with that firm for many years.（我们与那家公司有多年的生意往来。）

I never do business with dishonest people.（我从不跟不诚实的人做生意。）

We trade with people from many countries.（我们和许多国家的人进行贸易往来。）

2 解析重点2 **Thanks very much for your earnest efforts in taking care of this matter.**

这句话的意思是“非常感谢您能认真处理此事”。英文书信写作中有一个原则叫 Courtesy（礼貌）原则。此信中虽然是在催促对方赶快发货，本身包含着不满的情绪，但是表达时还是非常的礼貌客气。

1. We urge you to give your prompt attention to this matter.

我们强烈要求您立即对此事给予关注。

2. Please deliver the goods we ordered immediately.

请立即发送我们订购的货物。

3. I expect that you could *respond*[9] soon.

期望您快速做出回应。

4. The goods should have been delivered last week.

这批货物本应上周发货。

5. Please make prompt response and solve the problem.

请迅速回复并解决此事。

6. Thank you for turning your attention to this matter immediately.

感谢您立即关注此事。

7. The earliest possible delivery is greatly desired.

非常希望您能尽早发货。

必背关键单词

1. ***order*** [ˈɔːdə(r)] v. 订购
2. ***notification*** [ˌnəʊtɪfɪˈkeɪʃn] n. 通知；通知单
3. ***beginning*** [bɪˈɡɪnɪŋ] n. 起初；开始
4. ***specify*** [ˈspesɪfaɪ] v. 明确说明
5. ***delivery*** [dɪˈlɪvərɪ] n. 传送；交货；交付
6. ***reconsider*** [ˌriːkənˈsɪdə(r)] v. 重新考虑
7. ***track*** [træk] v. 跟踪；追踪
8. ***earnest*** [ˈɜːnɪst] adj. 真诚的；热心的
9. ***respond*** [rɪˈspɒnd] v. 回应

06 催促开立发票

这样写就对了

From paulcole@yahoo.com　Date August 20, 2008
To Ianmail@yahoo.com
Subject Please Send the Receipt Soon

Dear Mr. Ian,

I am afraid that we still have not received the ***receipt*** [1] for our ***payment*** [2] regarding ***invoice*** [3] No.00942926. Three months have passed since we made the payment in May.
This month, we will be **closing our books for the year**. If the receipt is not sent by the ***end*** [4] of the month, it will **hold up** our ***accounting*** [5] ***procedures*** [6].
Please turn you ***attention*** [7] to this problem at once, as I would not like to see something like this ***hinder*** [8] our business relationship.

Yours faithfully,
Paul

译文

亲爱的伊恩先生：

恐怕我们仍然没有收到发票号码为00942926的款项收据。自从5月份我们付款以来已经过了3个月。

本月我们要进行本年度的结算。如果月底的时候还收不到收据，将会耽搁我们账务处理程序。

请立即关注这个问题，因为我不想让此类的事情阻碍我们的业务关系。

保罗 谨上

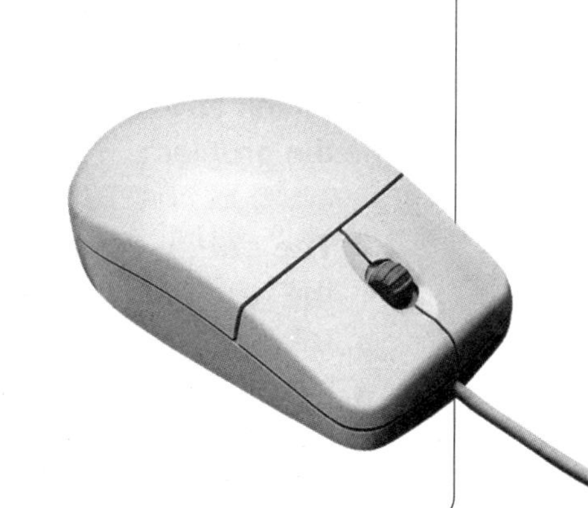

语法重点解析

1 解析重点1 close our books for the year

close our books for the year 的意思是“把这一年的账结清”，即“决算”。“决算期”可以写成 settlement term 或者 settlement period；“决算期末”可以用 account end；“决算日”可以用 accounting day 或者 closing day；“决算报告”可以用 financial statement；“会计年度”可以写成 fiscal year。

2 解析重点2 hold up

这个短语有“举起；支撑；耽搁；妨碍”的意思。这里是指“耽搁；延迟”。表达相同意思还可以用 delay, put off。请看以下例句：

I don't want to hold up your time.（我不想耽误你的时间。）

How long will the flight be delayed?（航班将延误多长时间？）

Why did you put off your visit?（你为什么拖延拜访日期？）

1. I hope you could give me the receipt for my payment at once.
我希望您立即为我的付款开具收据。

2. Please *address*[9] the problem immediately.
请立即处理这个问题。

3. I have called you several times to request the receipt.
我已经打了好几次电话向你索取收据。

4. I am afraid I still haven't got the receipt from you.
我恐怕仍然没有收到您的收据。

5. Please send me the invoice as soon as possible.
请尽快把收据寄给我。

6. Thank you for paying attention to the problem.
感谢您对这个问题予以关注。

7. The earliest possible delivery of the receipt is needed.
我需要您尽快寄出收据。

8. I really don't want this to affect our relationship.
我真的不想因这件事影响我们的关系。

必背关键单词

1. ***receipt*** [rɪˈsiːt] n. 收据
2. ***payment*** [ˈpeɪmənt] n. 付款
3. ***invoice*** [ˈɪnvɔɪs] n. 发票
4. ***end*** [end] n. 末尾；结束
5. ***accounting*** [əˈkaʊntɪŋ] n. 会计
6. ***procedure*** [prəˈsiːdʒə(r)] n. 程序；步骤
7. ***attention*** [əˈtenʃn] n. 注意
8. ***hinder*** [ˈhɪndə(r)] v. 阻碍；打扰
9. ***address*** [əˈdres] v. 处理；演说

07 催促 制订合同

这样写就对了

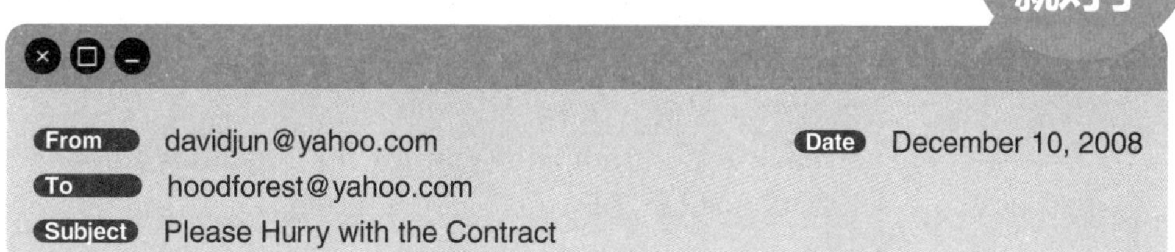

From davidjun@yahoo.com　Date December 10, 2008
To hoodforest@yahoo.com
Subject Please Hurry with the Contract

Dear Mr. Hood,

As we discussed before, we are ***ready***[1] to enter into a ***contract***[2] with you. We are waiting for you to send us the ***copies***[3] of the ***agreement***[4].
If you ***draft***[5] and send us the documents right away, we can commence ***construction***[6] next week. Any further delay in the ***conclusion***[7] of a contract will ***impede***[8] the construction schedule.

Please inform us in time if there have been some changes of the plan.

Sincerely yours,
David

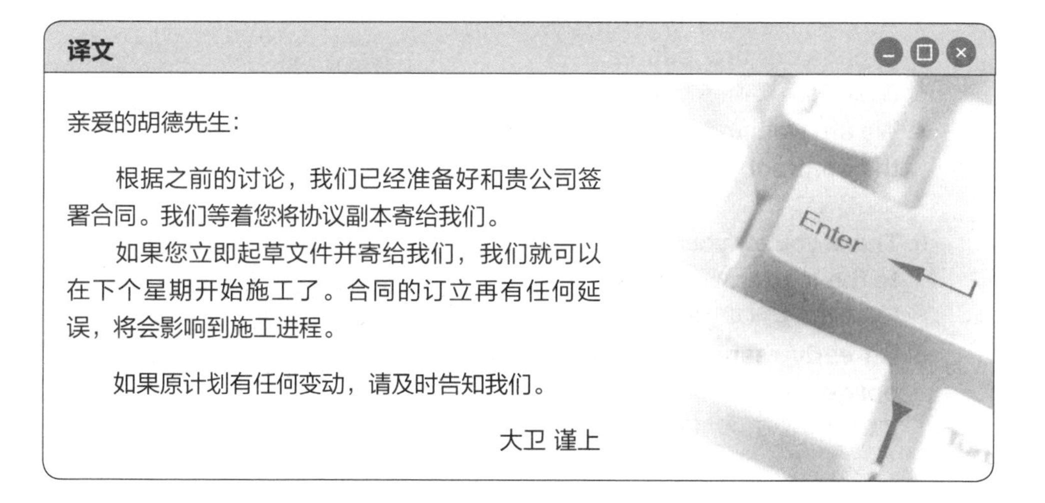

译文

亲爱的胡德先生：

　　根据之前的讨论，我们已经准备好和贵公司签署合同。我们等着您将协议副本寄给我们。

　　如果您立即起草文件并寄给我们，我们就可以在下个星期开始施工了。合同的订立再有任何延误，将会影响到施工进程。

　　如果原计划有任何变动，请及时告知我们。

大卫 谨上

语法重点解析

1 解析重点1 enter into a contract with

enter into a contract with sb 意为“与某人签合同”。这里的 enter into 不是“进入”的意思，而相当于 make，可理解为“制订；签署”。所以这个短语还可以用 make a contract with sb 来代替。请看以下例句：

We are glad to enter into a contract with you for 20-ton wool.（我们很高兴和贵方签订一份20吨羊毛的合同。）

I have come to make a contract with you for the business under discussion.（我来是为了与贵方签订正在洽淡的这笔业务的合约。）

2 解析重点2 commence

这个词的意思是“开始”，相当于 start, begin，但是 start 和 begin 都是非常普通的词，日常生活中经常会用到。commence 在日常英语交流中并不常见，多用于非常正式的场合。请看以下例句：

After the election, the new government commenced developing the roads.（选举过后，新政府开始修建道路。）

What time does the play start?（戏剧什么时候开始？）

The meeting is about to begin.（会议即将开始。）

1. **We are ready to make a contract with you.**
   我们已经准备好和贵公司签合约。
2. **Please send us the copies of agreement soon.**
   请把协议副本速寄至我公司。
3. **Any delay of the contract will impede our procedures.**
   合约的任何延误都将会阻碍我们的进程。
4. **We are waiting for you to enter into the contract.**
   我们等着与您签合约。
5. **Thank you for your prompt attention to the matter.**
   感谢您对此事予以即时关注。
6. **An early response will be greatly obliged.**
   如您能尽早回应，我将不胜感激。

必背关键单词 

1. ***ready*** [ˈredɪ] adj. 准备好的
2. ***contract*** [ˈkɒntrækt] n. 合约；合同
3. ***copy*** [ˈkɒpɪ] n. 副本
4. ***agreement*** [əˈgriːmənt] n. 协议
5. ***draft*** [drɑːft] v. 起草；草拟
6. ***construction*** [kənˈstrʌkʃn] n. 建筑
7. ***conclusion*** [kənˈkluːʒn] n. 结论；签订
8. ***impede*** [ɪmˈpiːd] v. 阻碍；妨碍

08 催促返还合同

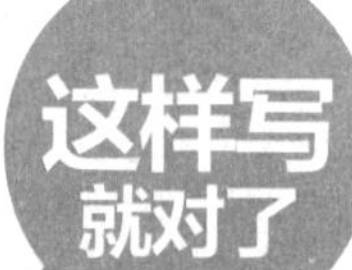

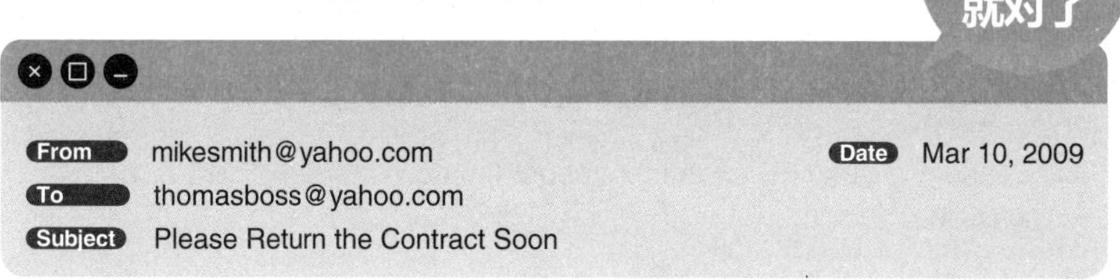
From mikesmith@yahoo.com
Date Mar 10, 2009
To thomasboss@yahoo.com
Subject Please Return the Contract Soon

Dear Mr. Thomas,

I am writing to ***enquire***[1] if you have ever had a ***chance***[2] to ***review***[3] the contract we sent. I ***refrained***[4] from mentioning this matter to you ***until***[5] today, as you ***instructed***[6] us to wait for your response.
Please ***sign***[7] and return one copy of the contract at your earliest convenience, because we really hope to start working on this project as soon as possible.

Thank you for your prompt attention!

Sincerely yours,
Mike

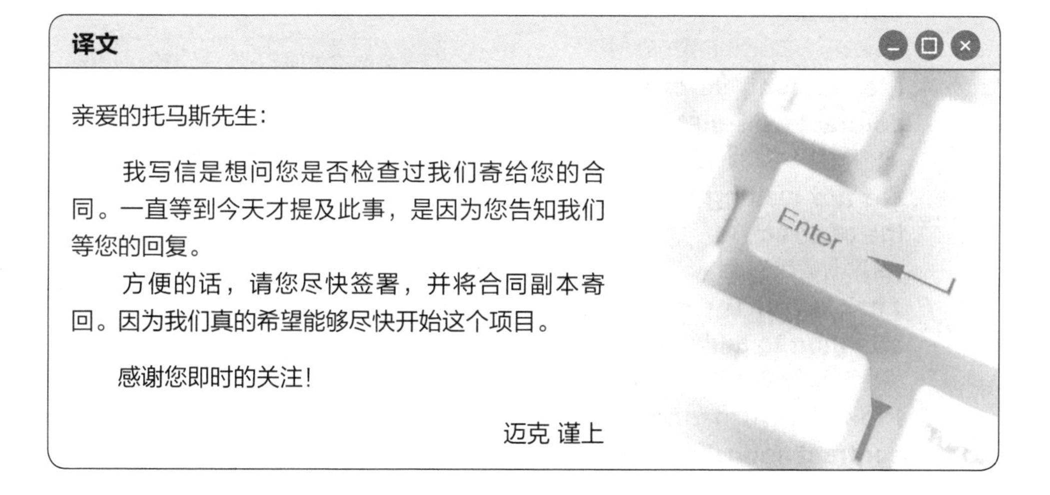
译文

亲爱的托马斯先生:

我写信是想问您是否检查过我们寄给您的合同。一直等到今天才提及此事，是因为您告知我们等您的回复。

方便的话，请您尽快签署，并将合同副本寄回。因为我们真的希望能够尽快开始这个项目。

感谢您即时的关注!

迈克 谨上

语法重点解析

1 解析重点1 refrain from

短语 refrain from doing sth 是“控制自己不做某事”的意思。表达“抑制；克制”还可以用 restrain。不过通常restrain 侧重指“控制……使不表露”，适用范围较小，一般用于感情、情绪方面。请对照以下例句：

I'm just trying to refrain from nodding off at work.（我只是试着不在工作时打瞌睡而已。）

I cannot restrain my excitement about the news.（听到这个消息，我无法控制我的兴奋之情。）

2 解析重点2 as you instructed

这个短语的意思是“按照您的指示”。表达“按照某人的指示”还可以说 as per one's instruction 或 by order of，其中 as per one's instruction 这个表达方法比较正式，所以一般适用于英文书信中。请看以下例句：

As per your instruction, we have asked the factory to improve the packing of the products.（根据您的指示，我们已要求工厂改进该产品的包装。）

The prisoner is removed by order of the court.（根据法院命令，该囚犯已被转移。）

高频例句

1. **We are ready to sign a contract with you.**
   我方已经准备好和贵公司签署合同。
2. **Have you ever reviewed our contract?**
   您已经检查过我们的合同了吗?
3. **The delay of the contract may put off our *schedule*[8].**
   合同的延误可能会阻碍我们的进程。
4. **Please send us one copy of the contract we sent to you.**
   请把我们寄给您的合同副本寄给我们。
5. **We are waiting for you to return the contract.**
   我们等着您返还合同。
6. **Thank you for your immediate attention to this matter.**
   感谢您对此事及时予以关注。
7. **Please return one copy of the contract soon.** 请尽快返还合同副本。

必背关键单词

1. ***enquire*** [ɪnˈkwaɪə(r)] v. 询问；打听
2. ***chance*** [tʃɑːns] n. 机会
3. ***review*** [rɪˈvjuː] v. 检查；评论
4. ***refrain*** [rɪˈfreɪn] v. 抑制；克制
5. ***until*** [ənˈtɪl] conj. 直到……才
6. ***instruct*** [ɪnˈstrʌkt] v. 指示；指导；命令
7. ***sign*** [saɪn] v. 签署
8. ***schedule*** [ˈʃedjuːl] n. 计划；安排

09 催促开立信用证

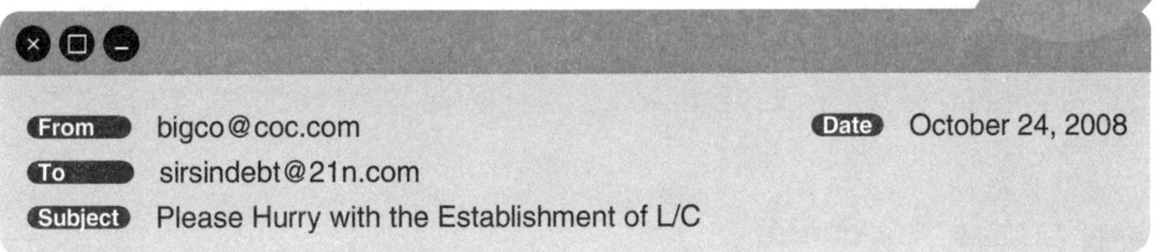
From bigco@coc.com Date October 24, 2008
To sirsindebt@21n.com
Subject Please Hurry with the Establishment of L/C

Dear Sirs,

With regard to your order No. AC178 for 200 tons of ***cotton***[1], we ***regret***[2] to tell you that up to this date we have received ***neither***[3] the required credit nor any further information from you.
Please note that we agreed that the payment for the above order is to be paid via sight letter of credit, and it must be established within 2 weeks upon the arrival of our Sales Confirmation.
We hereby request you to open **by** ***cable***[4] an ***irrevocable***[5] sight letter of credit for the ***amount***[6] of one ***million***[7] US dollars in our favor, with which we can ***execute***[8] the above order according to the original schedule.

Yours truly,
COC Co.

译文  

尊敬的先生们：

关于贵方订单号为AC178的200吨棉花订单事宜，我们很遗憾地告诉您，我们至今尚未收到要求开立的信用证，也未收到贵方任何进一步的消息。

请注意，上述订单的货款经双方同意是以即期信用证的方式支付，且信用证必须在收到我们销售确认书后的两个星期内开出。

我方在此恳请贵方以电报开立金额为一百万美元，以我方为受益人的，不可撤销的即期信用证，使我方得以按原定计划执行上述订单。

COC公司 谨上

语法重点解析

1 解析重点1 **with regard to**

这个短语的意思是“关于；至于”。表达“关于”还可以说：in regard of, in regard to, regarding, about 等。请看以下例句：

I would like to talk to you with regard to the letter you sent me.（我想就你写给我的信与你谈谈。）

In regard to his work, we have no complaints.（关于他的工作，我们没有什么可抱怨的。）

2 解析重点2 **by cable**

cable 是“电缆”的意思，by cable 就是“通过电报”的意思。在表达通过什么方式的时候，一般都会到用到介词 by，如：by rail（乘火车），by land（由陆路），by pipeline（通过管道）等。请看以下例句：

The major modes of transportation for merchandise are by rail, by land, by pipeline, by water and by air.（商品通过铁路、公路、管道、水路和空运等主要方式运输。）

1. **The goods for your order No.123 have been ready for shipment for quite some time.**
   贵方第123号订单的货物已备妥待运有相当长时间了。
2. **It is *imperative*[9] that you take immediate action to have the covering credit established as soon as possible.**
   贵方必须立即行动，尽快开出信用证。
3. **We repeatedly requested you by fax to expedite the opening of the relative letter of credit.**
   我们已经多次传真要求贵方尽快开立相关的信用证。
4. **We have effected shipment for the above-mentioned order.**
   上述订单已出货。
5. **We have not yet received the covering L/C after the lapse of three months.**
   三个月过去了，我们仍未收到相关的信用证。

必背关键单词

1. ***cotton*** [ˈkɒtn] n. 棉花
2. ***regret*** [rɪˈgret] v. 遗憾；后悔
3. ***neither*** [ˈnaɪðə(r)] conj. 两者都不
4. ***cable*** [ˈkeɪbl] n. 电缆
5. ***irrevocable*** [ɪˈrevəkəbl] adj. 不可撤销的；不可挽回的
6. ***amount*** [əˈmaʊnt] n. 数量；总额
7. ***million*** [ˈmɪljən] n. 百万
8. ***execute*** [ˈeksɪkjuːt] v. 执行；完成
9. ***imperative*** [ɪmˈperətɪv] adj. 必要的

10 催促 支付货款

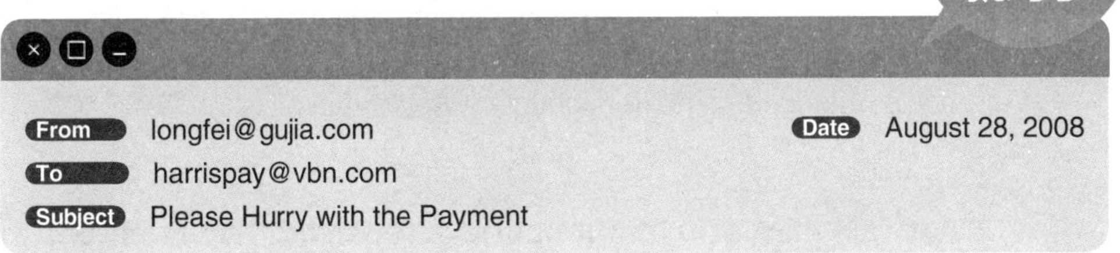
From longfei@gujia.com Date August 28, 2008
To harrispay@vbn.com
Subject Please Hurry with the Payment

Dear Mr. Harris,

This is about your ***Account***[1] No.7658.
As you are usually very prompt in ***settling***[2] **your accounts**, we ***wonder***[3] whether there is any ***special***[4] reason why we have not received payment for the above account, already a month ***overdue***[5].
We think you might not have received the *statement*[6] **of account** we sent you on July 25th showing the ***balance***[7] of US$60,000 you owe. We have sent you another copy today and hope it may have your early attention.

Yours faithfully,
Longmans Co.

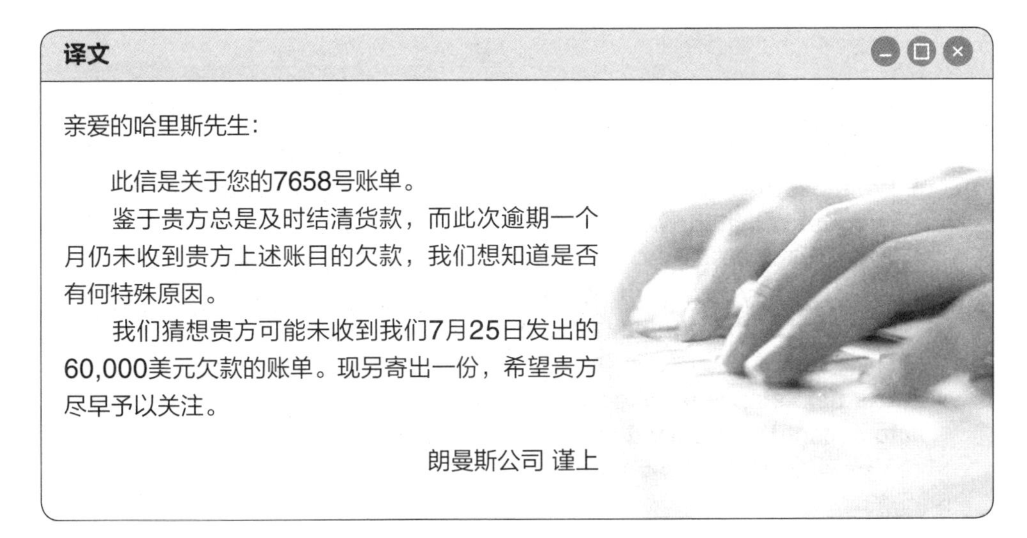
译文

亲爱的哈里斯先生：

此信是关于您的7658号账单。

鉴于贵方总是及时结清货款，而此次逾期一个月仍未收到贵方上述账目的欠款，我们想知道是否有何特殊原因。

我们猜想贵方可能未收到我们7月25日发出的60,000美元欠款的账单。现另寄出一份，希望贵方尽早予以关注。

朗曼斯公司 谨上

语法重点解析

1 解析重点1 settle your accounts

短语 settle account 是“结账；清算；支付”的意思。表达“结账；支付”含义的短语还有 square account。请看以下例句：

Could you give us another month to settle the account?（你能再宽限我们一个月的结账期吗？）

We should manage to square accounts with the bank.（我们应该设法与银行结清帐目。）

2 解析重点2 We think you might not have received the statement of account.

这句话的意思是“我们猜想贵方可能未收到账单”。一般来说，对方未能及时支付货款的原因可能有两种：一种是客观上确实出现了不可预见的状况；另外一种就是对方恶意拖延付款。不管出于哪种原因，写信者在催款的时候还是要表现出应有的礼貌。如这句话，写信者就为对方找了个借口，即“猜想贵方可能未收到账单”，从而使对方不至于太难堪，以此达到让对方尽快支付货款的目的，所以一定要注意措辞的艺术。

1. **The following items totaling US$4,000 are still *open*[8] on your account.** 您的欠款总额仍有4,000美元。
2. **It has been several weeks since we sent you our first invoice and we have not yet received your payment.**
   我们的第一份发票已经寄出好几个星期了，但我们尚未收到您的款项。

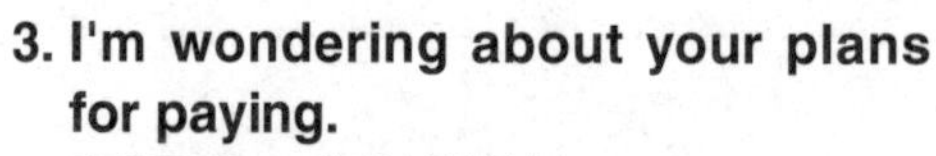

3. **I'm wondering about your plans for paying.**
   我想了解一下您的付款计划。
4. **We must now ask you to settle this account within the next few days.**
   请你务必在接下来的几日内结清这笔款项。
5. **We hope you could send the payment within the next five days.**
   我希望您能在接下来的5日内付款。
6. **Our next step is to take legal action to collect the money due to us.**
   下一步我们只能采取法律行动索取欠我们的款项了。

必背关键单词

1. ***account*** [əˈkaʊnt] n. 账目；账户
2. ***settle*** [ˈsetl] v. 解决；安排
3. ***wonder*** [ˈwʌndə(r)] v. 想知道；惊奇
4. ***special*** [ˈspeʃl] adj. 特别的
5. ***overdue*** [ˌəʊvəˈdjuː] adj. 过期的；到期未付的
6. ***statement*** [ˈsteɪtmənt] n. 声明；陈述；报告
7. ***balance*** [ˈbæləns] n. 余额；结存
8. ***open*** [ˈəʊpən] adj. 悬而未决的

Unit 11 投诉 Complaint

01 投诉货品错误

这样写就对了

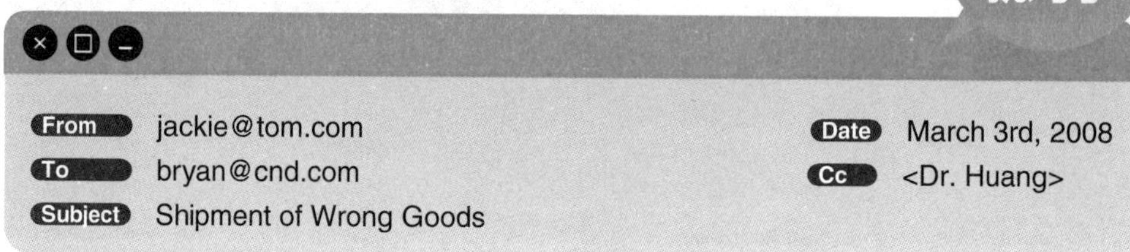

From	jackie@tom.com	Date	March 3rd, 2008
To	bryan@cnd.com	Cc	<Dr. Huang>
Subject	Shipment of Wrong Goods		

Dear Mr. Bryan,

I ***received***[1] a ***consignment***[2] of order No. 201314 from you yesterday, but the ***model***[3] is wrong. I ordered Model SP-520 **instead of** SP-502.
Attached[4] please find our order No. 201314 **for your reference**. I will ***return***[5] this shipment and the ***freight***[6] is at your cost.

Looking forward to receiving the correct shipment ***ASAP***[7].

Sincerely yours,
Jackie Black

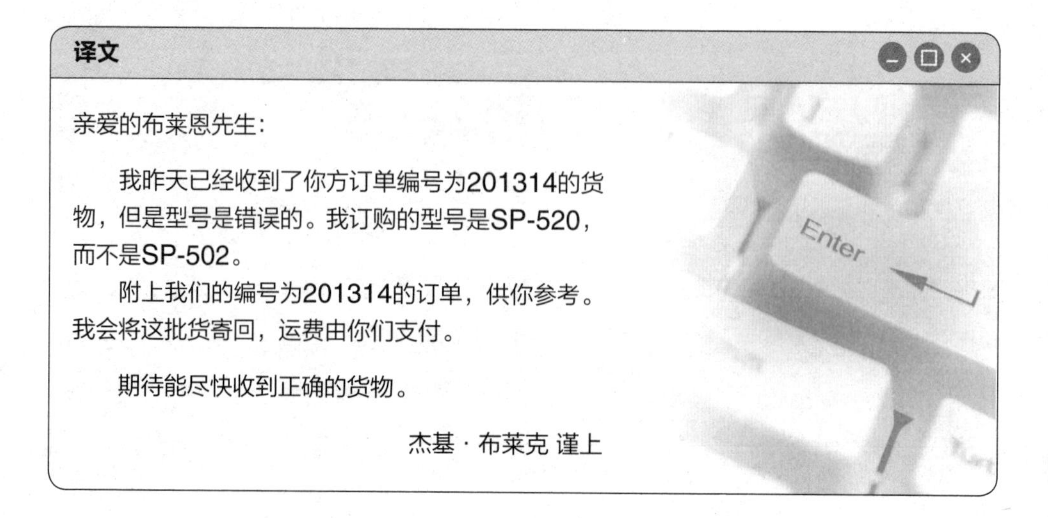

译文

亲爱的布莱恩先生：

我昨天已经收到了你方订单编号为201314的货物，但是型号是错误的。我订购的型号是SP-520，而不是SP-502。

附上我们的编号为201314的订单，供你参考。我会将这批货寄回，运费由你们支付。

期待能尽快收到正确的货物。

杰基 · 布莱克 谨上

语法重点解析

1 解析重点1 A instead of B

A instead of B 的意思是“是 A 而不是 B ”。一般使用这个短语来表示 A 与 B 中只能选一个，选择了 A 而不是 B。

请对照以下例句：

He wanted to play volleyball instead of going for a walk.（他想打排球而不是去散步。）

He sent his ***neighbor*** [8] instead of coming himself.（他不亲自来，而是叫他的邻居来。）

2 解析重点2 for your reference

reference 通常译为“提及；涉及”，而此处的语意比较接近“参考”。这个短语的意思是“供你参考”，亦用 study 代替 reference，但语意稍有不同。请对照以下例句：

This contract is only for your reference.（这份合同仅供您参考。）

Attached is the copy of our ***purchase*** [9] order for your study. Please check.（附上我们的采购单副本供您检阅，请查实。）

高频例句

1. **Please check the purchase order and send the right goods to me ASAP.**
   请核查采购单，并尽快将正确的货物寄给我。
2. **As to the compensation, we would like to know your opinion.**
   至于赔偿，我们想听听你们的意见。
3. **I believe that it is due to a *clerical* [10] error on your side.**
   我相信那是由于你们的笔误所造成的。
4. **Do you want me to send it back to your factory via freight collect?**
   你希望我以运费到付的方式寄回工厂吗？
5. **The freight should be at your cost.**
   运费应该由你方支付。
6. **We are looking forward to your earliest reply.**
   我们期待您尽早回复。
7. **Attached please find the latest delivery copy for your information.**
   附件是最新的送货单供您参考，请查收。

必背关键单词

1. ***receive*** [rɪˈsiːv] v. 收到
2. ***consignment*** [kənˈsaɪnmənt] n. 运送；装运的货物
3. ***model*** [ˈmɒdl] n. 模型；型号
4. ***attach*** [əˈtætʃ] v. 附加
5. ***return*** [rɪˈtɜːn] v. 返回；退回
6. ***freight*** [freɪt] n. 运费
7. ***ASAP = as soon as possible*** abbr. 尽快
8. ***neighbor*** [ˈneɪbə] n. 邻居
9. ***purchase*** [ˈpɜːtʃəs] n. 购买
10. ***clerical*** [ˈklerɪkl] adj. 文书工作的；文职人员的

02 投诉货品数量错误

这样写就对了

From christina@tom.com Date October 2nd, 2008
To jacobi@yahoo.com
Subject Shipment of Wrong Quantity

Dear Mr. Jacobi,

Your ***delivery***[1] of order No. 52099 has just ***arrived***[2] at our company a few hours ago. However, we found a ***shortage***[3] of the quantity because only 400 pieces were received. Please ***explain***[4] this situation, and tell us when we can receive the rest of the goods. And the ***freight***[5] ***caused***[6] will be on your charge.

Looking forward to receiving the ***remaining***[7] ***shipment***[8] ASAP.

Sincerely yours,
Christina Chen

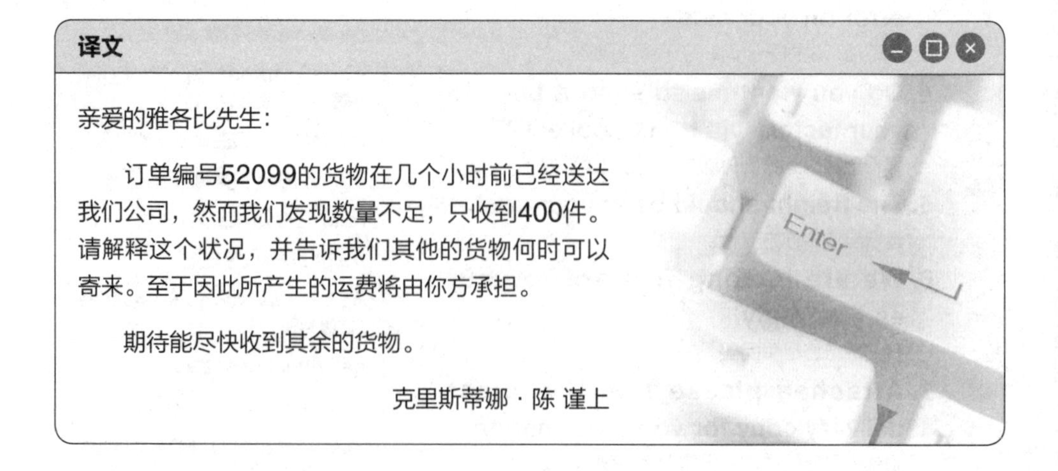

译文

亲爱的雅各比先生：

订单编号52099的货物在几个小时前已经送达我们公司，然而我们发现数量不足，只收到400件。请解释这个状况，并告诉我们其他的货物何时可以寄来。至于因此所产生的运费将由你方承担。

期待能尽快收到其余的货物。

克里斯蒂娜 · 陈 谨上

1 解析重点1 however

however 一般用于转折，通常译为“然而；但是；无论”。请对比下面的句子：

I feel a little tired. However, I can hold on.（我有点累了，但是我能坚持。）

However hot it is, he will not take off his coat.（无论多热，他也不愿脱掉外套。）

2 解析重点2 on your charge

charge 在此处意为“索价；收费”，on your charge 意为“由你付费”。同样，我们也可以换成 on my charge（由我付费）。请对比下面的句子：

The damage of this shipment will be on your charge.（这批货物的毁损将由你方付费。）

Open a bottle of wine for celebrating, and that will be on my charge.（开一瓶红酒庆祝吧！由我付费。）

1. **The shortage of this shipment caused delayed delivery to our customer.**
   这批货物的短缺导致我们延迟交货给顾客。
2. **Regarding to the shortage, we want to know how you will compensate our *loss*[9].**
   关于货物短缺，我们想知道你们将如何补偿我们的损失。
3. **We will deliver the remaining shipment immediately, and the freight charge will be on our account.**
   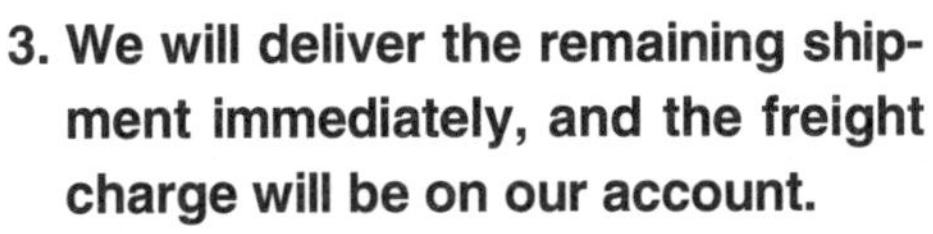
   我们会将余下的货物马上寄出，运费由我方支付。
4. **The *total*[10] amount of shortage is 350 kgs.**
   货物总计缺少了350公斤。
5. **The shortage of this cargo is caused by our employee's oversight.**
   这批货物的短缺是由于我方员工的疏忽造成的。

必背关键单词 

1. ***delivery*** [dɪˈlɪvərɪ] n. 投递；交货
2. ***arrive*** [əˈraɪv] v. 抵达
3. ***shortage*** [ˈʃɔːtɪdʒ] n. 缺少；不足
4. ***explain*** [ɪkˈspleɪn] v. 解释；说明
5. ***freight*** [freɪt] n. 运费
6. ***cause*** [kɔːz] v. 使发生；引起
7. ***remaining*** [rɪˈmeɪnɪŋ] adj. 剩余的
8. ***shipment*** [ˈʃɪpmənt] n. 装运；装货
9. ***loss*** [lɒs] n. 损失
10. ***total*** [ˈtəʊtl] adj. 全部的

03 投诉货品瑕疵

From bill@tom.com　　Date June 20th, 2008
To bob@yahoo.com
Subject Shipment of Defective Merchandise

Dear Mr. Bob,

We are glad to ***inform***[1] you that your delivery arrived at our company **in good condition** on June 17th, 2008. ***On the other hand***[2], we also found some ***defective***[3] ***merchandise***[4]. The ***system***[5] of this merchandise will ***shut down***[6] ***automatically***[7] while ***instructing***[8] some commands. Attached are some pictures for your study, and please ***transfer***[9] these information to your NBRMA.

Your **prompt** reply will be highly appreciated.

With Best Regards,
Bill Peterson

译文

亲爱的鲍伯先生:

我们很高兴地通知您，您发出的货物已经于2008年6月17日完好无缺地抵达我们公司了。另一方面，我们也发现了一些有瑕疵的商品。这个商品的系统在下达某些指令时会自动关机。附上一些图供您查阅，并请将这些信息转交给贵公司售后服务部门。

若您能尽快回复，则不胜感激。

比尔 · 彼得森 谨上

语法重点解析

1 解析重点1 in good condition

condition 的意思为“形势；状态”，in good condition 常被用在商业书信中形容货物接收时的状况良好。请对比下面的句子：

Conditions were favorable for business at that time.（那时的形势有利于经商。）

That cargo are received in good condition.（那批货物的接收状况良好。）

2 解析重点2 prompt

prompt 的意思是“及时的；迅速的；立即的”。一般我们都会用 quickly 来表示“快速地”，但商业书信上较常使用 prompt，请对比下面的句子：

Please come to my office quickly.（请快点到我的办公室。）

He is prompt in paying his bill.（他支付账单从不拖延。）

1. **Our *model*[10] SP-250 may have one possible defect.**
   我们的型号SP-250可能有一个缺陷。
2. **You can find a complete solution on our website first if there is any defect on our products.**
   如果我们的产品有任何缺陷，你可以先在我们的网站上寻找全方位的解答。
3. **Please send back all the defective products to our RMA dept.**
   请将所有的瑕疵产品寄回我们的维修部门。
4. **Can you accept that we supply the defect parts and you assemble them in your factory if needed?**
   你们能否接受我们提供有瑕疵的零件？如有需要，由你们工厂自行组装。
5. **There are some light spots on the surface of your product IR-150.**
   你们的IR-150产品表面有淡淡的斑点。

必背关键单词

1. ***inform*** [ɪnˈfɔːm] v. 通知；告知
2. ***on the other hand*** [ɒn ðə ˈʌðə hænd] phr. 另一方面
3. ***defective*** [dɪˈfektɪv] adj. 瑕疵的
4. ***merchandise*** [ˈmɜːtʃəndaɪs] n. 商品
5. ***system*** [ˈsɪstəm] n. 系统
6. ***shut down*** [ʃʌt daʊn] phr. 停工；关闭
7. ***automatically*** [ˌɔːtəˈmætɪklɪ] adv. 自动地
8. ***instruct*** [ɪnˈstrʌkt] v. 指示；命令
9. ***transfer*** [trænsˈfɜː(r)] v. 转换；转交
10. ***model*** [ˈmɒdl] n. 型号

04 投诉货品毁损

这样写就对了

From john@tom.com
To jerry@yahoo.com
Subject Shipment of Damaged Merchandise
Date June 20th, 2008
Cc matthew@yahoo.com

Dear Mr. Jerry,

I regret to inform you that your ***damaged***[1] shipment arrived yesterday. ***Several***[2] ***corners***[3] of your products are ***broken***[4], which should be caused by ***careless***[5] delivery. We hope that you have bought ***insurance***[6] for this shipment. ***Anyway***[7], I will send the damaged parts to you COD. Please let Matthew know when they arrive, and he will ***take*** it ***over***[8] from here.

If there are any other questions, please contact Matthew directly.

With Best Regards,
John Lin

译文

亲爱的杰瑞先生:

我很遗憾地通知您昨天送达的货物有毁损情况。

您的产品好几个角都有破损，应该是发货不小心造成的。我们希望你为这批货购买了保险。不管怎样，我将会把损坏的部分以货到付款的方式寄回。收到这批货后请让马修知道，他会从这里开始接手。

如果有任何其他的问题，请直接联络马修。

约翰 · 林 谨上

语法重点解析

1 解析重点1 be caused by

cause 意为“引起；造成”，be caused by 意为“由……引起”，类似的表达还有arise，result，from，come of等。请对照下面的句子：

His disability is caused by a fever.（他的残疾是因发高烧而引起的。）

The problem just resulted from different opinions.（这个问题正是由于不同的观点产生的。）

2 解析重点2 COD = collect / cash on delivery

COD 属于贸易常用语，意为“货到付款”。collect on delivery 为美式英语。请对比下面的句子：

I want to order an urgent order of 30 PCs ER-125, and please send to my office COD.（我要下一个包含30件ER-125的紧急订单，并请送到我的办公室，货到付款。）

If the label won't stick tight or the product can't function well, I will send them back to you COD.（如果标签没有粘牢，或产品无法正常使用，我会以货到付款的方式寄回给你。）

高频例句

1. **Our transportation agent just informed us that your container has been destroyed by a fire on June 30th, 2008.**
   我们的运输代理商刚刚通知我们，你们的集装箱在2008年6月30日被一场大火烧毁了。
2. **A few of your glass pots has a small *crack* [9] when they arrived at our company, do you need us to send them back for replacement?**
   少数的玻璃罐在送达我们公司时有一个小裂缝，你是否需要我们寄回换新？
3. **Your machine has arrived, but it's a mess.**
   你的机器已送达，但它一塌糊涂。

必背关键单词

1. ***damaged*** ['dæmɪdʒd] adj. 毁损的
2. ***several*** ['sevrəl] adj. 几个的；各自的
3. ***corner*** ['kɔːnə(r)] n. 角；角落
4. ***break*** [breɪk] v. 打破；破裂
5. ***careless*** ['keələs] adj. 粗心的；草率的
6. ***insurance*** [ɪn'ʃʊərəns] n. 保险
7. ***anyway*** ['enɪweɪ] adv. 无论如何；不管怎样
8. ***take over*** [teɪk 'əʊvə] phr. 接管
9. ***crack*** [kræk] n. 裂缝

05 投诉货品与说明不符

这样写就对了

From jackson@tom.com
Date June 20th, 2008
To jane@yahoo.com
Subject Shipment of Wrong Merchandise

Dear Ms. Jane,

Please be informed that your ***software***[1] of IT-250 (***Website***[2] No. IT-01001250) comes with a wrong ***statement***[3]. This software has been ordered on June 25th, 2008 ***via***[4] your ***online***[5] website, which ***stated***[6] that this software can **be applied to** Windows XP. After I ***installed***[7] it, **it couldn't work at all.** I ***simply***[8] want to know if I need a ***replacement***[9] ?

Looking forward to hearing from you very soon.

Sincerely Yours,
Jackson Woods

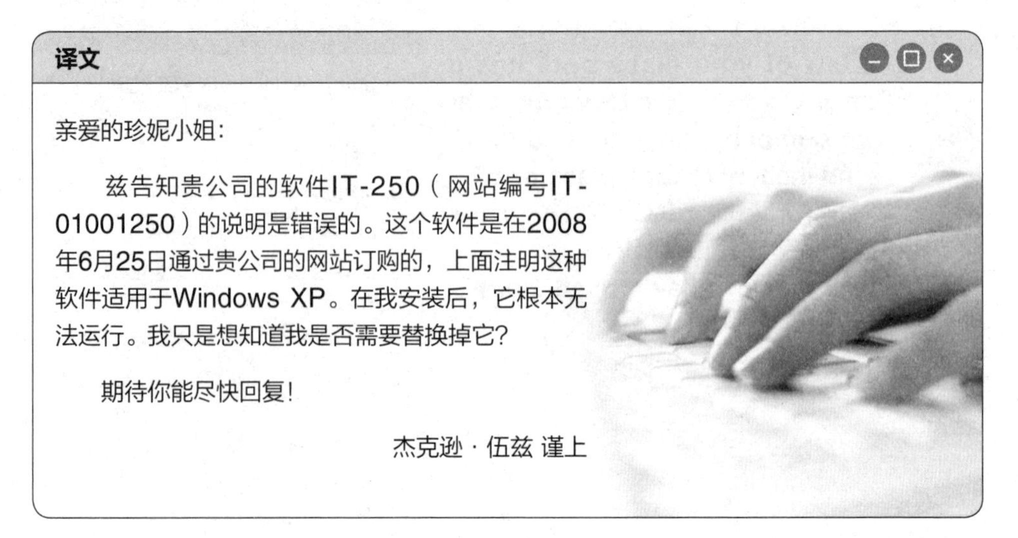

译文

亲爱的珍妮小姐：

兹告知贵公司的软件IT-250（网站编号IT-01001250）的说明是错误的。这个软件是在2008年6月25日通过贵公司的网站订购的，上面注明这种软件适用于Windows XP。在我安装后，它根本无法运行。我只是想知道我是否需要替换掉它?

期待你能尽快回复!

杰克逊·伍兹 谨上

1 解析重点1 be applied to

apply 意为“申请；适用”，be applied to 的意思为“应用于；施加于；试用于”。请对比下面的句子：

I applied to the Consul for Application of Certificate of Origin.（我向领事提出原产地证明书的申请。）

He doesn't know how steam can be applied to navigation.（他不知道蒸汽如何应用于航行。）

2 解析重点2 It couldn't work at all.

work 的原意为“工作”，在此指“运作”。请比对下面的句子：

The marker doesn't work.（这支记号笔不能用。）

My MP3 player doesn't work at all.（我的 MP3播放器坏了。）

1. **I just received your product of SP-125, but its color is blue instead of green.**
   我刚收到你寄来的SP-125产品，但颜色却是蓝色而不是绿色。
2. **Your crystals have just arrived at our company, but the purity is completely wrong.**
   你（发出）的水晶刚刚送达我们公司，但是成色却完全错了。
3. **I just installed your software of XR-100, but found that you sent me a wrong version.**
   我刚安装了贵公司的XR-100软件，但是发现寄给我的版本错了。
4. **Your courier just delivered my order of PO0903001, but the cables should be USB port.**
   贵公司的配送员刚刚送达了我订单号为PO0903001的货物，但是连接线应该是USB插头的。
5. **Your printer just arrived in good condition, but the installation software is wrong.**
   贵公司的打印机刚刚完好无损地抵达了，但安装软件却错了。
6. **Please inform your R.D. that your router can't be applied to my computer.**
   请通知贵公司的研发人员，你们的路由器无法与我的电脑兼容。

必背关键单词

1. ***software*** [ˈsɒftweə(r)] n. 软件
2. ***website*** [ˈwebsaɪt] n. 网站
3. ***statement*** [ˈsteɪtmənt] n. 陈述；说明
4. ***via*** [ˈvaɪə] prep. 经由
5. ***online*** [ˌɒnˈlaɪn] adj. 在线的
6. ***state*** [steɪt] v. 陈述；说明
7. ***install*** [ɪnˈstɔːl] v. 安置；安装
8. ***simply*** [ˈsɪmplɪ] adv. 简单地；仅仅
9. ***replacement*** [rɪˈpleɪsmənt] n. 替代品；更换

06 投诉货品与样品不符

这样写就对了

From ramond@tom.com **Date** October 2nd, 2008
To jimmy@yahoo.com
Subject Merchandise Different from Samples

Dear Mr. Jimmy,

I **regret to** inform you that your products of SP-125 just arrived at my ***office***[1], and the color is ***apparently***[2] not ***correct***[3]. Our order of these ***products***[4] should be ***light***[5] green, instead of ***dark***[6] green. Please **re-check** this order and ***send***[7] me the right products ***immediately***[8].

If there are any other ***questions***[9], please feel free to let me know.

Sincerely Yours,
Ramond Wang

译文

亲爱的杰米先生：

我很遗憾地通知您，贵公司的产品SP-125刚刚送达我的办公室，但是颜色很明显不对。我们订的这批货应该是淡绿色，而不是深绿色。请再次确认订单，并立刻寄给我正确的产品。

如果还有任何问题，请尽管告诉我。

雷蒙德 · 王 谨上

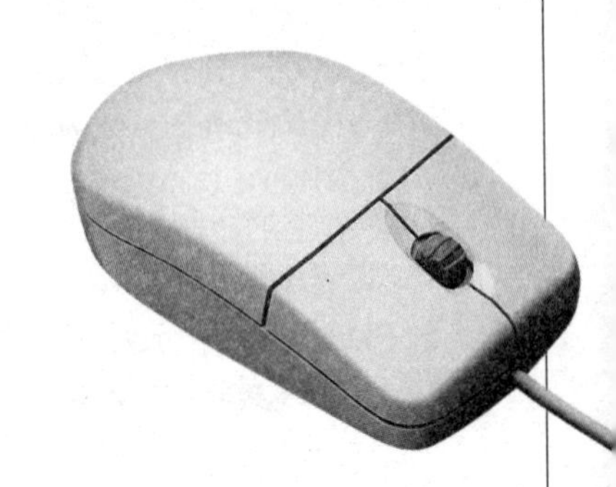

语法重点解析

1 解析重点1 regret to

regret to do sth 意为“遗憾做某事”，有可惜的意味。regret 除了做动词外，还可以当作名词使用，例如：

She expressed regret over the sad incident.（她对这件不幸的事表示遗憾。）

Don't you feel any regret over failing the exam?（没有通过考试，你不觉得懊悔吗？）

2 解析重点2 re-check

check 意为“检查”，而前缀 re- 有“再次；重新”的意思，一般的动词前都可以加前缀 re-，表示重复这个动作。请对比下面的句子：

Please re-do this work, it's completely wrong.（请重新做这项工作，全做错了。）

That machine should be re-checked today, because it will be delivered to our buyer tomorrow.（那台机器今天要重新检查，因为明天会送到我们的买家手中。）

1. **Your printer has arrived at my office, but I can't find the cables which should be included in the package.**
   贵公司的打印机已经送达我的办公室，但我找不到本应该在包装内的连接线。
2. **Some of your diamonds are with different classes of purity.**
   你们的一些钻石有不同等级的纯度。
3. **The delivery of SP-125 has been packed in well condition, but the sticker is *different*[10] from the sample.**
   SP-125产品包装得很好，但是贴纸和样品不一样。
4. **Unlike your samples, these mice don't work with PS2 ports.**
   和样品不一样，这些鼠标不适用于PS2端口。

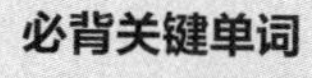

1. ***office*** [ˈɒfɪs] n. 办公室
2. ***apparently*** [əˈpærəntlɪ] adv. 表面上；显然地
3. ***correct*** [kəˈrekt] adj. 正确的
4. ***product*** [ˈprɒdʌkt] n. 产品
5. ***light*** [laɪt] adj. 明亮的；浅色的
6. ***dark*** [dɑːk] adj. 黑暗的；深色的
7. ***send*** [send] v. 发送；寄
8. ***immediately*** [ɪˈmiːdɪətlɪ] adv. 直接地；马上
9. ***question*** [ˈkwestʃən] n. 问题
10. ***different*** [ˈdɪfrənt] adj. 不同的

07 投诉货品问题并要求取消订单

这样写就对了

From richard@tom.com
Date July 1st, 2008
To josephine@yahoo.com
Cc jenny@yahoo.com; John@tom.com
Subject Complaint about Canceled Order

Dear Ms. Josephine,

We regret to inform you that we have to ***cancel***[1] our order No. PO0907020. This ***shipment***[2] was ***completely***[3] ***wet***[4] when they arrived on May 8th, 2008, and their system could **not work at all**. Please understand that I already **did my best** to ***avoid***[5] this ***outcome***[6].

Thanks in advance.

Sincerely Yours,
Richard Hofmann

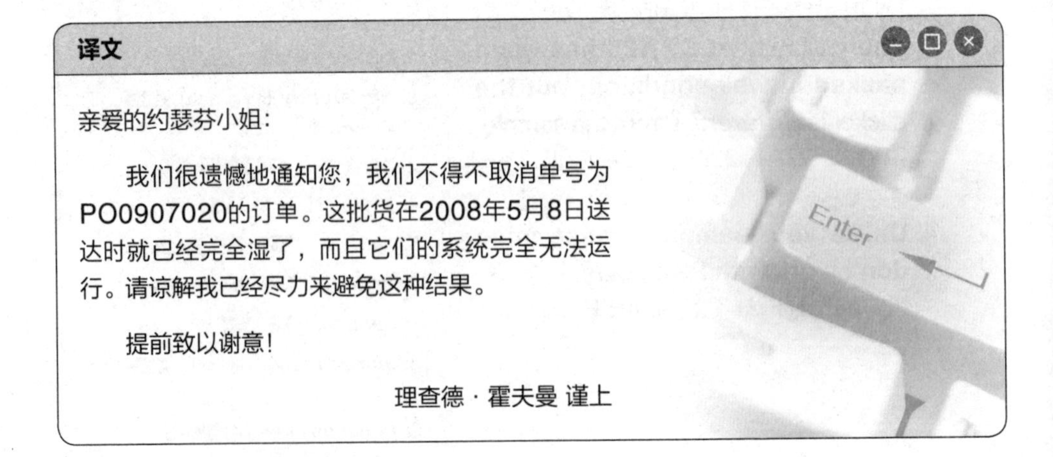

译文

亲爱的约瑟芬小姐：

我们很遗憾地通知您，我们不得不取消单号为PO0907020的订单。这批货在2008年5月8日送达时就已经完全湿了，而且它们的系统完全无法运行。请谅解我已经尽力来避免这种结果。

提前致以谢意！

理查德 · 霍夫曼 谨上

语法重点解析

1 解析重点1 not...at all

此短语意为“一点也不……”，not work at all 意为“一点反应也没有；完全没有作用”。not at all 也可以单独使用，请看下面的句子：

Thank you very much.（非常谢谢你！）

Not at all! It was the least I can do.（一点也不！这是我最起码能做的。）

2 解析重点2 do one's best

此短语的意思为“竭尽某人全力做……”，而结果通常没有成功。表示所有能做的事都做过了，但结果仍是如此，已经无力挽回了。请看下面的句子：

I have done my best to please my mother-in-law, but she is still upset every day.（我已经竭尽全力来取悦我的婆婆，但她每天还是不快乐。）

Please do your best to ***prevent***[7] the worst situation.（请尽全力避免最糟的情况。）

高频例句

1. **I want to cancel the order of computer BN-2019, because you cannot deliver this shipment as scheduled.**
   我想要取消BN-2019电脑的订单，因为你们无法如期交货。
2. **If this problem cannot be solved well, I am afraid that we have to cancel this order.**
   如果这个问题无法妥善解决，我们恐怕只好取消这一订单。
3. **We have no *choice*[8] but cancel this order.**
   我们没有选择只好取消订单。

必背关键单词 

1. ***cancel*** ['kænsl] v. 取消
2. ***shipment*** ['ʃɪpmənt] n. 装运；装运物
3. ***completely*** [kəm'pli:tlɪ] adv. 完全地；彻底地
4. ***wet*** [wet] adj. 湿的
5. ***avoid*** [ə'vɔɪd] v. 避免
6. ***outcome*** ['aʊtkʌm] n. 后果
7. ***prevent*** [prɪ'vent] v. 避免
8. ***choice*** [tʃɔɪs] n. 选择

08 投诉请款金额错误

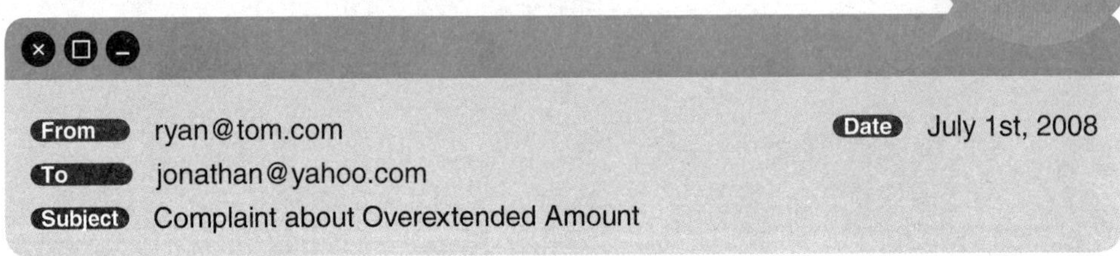

From ryan@tom.com
Date July 1st, 2008
To jonathan@yahoo.com
Subject Complaint about Overextended Amount

Dear Mr. Jonathan,

I was surprised to receive your bill of the last order. According to our ***agreement***[1], your ***commission***[2] should be 5% per order. But the ***unit***[3] price of ***latest***[4] orders already ***included***[5] your commission, and you added the commission again in the total amount. ***Therefore***[6], your commission ***became***[7] 5.25% on latest orders. Please re-calculate and send another bill to us as soon as you can.

Looking forward to receiving another bill very soon.

Sincerely Yours,
Ryan Mill

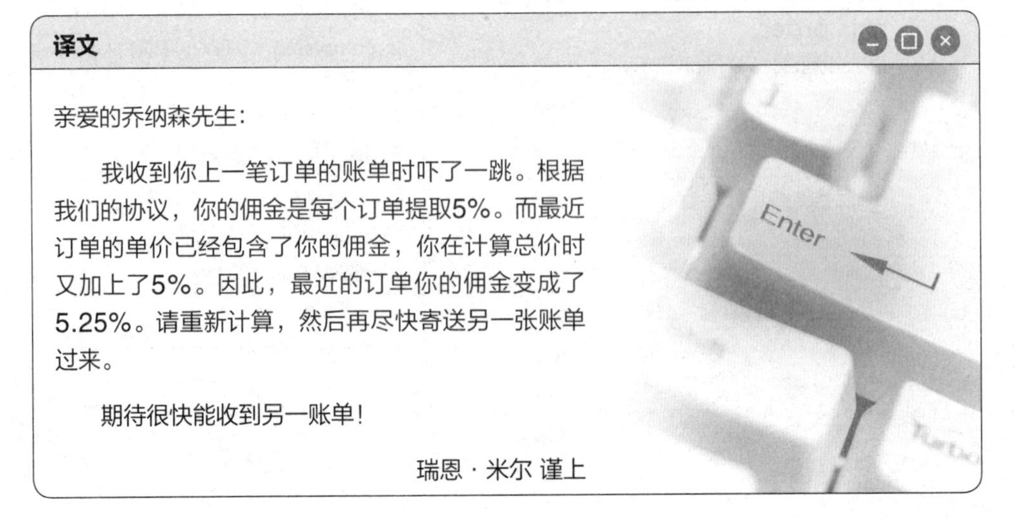

译文

亲爱的乔纳森先生：

我收到你上一笔订单的账单时吓了一跳。根据我们的协议，你的佣金是每个订单提取5%。而最近订单的单价已经包含了你的佣金，你在计算总价时又加上了5%。因此，最近的订单你的佣金变成了5.25%。请重新计算，然后再尽快寄送另一张账单过来。

期待很快能收到另一账单！

瑞恩 · 米尔 谨上

语法重点解析

1 解析重点1 per order

per 可以翻译为“经由；每一；按照”，per order 在商业书信中通常翻译为“每一个订单”。请对比下面的句子：

The ***buffet***[8] of this restaurant costs 200 yuan per person.（这间餐厅的自助餐每人200元。）

That hand-made table was delivered per rail.（那张手工制的桌子是由铁路运输的。）

2 解析重点2 bill

bill 这个单词原来的意思为“账单”，但后来引申为“目录；票据；请款单”等。在美国甚至连钞票都可以称为 bill。请对比下面的句子：

That store ***collected***[9] me a bill for US$25.（那家商店给了我一张25美元的账单。）

Please send me a bill of exchange as the payment of that order.（请给我一张汇票支付那笔订单。）

1. **I'm afraid that this order could be canceled because of overextended total amount.**
   恐怕这个订单要取消了，因为超出了我们的预算。
2. **If you keep on extending this order, it will be very tough for our business in the future.**
   如果你再继续这样延期订单，我们未来的生意将会很难做。
3. **This order has been extended for too many times. Please place another order for replacement.**
   这张订单已经延期过太多次了，请重新下一张订单代替。
4. **I'd like to extend this quantity to 1,250 PCs, and also postpone the delivery to one month later.**
   我要将数量增加到1,250件，并且把发货时间推迟一个月。

必背关键单词

1. ***agreement*** [əˈgri:mənt] n. 协议
2. ***commission*** [kəˈmɪʃn] n. 佣金
3. ***unit*** [ˈju:nɪt] n. 单位；单元
4. ***latest*** [ˈleɪtɪst] adj. 最新的；最近的
5. ***include*** [ɪnˈklu:d] v. 包括；包含
6. ***therefore*** [ˈðeəfɔ:(r)] adv. 因此；所以
7. ***become*** [bɪˈkʌm] v. 变成
8. ***buffet*** [ˈbʊfeɪ] n. 自助餐
9. ***collect*** [kəˈlekt] v. 收账；收集

09 投诉未开发票

这样写就对了

From sam1112@tom.com
Date December 21st, 2008
To justin@yahoo.com
Subject Complaint for Not Issuing the Invoice

Dear Mr. Justin,

We are ***pleased***[1] to inform you that the goods of PO0811023 arrived at Shanghai, yesterday morning. But there is no formal **Commercial Invoice** attached. I have asked the courier ***agent***[2] to look for it ***throughout***[3] these ***cartons***[4] and nothing was found. Please mail the Invoice immediately for custom ***purpose***[5]. We sincerely ***hope***[6] that this situation will not ***happen***[7] again. It caused us a lot of trouble in **custom clearance**.

Looking forward to receiving your Commercial Invoice immediately.

Sincerely Yours,
Sam Collins

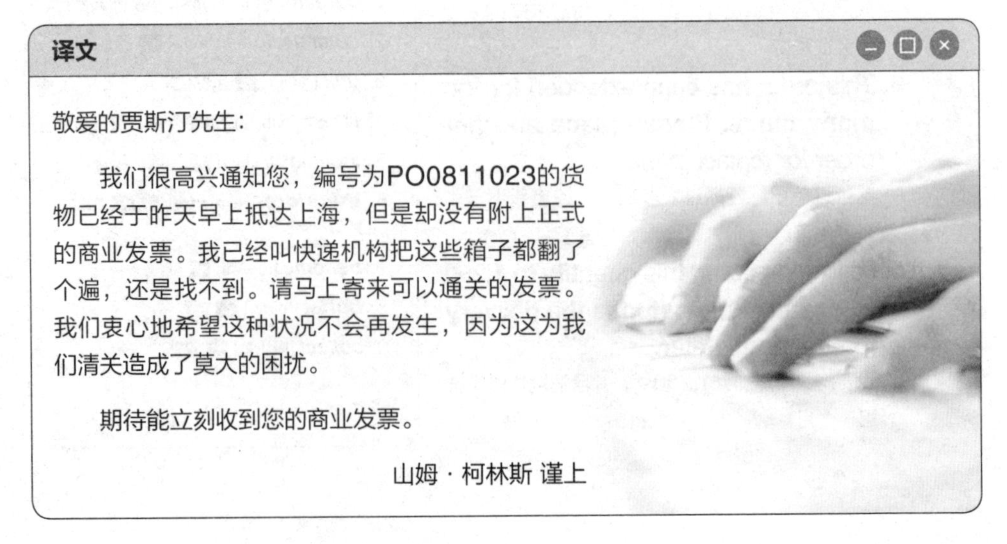

译文

敬爱的贾斯汀先生：

我们很高兴通知您，编号为PO0811023的货物已经于昨天早上抵达上海，但是却没有附上正式的商业发票。我已经叫快递机构把这些箱子都翻了个遍，还是找不到。请马上寄来可以通关的发票。我们衷心地希望这种状况不会再发生，因为这为我们清关造成了莫大的困扰。

期待能立刻收到您的商业发票。

山姆 · 柯林斯 谨上

语法重点解析

1 解析重点1 Commercial Invoice

Commercial Invoice 翻译为“商业发票”，此类单据的用途类似收据，主要用于通关。请看下面的句子：

Our Commercial Invoice is always ***attached***[8] on the envelope that ***stuck***[9] on the cartons.（我们的商业发票通常附在张贴在箱子上的信封里。）

Please send me the Commercial Invoice by e-mail first before delivery.（在装运前请先将商业发票用电子邮件寄给我。）

2 解析重点2 custom clearance

custom 意为“习俗；惯例；海关”，clearance 意为“清除”，custom clearance 意为“清关”。“清关”是指进出口或转运货物出入一国关境时，依法履行的手续。请对比下面的句子：

It is necessary for custom clearance to ***offer***[10] the Commercial Invoice, Packing List and so on.（货物清关必须要提供商业发票、装箱单等。）

Our custom officer asked for a Certificate of Origin for custom clearance.（我们的海关官员要求提供原产地证明以便清关。）

1. **Offering a commercial invoice for custom clearance is common sense.**
   通关时要提供商业发票是基本常识。
2. **An invoice is needed whenever we buy any merchandise.**
   我们买任何东西时都需要发票。
3. **Commercial Invoice is an invoice which states the unit price, total amount of the order, terms of trade, delivery and so on.**
   商业发票就是一张标明单价、订单总金额、贸易条款、运输等信息的发票。
4. **Please inform our courier agent that we will offer the Commercial Invoice to them immediately.**
   请通知我们的快递机构，我们会马上提供商业发票给他们。

必背关键单词 

1. ***pleased*** [pli:zd] adj. 高兴的
2. ***agent*** ['eɪdʒənt] n. 代理商
3. ***throughout*** [θru:'aʊt] adv. 处处；里里外外；贯穿
4. ***carton*** ['kɑ:tn] n. 纸盒；纸板箱
5. ***purpose*** ['pɜ:pəs] n. 目的
6. ***hope*** [həʊp] v. 期待；希望
7. ***happen*** ['hæpən] v. 发生
8. ***attach*** [ə'tætʃ] v. 附加；附带
9. ***stick*** [stɪk] v. 张贴；插入
10. ***offer*** ['ɒfə(r)] v. 提供；报价；提议

10 投诉商家取消订单

From vincent@yahoo.com

Date December 21st, 2008

To kenny@tom.com

Subject Complaint about Canceling Order

Dear Mr. Kenny,

I was **terribly** sorry to know that you want to cancel our order No. PO0907020 this morning. ***Several***[1] deliveries of the ***components***[2] of the products had been delivered to your company in these months. I already dispatched our engineers to solve the ***software***[3] problem in your ***factory***[4]. Please tell me **what else** I can do about this ***matter***[5].

Looking forward to hearing good news from you.

Sincerely Yours,
Vincent Lin

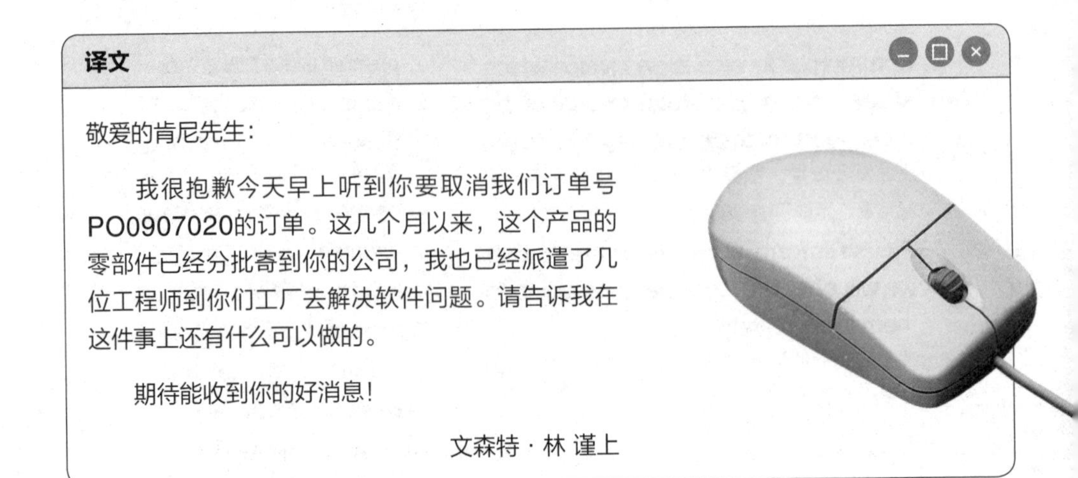

译文

敬爱的肯尼先生:

我很抱歉今天早上听到你要取消我们订单号PO0907020的订单。这几个月以来，这个产品的零部件已经分批寄到你的公司，我也已经派遣了几位工程师到你们工厂去解决软件问题。请告诉我在这件事上还有什么可以做的。

期待能收到你的好消息!

文森特 · 林 谨上

语法重点解析

1 解析重点1 terribly

terribly 的意思为“恐怖地；骇人听闻地”，口语上也可表示“很，非常”的意思，另外它的形容词为 terrible，解释为“可怕的”。请对比下面的句子：

I am terribly sorry to hear that your mother just ***passed away***[6]. （听到你母亲刚逝世这件事让我很遗憾。）

That prisoner died in terrible ***sufferings***[7]. （那个囚犯死的非常痛苦。）

2 解析重点2 what else

else 这个副词原来的意思为“除此之外；另外；否则”，常与 what，whose，somebody，somewhere，nothing，anyone 单词等连用。请对比下面的句子：

If you cannot find my ***chopsticks***[8], someone else's will do. （如果你找不到我的筷子，其他人的也可以。）

Except staying at home, we won't go anywhere else. （除了待在家里，我们哪儿也不去。）

高频例句

1. **Please be informed that we have to cancel this purchase order because of late delivery.**
   由于交货迟了，我们不得不取消这个采购订单，请知悉。
2. **If you keep on postponing the delivery of these machines, we will have to cancel this order.**
   如果你们一直拖延这些机器的交货日期，我们只好取消订单。
3. **The specifications of these mobile phones are totally different from our order, and we are considering canceling this order now.**
   这些手机的规格和我们的订单完全不同，现在我们正考虑取消这张订单。
4. **Please do not cancel this purchase order. We will do our best to meet your standard.**
   请不要取消订购单。我们会竭尽所能达到你们的标准。

必背关键单词

1. ***several*** [ˈsevrəl] adj. 几个的；数个的
2. ***component*** [kəmˈpəunənt] n. 零件；配件
3. ***software*** [ˈsɒftweə(r)] n. 软件
4. ***factory*** [ˈfæktrɪ] n. 工厂
5. ***matter*** [ˈmætə(r)] n. 事情；事件
6. ***pass away*** [pɑːs əˈweɪ] phr. 逝世
7. ***suffering*** [ˈsʌfərɪŋ] n. 痛苦；苦难
8. ***chopstick*** [ˈtʃɒpstɪk] n. 筷子

11 投诉违反合约

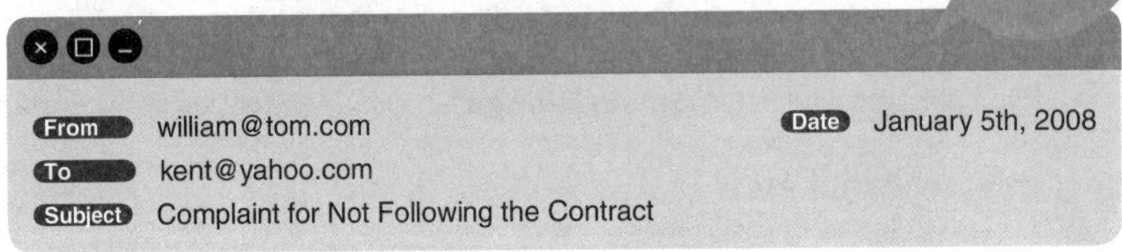
From william@tom.com
Date January 5th, 2008
To kent@yahoo.com
Subject Complaint for Not Following the Contract

Dear Mr. Kent,

Attached is our contract of ***Purchasing***[1] Order PO0801026 for your study. It ***seems***[2] that your company has not **followed** this ***contract***[3] for a long time, which has cost us a lot of ***inconvenience***[4] for several months. Our delivery should be ***once***[5] a month, and the quantity will be 20 dozens each time. Please do follow this contract, otherwise you will have to **bear** the ***consequences***[6].

If you have any other questions, please contact me directly.

Sincerely Yours,
William Yang

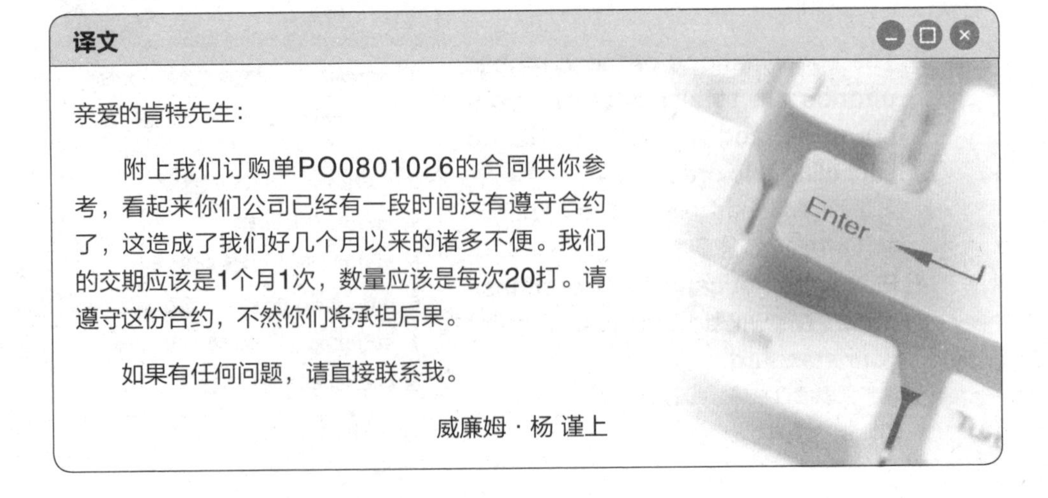
译文

亲爱的肯特先生:

附上我们订购单PO0801026的合同供你参考，看起来你们公司已经有一段时间没有遵守合约了，这造成了我们好几个月以来的诸多不便。我们的交期应该是1个月1次，数量应该是每次20打。请遵守这份合约，不然你们将承担后果。

如果有任何问题，请直接联系我。

威廉姆·杨 谨上

1 解析重点1 **follow**

follow 的原意为"跟随"，但在此处引申为"听从；领会；贯彻"的意思，请对比下面的句子：

Sonny will follow up these products from now on.（桑尼会从现在开始接手这些产品。）

Millions of families follow this TV soap operas ***devotedly***[7].（上百万的家庭非常热衷于收看这个电视肥皂剧。）

2 解析重点2 **bear**

bear 有"负担；支持；负荷"的意思，而在这里是"承担"的意思。其动词的变化形式为：过去式 bore；过去分词 born (borne)；现在分词 bearing。请看下列的例句：

I have to bear all the expenses.（我必须承担所有的费用。）

1. **According to our contract, your company needed to offer a warranty of these *scanners*[8].**
   根据我们的合同，贵公司必须提供这些扫描仪的保修单。
2. **Regarding the guarantee, our company considers that one year is a proper period.**
   关于保修期，我们公司认为一年期限为宜。
3. **As to the additional conditions of this contract, we appoint your company as our sole agent in France.**
   至于合同的附带条件，我们指定贵公司为我们在法国的独家代理商。
4. **Our courier agent has been appointed FedEx as our contract.**
   合同指定我们的快递机构为FedEx。
5. **Please do not violate our contract if possible.**
   请尽可能不要违反合同。
6. **I had tried my best to keep the contract, but failed.**
   我已经竭尽所能地信守合同，但是失败了。

必背关键单词

1. ***purchasing*** [ˈpɜːtʃəsɪŋ] n. 购买
2. ***seem*** [siːm] v. 看起来；似乎
3. ***contract*** [ˈkɒntrækt] n. 合同
4. ***inconvenience*** [ˌɪnkənˈviːnɪəns] n. 不便；麻烦
5. ***once*** [wʌns] adv. 一次；一回
6. ***consequence*** [ˈkɒnsɪkwəns] n. 结果；后果
7. ***devotedly*** [dɪˈvəʊtɪdlɪ] adv. 忠实地
8. ***scanner*** [ˈskænə(r)] n. 扫描仪

12 投诉延期交货

这样写就对了

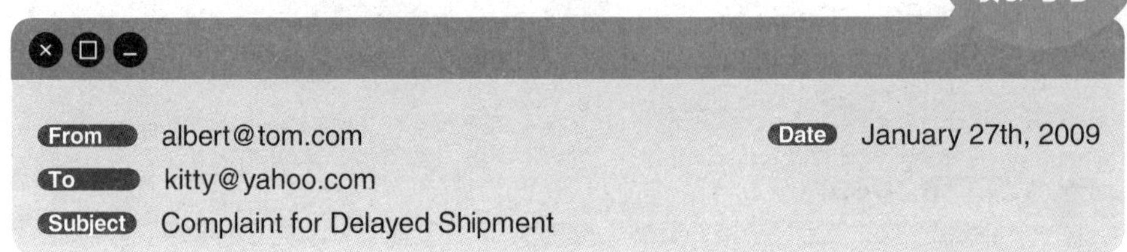

From albert@tom.com
Date January 27th, 2009
To kitty@yahoo.com
Subject Complaint for Delayed Shipment

Dear Ms. Kitty,

We won't ***accept***[1] any delayed shipment ***from now on***[2]. Your delivery of purchasing order PO0901023 was ***scheduled***[3] to be handled by UPS by the end of December. You ***postponed***[4] this shipment to January 6th, 2009, because of ***uncompleted***[5] ***accessories***[6]. On January 15th, 2009, this shipment was postponed again for bad ***weather***[7]. I don't care what's the **excuse** of your **delaying tactics** this time, but it won't be accepted anyway. Please ***proceed***[8] this shipment ***at once***[9], and let me know when it will arrive at our factory.

If you have any further questions, please do not ***hesitate***[10] to contact me now.

Sincerely Yours,
Albert Lee

译文

亲爱的基蒂小姐:

从现在开始我们不会再接受任何的装运延误，由UPS操作的订购单PO0901023原本安排在12月底。你因为配件未完成而将它延迟到2009年1月6日，而2009年1月15日，这批货又因为天气不佳而被耽搁了。不管你这次又有什么拖延的借口，但是我们无论如何都不会接受了。请立刻出货，并让我知道货品何时会抵达我们的工厂。

如果有任何其他的问题，现在就请不要迟疑地联络我。

艾伯特 · 李 谨上

语法重点解析

1 解析重点1 excuse

excuse 做动词意为“原谅；辩解”等，做名词意为“借口”。请对比下面的句子：

Excuse me! Would you please pass the pepper?（抱歉！能否请您将胡椒递过来？）

There is no excuse for this scandalous affair.（没理由做出这样的丑事。）

2 解析重点2 delaying tactics

delay 这个词的意思为“延缓；耽搁”，这里引申为“拖延”。这个短语的意思为“拖延战术”，请对比下面的句子：

Don't beat around the bush, otherwise their delaying tactics would work.（别再旁敲侧击了，否则他们的拖延战术就要奏效了。）

Delaying tactics won't work for an efficient company.（拖延战术对一个有效率的公司来说是无效的。）

1. **Please be informed that all of the delivery of sweaters will be postponed for one month due to popularity among the customers.**
   请知悉，由于客户的喜爱，所有毛衣的发货时间将推迟一个月。
2. **Regarding to the delivery of these orders, please don't postpone for over one week.**
   关于这些订单的交货，请不要延期超过一周。
3. **As to the additional conditions of this offer, please instruct that delivery cannot be postponed for three days.**
   至于这个报价的附带条件，请注明交货不能延期超过三天。
4. **UPS just informed us that these products will arrive at our company tomorrow.**
   UPS刚刚通知我们，这些产品明天才会送到我们的公司。

必背关键单词

1. ***accept*** [əkˈsept] v. 接受
2. ***from now on*** [frɒm naʊ ɒn] phr. 从现在开始
3. ***schedule*** [ˈʃedju:l] v. 安排；预定
4. ***postpone*** [pəˈspəʊn] v. 延迟；延后
5. ***uncompleted*** [ˌʌnkəmˈpli:tɪd] adj. 未完成的
6. ***accessory*** [əkˈsesərɪ] n. 配件
7. ***weather*** [ˈweðə(r)] n. 天气
8. ***proceed*** [prəˈsi:d] v. 着手；进行
9. ***at once*** [æt wʌns] phr. 立刻
10. ***hesitate*** [ˈhezɪteɪt] v. 犹豫；踌躇

13 投诉货品的残次问题

这样写就对了

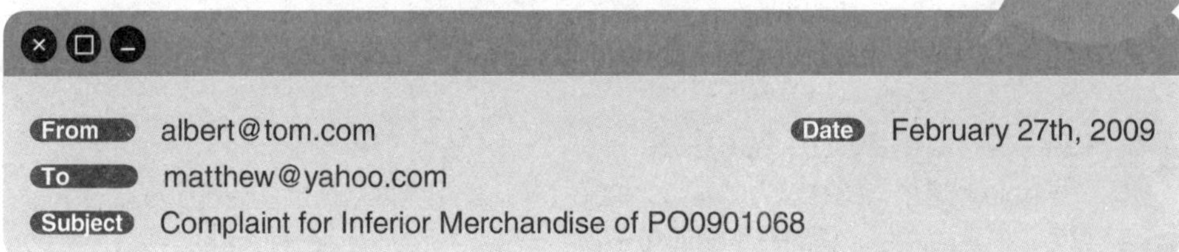
From albert@tom.com Date February 27th, 2009
To matthew@yahoo.com
Subject Complaint for Inferior Merchandise of PO0901068

Dear Mr. Matthew,

I regret to inform you that the delivery of these products was ***classified***[1] as ***inferior***[2]. Please take care of this matter very ***carefully***[3]. These products arrived at our factory in good condition, but we found that the software was not ***stable***[4]. As a ***result***[5] of it, we have to ask your company to ***compensate***[6]. Maybe you can give us a discount or **something like that** on our next purchasing order.

Please send my ***greetings***[7] to your new family member, and contact me **if needed**.

Sincerely Yours,
Albert Lai

译文

敬爱的马修先生：

很遗憾地通知你这些出货的产品被定位为次级品，在处理这件事时请多加注意。这些产品到我们工厂时状况良好，但我们发现软件不稳定。因此，我们必须向贵公司索赔，也许是在我们下次订购时打折之类的赔偿。

请代我向你家的新成员问好，如有需要请联系我。

艾伯特 · 赖 谨上

语法重点解析

1 解析重点1 something like that

something 的意思为“某事；某物”，此句意为“那样的东西”的意思。在美语中是比较口语化的说法，请对比下面的句子：

Do you have any ***disease***[8] like hydrophobia, AIDS, diabetes or something like that?（你有没有诸如狂犬病、艾滋病、糖尿病等类似的疾病？）

Do you have ***knives***[9], blades or something like that?（你有没有刀或刀片之类的东西？）

2 解析重点2 if needed

need 意为“需要；有必要”。请对照下面的句子：

The premium will be paid for insuring property if needed.（如有需要，保险费将用于财产保险。）

A friend in need is a friend indeed.（患难见真情。）

1. **Please be informed that all the T-Shirts will be returned because of the obviously wrong pattern.**
   请知悉，所有的短袖运动衫都将因明显的花样错误而被退回。
2. **All the binding of these books for PO0902010 are wrong, and they will be returned to your factory.**
   PO0902010订单的所有书都装订错误，它们将会被退回至贵工厂。
3. **Your design attracts us very much, but the irregular size of your sport swears are unacceptable.**
   贵公司设计的运动衫款式相当吸引人，但是尺码不规范却是无法接受的。
4. **Some of the wine on this delivery turned into vinegar.**
   这次货运的红酒有些变成了醋。
5. **Your Pen Drive cannot be detected by our computer.**
   你们的随身存储器无法被我们的电脑识别。

必背关键单词

1. ***classify*** [ˈklæsɪfaɪ] v. 分类
2. ***inferior*** [ɪnˈfɪərɪə(r)] adj. 次级的；低劣的
3. ***carefully*** [ˈkeəfəlɪ] adv. 小心谨慎地；仔细地
4. ***stable*** [ˈsteɪbl] adj. 稳定的
5. ***result*** [rɪˈzʌlt] n. 结果
6. ***compensate*** [ˈkɒmpenseɪt] v. 补偿；赔偿
7. ***greeting*** [ˈgriːtɪŋ] n. 问候
8. ***disease*** [dɪˈziːz] n. 疾病
9. ***knife*** [naɪf] n. 刀

14 投诉售后服务不佳

这样写就对了

From anthony@tom.com
Date May 7th, 2008
To martin@yahoo.com
Subject Complaint for Bad After-sales Service

Dear Mr. Martin,

We delivered our products to your ***RMA Dept.***[1] **at the end of** March, which should be returned our company in ***early***[2] April. But we still have not received them yet. **After all**, we have been your customer for five years, but I don't think your ***after-sales service***[3] is satisfactory this time. Please let me know your company's ***maintaining***[4] schedule.

Your earliest reply will be ***highly***[5] ***appreciated***[6].

Sincerely Yours,
Anthony Johnson

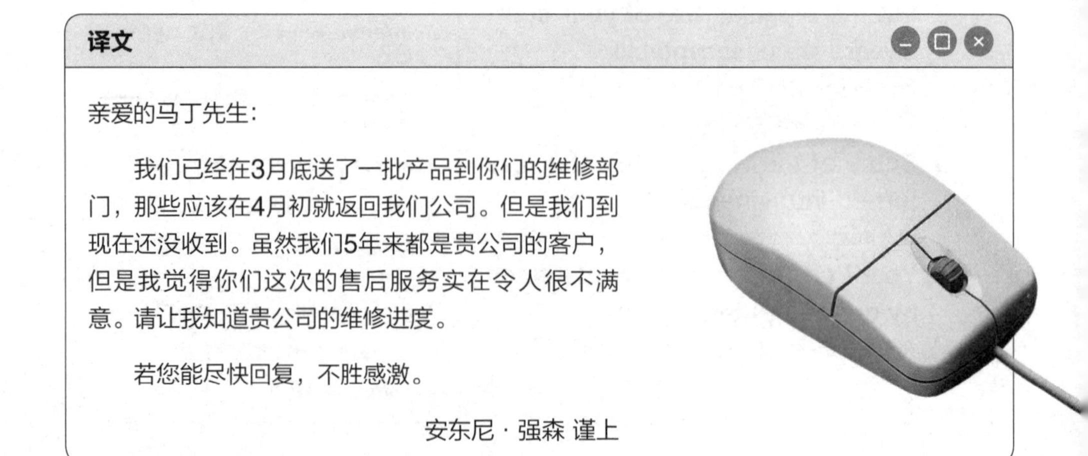

译文

亲爱的马丁先生:

我们已经在3月底送了一批产品到你们的维修部门，那些应该在4月初就返回我们公司。但是我们到现在还没收到。虽然我们5年来都是贵公司的客户，但是我觉得你们这次的售后服务实在令人很不满意。请让我知道贵公司的维修进度。

若您能尽快回复，不胜感激。

安东尼 · 强森 谨上

语法重点解析

1 解析重点1 at the end of

end 的原意为“终了；末端”等，at the end of 的意思为“在……的末期”，一般是指一段时间或时期即将结束的时候，请看下面的句子：

Taking a hot bath in the ***bathtub***[7] is the greatest pleasure at the end of a day.（在一天结束之后，在浴缸里泡个热水澡是最大的享受。）

2 解析重点2 after all

after all 可翻译为“虽然；尽管；毕竟；终究”。这个短语翻译为“虽然；尽管”时有表示结果的意味，当翻译为“毕竟；终究”时则有经过全方位的考虑之后的意思。请看下面的句子：

After all, you are still my friends.（毕竟你们还是我的朋友。）

1. **Please pay attention to your *attitude*[8], which may cause our huge loss.**
   请注意你的态度，那可能会带给我们极大的损失。
2. **These dictionaries have been *bound*[9] upside down, and your factory refuses to re-do them.**
   这些词典被装订得上下颠倒，而你的工厂拒绝重订。
3. **The shortcut in your mobile phone disabled some functions.**
   你手机中的快捷方式禁用了一些功能。
4. **Maintenance is very important when the economy is depressed.**
   经济萧条时，维护是非常重要的。
5. **Your factory suggested a replacement because these ethernet switches are too hard to maintain.**
   因为这些以太网交换机太难维护，你的工厂建议将它们更换掉。

必背关键单词

1. ***RMA Dept.= Return Merchandise Authorization Department*** phr. 维修部门
2. ***early*** [ˈɜ:lɪ] adj. 早期的；早的
3. ***after-sales service*** [ˈæftəˈselz ˈsɜ:vɪs] phr. 售后服务
4. ***maintain*** [meɪnˈteɪn] v. 维修
5. ***highly*** [ˈhaɪlɪ] adv. 非常地；高度地
6. ***appreciate*** [əˈpri:ʃieɪt] v. 感激
7. ***bathtub*** [ˈbɑ:θtʌb] n. 浴缸
8. ***attitude*** [ˈætɪtju:d] n. 态度
9. ***bind*** [baɪnd] v. 装订；捆；绑

Unit 12 拒绝 Refusal

01 因库存短缺而拒绝订单

这样写就对了

From "Ben Affleck" (baffleck@okey.com) Date Wed., October 1, 2008
To "Keanu Reeves" (kreeves@choose.com)
Subject Out of Stock

Dear Mr. Reeves,

We thank you for your Order No. 666 for ***T-shirts***[1] today, **but regret to *disappoint***[2] **you.**
At present, we are out of stock and **we need another two *months***[3] **before we can replenish stocks**. So we suggest you obtain them ***elsewhere***[4].

Yours sincerely,
Ben Affleck

译文 

亲爱的里夫斯先生：

我们很感谢今天收到贵公司订单号为666的T恤的订购单。但很抱歉要让您失望了。

目前，我们没有T恤的存货，要在两个月之后才有新货供应。因此，我们建议您到别处购买。

本·阿弗莱克 谨上

语法重点解析

1 解析重点1 **..., but regret to disappoint you.**

因为库存短缺而需要向客户说明情况的时候，我们一般会想到用 We very much regret that...（很抱歉……）或者是 we are very sorry that...（很抱歉……）的句型。此外，我们还可以用 but regret to disappoint you（但很抱歉可能要让您失望了）的句型。

2 解析重点2 **We need another two months before we can replenish stocks.**

一般库存短缺的时候，我们还需要告知订货方何时候才会有货，也好协商一下是取消订单还是继续等待。除了邮件内文的表达，我们还可以用 It will be another two months before we can replenish stocks（要在两个月之后才有新货供应）表示对方需要等待的时间或某件事情所需要花费的时间。

1. **We regret to inform you that we can't accept your order.**
   很抱歉通知您，我们无法接受您的订单。
2. **We *apologize*[5] for any inconvenience this may have *caused*[6].**
   对给您造成的任何不便，我们深表歉意。
3. **Shoes of this kind are not available at the moment.**
   这种样式的鞋子目前缺货。
4. **We don't know when this item will be back in stock.**
   我们不知道这项产品何时才会有货。
5. **The jeans of this *style*[7] are in short supply now.**
   这种式样的牛仔裤目前缺货。
6. **Owing to short supply, we cannot make you an offer at present.**
   由于缺货，目前我们不能给贵方报价。
7. **We have decided not to accept your order because of the current shortage of the goods.**
   由于目前此项商品缺货，所以我们决定不接受您的订单。
8. **There is a serious shortage of this *commodity*[8] at present.**
   这款商品目前严重缺货。

1. ***shirt*** [ʃɜːt] n. 衬衫
2. ***disappoint*** [ˌdɪsəˈpɔɪnt] v. 使失望
3. ***month*** [mʌnθ] n. 月
4. ***elsewhere*** [ˌelsˈweə(r)] adv. 在别处
5. ***apologize*** [əˈpɒlədʒaɪz] v. 道歉；认错
6. ***cause*** [kɔːz] v. 引起
7. ***style*** [staɪl] n. 风格；时尚；式样
8. ***commodity*** [kəˈmɒdətɪ] n. 商品

02 婉拒报价

这样写就对了

From "Will Smith" (wsmith@ilook.com)　Date Thurs., October 2, 2008
To "Orlando Bloom" (obloom@neat.com)
Subject Declining your Offer

Dear Mr. Bloom,

Thank you for your offer.
Unfortunately, we must ***decline*** [1] your offer at this time, as **we can obtain a *price*** [2] **of US$50 *per*** [3] **item with another *firm*** [4]. **This is US$5 per item lower than your price.**

Please accept our ***sincere*** [5] regrets.

Yours sincerely,
Will Smith

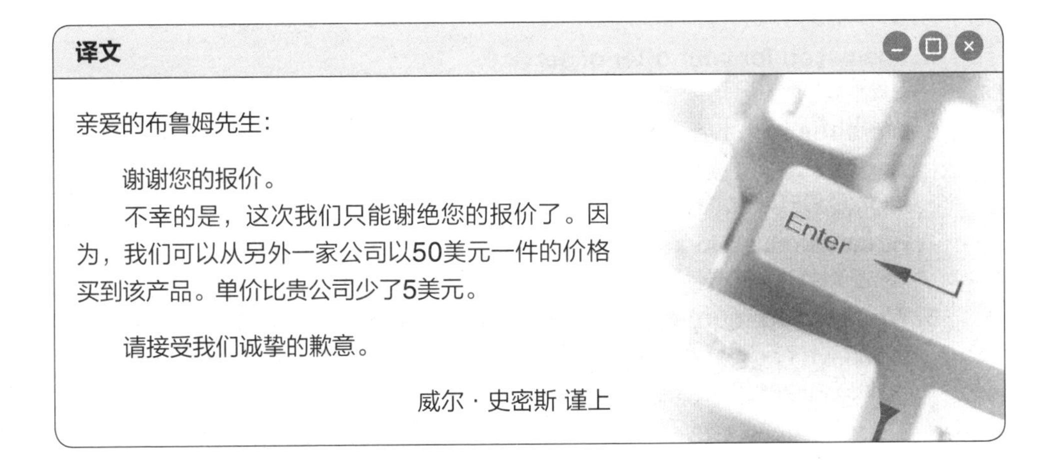

译文

亲爱的布鲁姆先生：

谢谢您的报价。

不幸的是，这次我们只能谢绝您的报价了。因为，我们可以从另外一家公司以50美元一件的价格买到该产品。单价比贵公司少了5美元。

请接受我们诚挚的歉意。

威尔 · 史密斯 谨上

语法重点解析

1 解析重点1 We can obtain a price of US$50 per item with another firm.

由于其他厂商给予我们比较低的价格，所以我们可能需要婉拒出价较高的一方。因此，不管出价方是不是会降价，我们还是要让他了解目前的情况。正如邮件中所述：We can obtain a price of US$50 per item with another firm.（我们可以从另外一家公司以50美元一件的价格买到产品。）

2 解析重点2 This is US$5 per item lower than your price.

只有货比三家，我们才会得到一个对自己比较有利的价格。如果我们想对商品供应方说，你的价格比别人高了或是别人的价格比你低，我们可以像邮件中所说的那样来表达：This is US$5 per item lower than your price.（单价比你们的少了5美元。）This is...lower than... 这个句型就是表示前者比后者数字小。

1. **We have to *turn*[6] down your offer, as other suppliers are under-quoting you.**
   我们不得不谢绝你方的报价，因为其他供应商报价比你们低。
2. **I think I should turn down your offer.**
   我认为我应该拒绝您的报价。
3. **His quotation is much lower than yours.**
   他的报价远低于你方的报价。
4. **We are writing to *reject*[7] your quotation.**
   我们写信是为了拒绝您的报价。
5. **Thank you for your offer of service.**
   非常感谢您的报价。
6. **We already have a partner in the *region*[8].**
   我们在本区已经有合作伙伴了。
7. **He can give us a more favorable price.**
   他能提供给我们一个更优惠的价格。
8. **We find your quotation higher than those we can get elsewhere.**
   我们发现你们的报价比别处的要高。

必背关键单词

1. ***decline*** [dɪˈklaɪn] v. 下降；衰败；拒绝
2. ***price*** [praɪs] n. 价格；代价
3. ***per*** [pə(r)] adj. 每个的
4. ***firm*** [fɜːm] n. 商号；商行；公司
5. ***sincere*** [sɪnˈsɪə(r)] adj. 真挚的；诚挚的
6. ***turn*** [tɜːn] v. 旋转；转动
7. ***reject*** [rɪˈdʒekt] v. 拒绝
8. ***region*** [ˈriːdʒən] n. 区域

03 拒绝降价请求

这样写就对了

From "Will Smith" (wsmith@aigo.com)　**Date** Fri., October 3, 2009
To "Keanu Reeves" (kreeves@cool.com)
Subject Declining the Request for a Discount

Dear Mr. Reeves,

Thank you for your e-mail of September 28.
Unfortunately, we must say **we can't make a better offer than the one we suggested to you**. **We feel that the offer is the most favorable one presently**.

We hope you will be able to accept our offer after ***reconsideration***[1].

Yours sincerely,
Will Smith

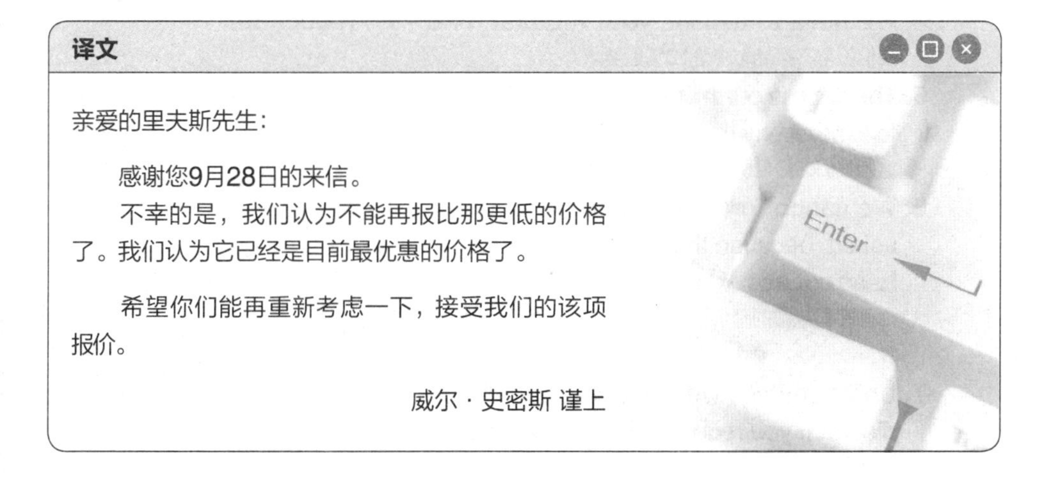

译文

亲爱的里夫斯先生:

感谢您9月28日的来信。

不幸的是，我们认为不能再报比那更低的价格了。我们认为它已经是目前最优惠的价格了。

希望你们能再重新考虑一下，接受我们的该项报价。

威尔 · 史密斯 谨上

语法重点解析

1 解析重点1 **We can't make a better offer than the one we suggested to you.**

当对方要求我们再次降价，而我们又不能再降价的时候，我们可以这样来说：We can't make a better offer（我们报不出更低的价格了），此外，还可以更具体详细地说明情况：We can't make a better offer than the one we suggested to you（我们报不出来比那更低的价格了）。

2 解析重点2 **We feel that the offer is the most favorable one presently.**

在说明自己的报价已经是最好或最优惠时，我们可以这样说：We feel that the offer is the most favorable one presently（我们认为已经是目前最优惠的价格了），注意单词 favorable 的使用，它是用来表示有利的价格或条件。

1. **We can't *reduce*[2] the price as you *required*[3].**
   我们无法按您的要求降价。
2. **We have already *marked*[4] all prices down by 10%.**
   我们所有的商品都已经降价10%了。
3. **We cannot do more than a 5% *reduction*[5].**
   我们只能降价5%，不能再多了。
4. **We have to decline your request for a 10% reduction.**
   我们不得不拒绝您降价10%的要求。
5. **The price is our *minimum*[6]; we refuse to lower it any more.**
   这是我们的最低价，我们拒绝再次降价。
6. **We cannot grant the reduction you asked, because the price has already been cut as *far*[7] as possible.**
   我们不同意您的降价要求，因为价格已尽可能地降至最低点。
7. **Unfortunately, we can't give you the discount you requested for the goods.**
   很遗憾，我们不能给你方所要求的折扣。

必背关键单词

1. ***reconsideration*** 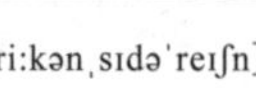 [ˌriːkənˌsɪdəˈreɪʃn] n. 再考虑；再审议
2. ***reduce*** [rɪˈdjuːs] v. 减少
3. ***require*** [rɪˈkwaɪə(r)] v. 需要
4. ***mark*** [mɑːk] v. 作记号；标出
5. ***reduction*** [rɪˈdʌkʃn] n. 减少
6. ***minimum*** [ˈmɪnɪməm] n. 最小量
7. ***far*** [fɑː(r)] adv. 远地；很大程度

04 拒绝变更交易条件

From "Chris Evans" (cevans@gome.com)　Date Sat., October 4, 2009
To "Wentworth Miller" (wmiller@choice.com)
Subject About Your Request for Change of Conditions

Dear Mr. Miller,

We are sorry to inform you that we have to decline your request to ***alter***[1] our conditions.
If you find that **our terms are not in *accord*[2] with your requirements, we might suggest you try seeking other *suppliers*[3]**.

Yours sincerely,
Chris Evans

译文

亲爱的米勒先生：

非常抱歉地告诉您，我们不得不拒绝您想要修改条件的要求。
如果您觉得我们的条款和您的要求不一致，我们建议您试着再找找其他的供应商。

克里斯 · 埃文斯 谨上

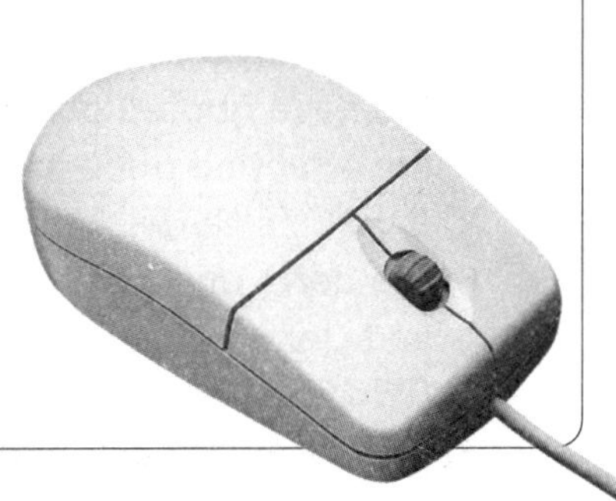

语法重点解析

1 解析重点1 **Our terms are not in accord with your requirements.**

有时候客户不是很满意我们的交易条件，要求更改交易条件。但如果我们无法更改条件时，该如何委婉回复客户呢？我们可以说：Our terms are not in accord with your requirements（我们的条款和您要求的不一致），这种表达方式把原因说成是我们无法满足客户的要求，而不是说客户的要求太过份，显得更礼貌（Courtesy）。

2 解析重点2 **We might suggest you try seeking other suppliers.**

如果我们实在满足不了客户的要求，无法变更交易条件，那么，我们便可以礼貌（Courtesy）而又体贴（Consideration）地建议客户寻找其他合适的供应商。We might suggest you try seeking other suppliers（我们建议您再找找其他的供应商看看），might 在这个句子中的使用，会使得整个句子既礼貌又委婉。

1. **We can't *relax*[4] the *basic*[5] terms of the transaction.**
   我们不能放宽交易的基本条件。
2. **Please inform us if you can reconsider these terms.**
   请告知我方，贵方是否能重新考虑这些交易条件。
3. **You said that you would like to *negotiate*[6] with us on prices, delivery schedule, and purchase terms.**
   您说需要与我们协商一下价格、交货时间和购买条款问题。
4. **You told us you should get the best possible *deal*[7].**
   您告诉我们您应该获得更好的交易条件。
5. **We have quoted our best terms in the attached price list.**
   我方已在所附价格表中报出最优惠价。
6. **I've done my best to negotiate the trade terms with the supplier for you.**
   我已经尽力为您与供应商协商贸易条款。
7. **Our products are very *popular*[8] in the area.**
   我们的产品在这个地方很受欢迎。

必背关键单词  

1. ***alter*** [ˈɔ:ltə(r)] v. 更改；改变
2. ***accord*** [əˈkɔ:d] n. 一致；和谐
3. ***supplier*** [səˈplaɪə(r)] n. 供应商
4. ***relax*** [rɪˈlæks] v. 放松
5. ***basic*** [ˈbeɪsɪk] adj. 基本的
6. ***negotiate*** [nɪˈgəʊʃieɪt] v. 商议；谈判
7. ***deal*** [di:l] n. 买卖；交易
8. ***popular*** [ˈpɒpjələ(r)] adj. 流行的；受欢迎的

05 拒绝接受退货

From "Tobey Maguire" (tmaguire@ilike.com) Date Sun., October 5, 2009
To "Paul Walker" (pwalker@good.com)
Subject Declining the Return of Goods

Dear Mr. Walker,

Thank you for your e-mail of October 4 requesting to return all the 200 items. **We are *unable*[1] to accept the return of goods**, as **we have informed you to check before you *sign*[2]** for them.

Please understand.

Yours sincerely,
Tobey Maguire

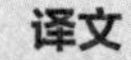

译文

亲爱的沃克先生:

感谢您10月4日的来信，信中要求退回所有200件货品。

我们无法接受您的退货请求，因为我们已经告知您检查后再签收。

请谅解!

托比 · 马奎尔 谨上

语法重点解析

1 解析重点1 **We are unable to accept the return of goods.**

当我们要向客户说明我方不接受退货时，我们可以这样说：We are unable to accept the return of goods（我们无法接受您的退货请求）。我们还可以用 can 来表达：We can't accept the return of goods（我们不能接受您的退货请求）。在这里，后一个句子比前一个句子生硬，因此，前一个句子比较恰当。

2 解析重点2 **We have informed you to check before you sign for them.**

退货条件一般都是固定的，例如，若超出了退货期限就不予退货；不当使用造成的损坏不能退货；没确认好就签收的产品也不予退货。We have informed you to check before you sign for them（因为我们已经告知您请确认好之后再签收），因此，在顾客购买物品的时候，厂家就要事先说明相关事项，以免造成不必要的麻烦。

高频例句

1. **Customers are supposed to check all goods thoroughly before purchase as the shop cannot give *refunds*[3].**
   因为本店不接受退货，望顾客在购买前彻底检查所有商品。
2. **Our company won't take the *china*[4] back if it *breaks*[5].**
   瓷器如有破损，我们公司将不予退货。
3. **We are sorry that we can't accept your request to return these goods.**
   很抱歉，我们不能接受您的退货要求。
4. **Several customers have returned a large quantity of them to us.**
   有几位顾客已大量退货给我们。
5. **We don't take returns on sale items.**
   我们不接受特价商品的退货。
6. **These products are *refundable*[6] in specific *period*[7].**
   这些产品只能在某个限期内退货。

必背关键单词

1. ***unable*** [ʌnˈeɪbl] adj. 不能的；不会的
2. ***sign*** [saɪn] v. 签名；签署
3. ***refund*** [ˈriːfʌnd] n. 归还；偿还
4. ***china*** [ˈtʃaɪnə] n. 瓷器
5. ***break*** [breɪk] v. 打破；弄破；弄坏
6. ***refundable*** [rɪˈfʌndəbl] adj. 可退还的；可偿还的
7. ***period*** [ˈpɪərɪəd] n. 时期；期间

06 无法取消订单

这样写就对了

From "Ronan Keating" (rkeating@ilook.com) Date Mon., October 6, 2008
To "Aaron Carter" (acarter@choose.com)
Subject We Can't Cancel Your Order

Dear Mr. Carter,

We have got your e-mail of October 5 requesting to cancel the order of No. 8888. However, we are unable to accept such a ***cancellation*** [1], as **we have *already*** [2] **had the goods *shipped*** [3]. **We most sincerely hope you will *afford*** [4] **us your understanding**.

Yours sincerely,
Ronan Keating

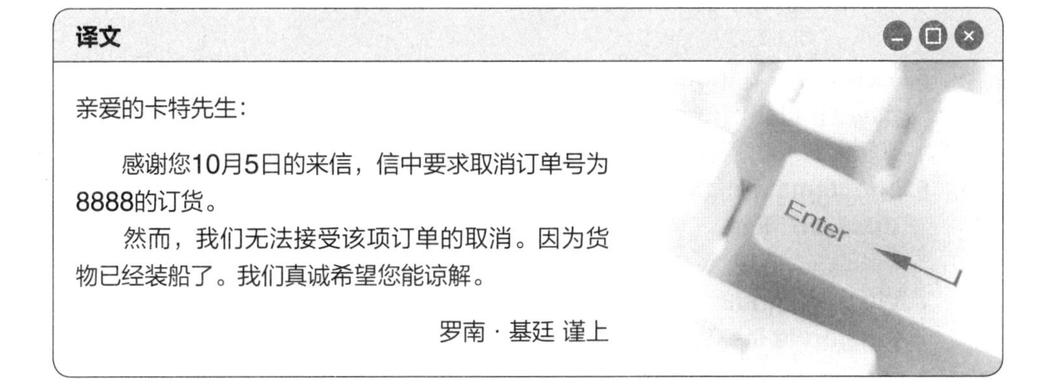

译文

亲爱的卡特先生:

感谢您10月5日的来信，信中要求取消订单号为8888的订货。

然而，我们无法接受该项订单的取消。因为货物已经装船了。我们真诚希望您能谅解。

罗南 · 基廷 谨上

语法重点解析

1 解析重点1 We have already had the goods shipped.

已经做了某事时，要使用现在完成时。如果我们要告知对方货物已经装船了，我们可以这样说：

We have already had the goods shipped.（货物已经装船了。）

2 解析重点2 We most sincerely hope you will afford us your understanding.

不能满足客户要求的时候，我们一定要很真诚地请求对方的谅解。我们可以将 most sincerely 放在 hope 之前来表示我们的诚意。另外，...you will afford us your understanding（敬请谅解）是 thank you for your understanding（敬请理解）的另一种说法。

1. **The goods were shipped yesterday as you required.**
   货物已在昨天按您的要求发货了。
2. **To my *deep*[5] regret, we can't cancel the order.**
   非常抱歉，我们无法撤销订单。
3. **After receiving your e-mail, we've executed orders promptly.**
   收到您的邮件后，我们就立即执行订单了.
4. **To cancel the order will *violate*[6] our contract.**
   取消订单将违反我们的合约。
5. **You can't cancel our order and place it elsewhere.**
   贵公司不能取消订单而转向他处订购。
6. **The items are already on the *production*[7] line.**
   产品已经在生产线上了。
7. **We were under contract to deliver the goods last week.**
   依照合同，我们已于上周发货。
8. **The order cannot be cancelled for the goods are on the *way*[8].**
   订单无法取消，因为已经在发货途中了。

必背关键单词 

1. ***cancellation*** [ˌkænsəˈleɪʃn] n. 取消
2. ***already*** [ɔːlˈredɪ] adv. 已经
3. ***ship*** [ʃɪp] v. 运送；装船
4. ***afford*** [əˈfɔːd] v. 给予；供给
5. ***deep*** [diːp] adj. 深深的
6. ***violate*** [ˈvaɪəleɪt] v. 妨害；违反
7. ***production*** [prəˈdʌkʃn] n. 制造；生产
8. ***way*** [weɪ] n. 路；道路

07 婉拒提议

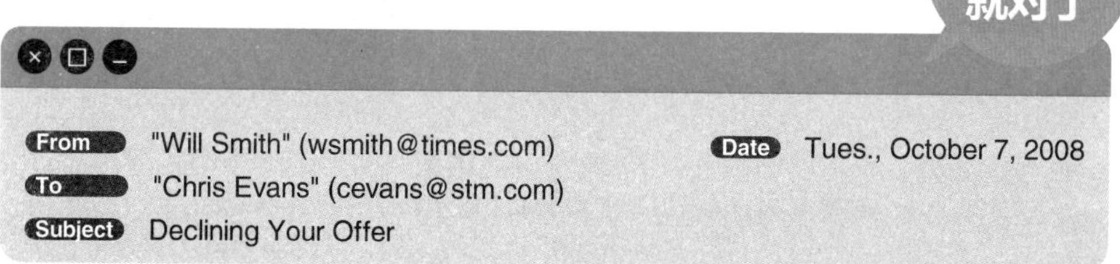

From "Will Smith" (wsmith@times.com)
Date Tues., October 7, 2008
To "Chris Evans" (cevans@stm.com)
Subject Declining Your Offer

Dear Mr. Evans,

Thank you for your offer to ***visit*** [1] our company next week.
We very much regret, however, that our project is still in its ***infancy*** [2]. We will ***record*** [3] your company ***name*** [4] and contact you if there is a need.

Yours sincerely,
Will Smith

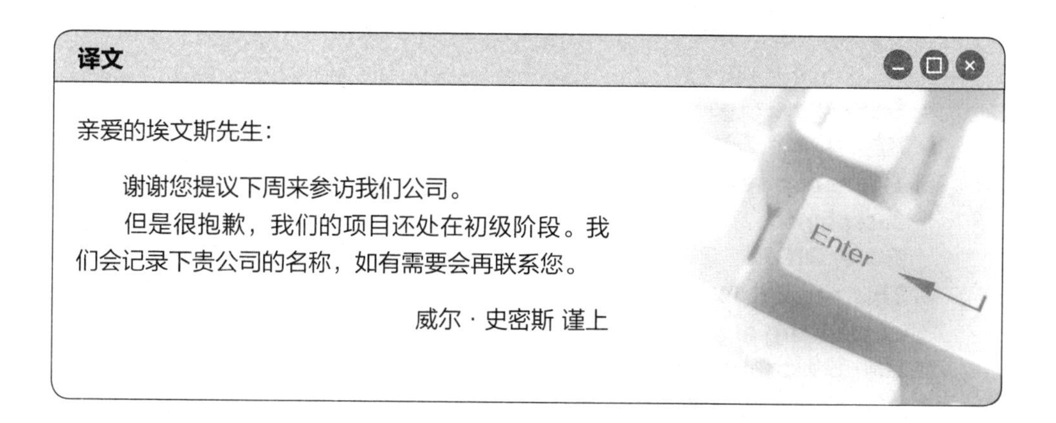

译文

亲爱的埃文斯先生：

谢谢您提议下周来参访我们公司。

但是很抱歉，我们的项目还处在初级阶段。我们会记录下贵公司的名称，如有需要会再联系您。

威尔 · 史密斯 谨上

语法重点解析

1 解析重点1 Our project is still in its infancy.

项目还不成熟，还没到寻找合作伙伴的时候或是还在策划阶段，而某些客户又向我们寻求合作，我们就需要说明目前的情况，并婉拒他们：Our project is still in its infancy（我们的项目还处在初级阶段），对于这句话，我们还可以换个表达方式，例如：Our project is still in its initial stage.（我们的项目还处在初级阶段。）

2 解析重点2 We will record your company name and contact you if there is a need.

虽然现在合作不成，但是等到项目成熟的时候，也许双方还有可能一起合作，所以，在这个时候，我们还是应该细心地记下对方公司的名称、电话之类的信息，并且告诉对方如果以后有合作的机会，会及时通知他们。We will record your company name and contact you if there is a need（我们会记录下贵公司名称，如有需要会再联系您），这样也可以算是对询问方一种比较礼貌（Courtesy）的回答吧！

1. **I have to refuse your offer about this *investment*[5].**
   我不得不拒绝您关于这次投资的提议。
2. **Thank you for your offer to visit us on your next *trip*[6].**
   谢谢您提议在下次旅行的时候来拜访我们。
3. **We are not in need of an engineer at this time.**
   我们现在不需要工程师。
4. **Sorry, I have to decline your *proposal*[7].**
   抱歉，我得拒绝您的提议了。
5. **The proposal didn't find wide acceptance.**
   这项提议没能得到广泛的认可。
6. **This project is still in the *stage*[8] of planning at present.**
   目前，该项目还在计划阶段。
7. **Please don't feel annoyed at our rejection of your offer.**
   请不要为我们拒绝您的提议而感到不快。
8. **I cannot but decline his offer.**
   我不得不拒绝他的提议。

必背关键单词

1. ***visit*** [ˈvɪzɪt] v. 访问；拜访
2. ***infancy*** [ˈɪnfənsɪ] n. 初期；未发达阶段
3. ***record*** [rɪˈkɔːd] v. 记录
4. ***name*** [neɪm] n. 名字；姓名；名称；名义
5. ***investment*** [ɪnˈvestmənt] n. 投资额；投资
6. ***trip*** [trɪp] n. 旅行；出行
7. ***proposal*** [prəˈpəʊzl] n. 提议
8. ***stage*** [steɪdʒ] n. 舞台；阶段

08 无法提早交货

这样写就对了

From "Jared Leto" (jleto@asset.com)
Date Wed., October 8, 2008
To "Ryan Reynolds" (rreynolds@excel.com)
Subject About Your Request for Earlier Delivery

Dear Mr. Reynolds,

Thank you for your e-mail requesting an earlier delivery of the goods.
We have checked our delivery schedule, but found that there is no ***possibility***[1] to ***advance***[2] your delivery. It usually takes two weeks to ***finish***[3] the ***whole***[4] ***process***[5].

Thank you for your understanding.

Yours sincerely,
Jared Leto

译文

亲爱的雷诺兹先生：

感谢您的来信，信中您要求提前发货。

我们已查阅了发货进度，但是发现没有办法提前供货给您，因为通常需要两周的时间来完成整个流程。

敬请谅解！

贾里德·莱托 谨上

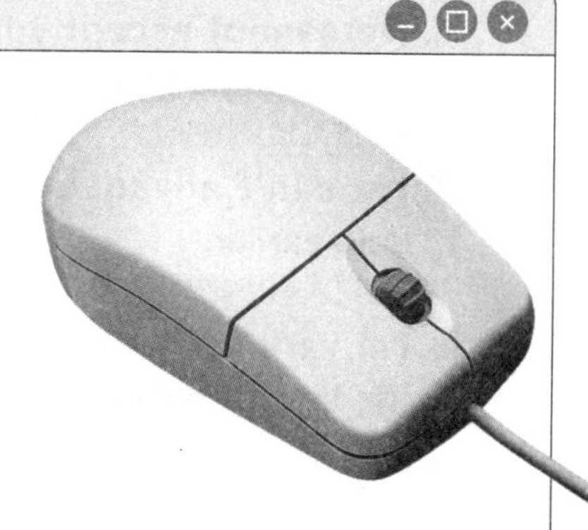

语法重点解析

1 解析重点1 **We have checked our delivery schedule.**

在客户要求提早交货的时候，我们一般要查看货物的进度，全方面考虑清楚事情的安排。如不能提早交货，在回复客户时，就要清楚明白地告知客户实在无法提前交货，从而寻求客户的谅解。

2 解析重点2 **There is no possibility to advance your delivery.**

当我们无法提前交货，我们应该回复：There is no possibility to advance your delivery（实在没办法提前供货给您）。此外，我们还可以这样说：It's impossible for us to advance your delivery.（我们实在没办法提前供货给您。）

高频例句

1. **We cannot advance the time of delivery.**
我们无法提前交货。
2. **I see no way of moving up your delivery.**
我们没办法提前交货给您。
3. **I have checked our production schedule.**
我已经查看了我们的生产进度。
4. **We can't *expedite*[6] production to advance the date of delivery.**
我们无法加速生产，以提前交货。
5. **We have done our best to deliver the goods.**
我们已经尽最大努力交货了。
6. **We cannot accept your request for making an earlier *shipment*[7].**
我们无法接受您的要求，提前交货。
7. **We can't advance the delivery date earlier.**
我们不能再提前交货日期了。
8. **I'm very sorry we can't advance the time of delivery.**
非常抱歉，我们不能提前交货。

必背关键单词

1. ***possibility*** [ˌpɒsəˈbɪləti] n. 可能性

2. ***advance*** [ədˈvɑːns] v. 提前
3. ***finish*** [ˈfɪnɪʃ] v. 完成；结束
4. ***whole*** [həʊl] adj. 全部的；整个的
5. ***process*** [ˈprəʊses] n. 过程；流程
6. ***expedite*** [ˈekspədaɪt] v. 迅速完成；加速进展
7. ***shipment*** [ˈʃɪpmənt] n. 装运

09 无法提供协助

 "Hugh Jackman" (hjackman@ipod.com) 

From "Hugh Jackman" (hjackman@ipod.com)
To "Ryan Reynolds" (rreynolds@travel.com)
Subject I Can't Accompany You to New York
Date Thurs., October 9, 2008

Dear Mr. Reynolds,

I am sorry that I can't ***accompany***[1] you to New York for the ***exhibition***[2] this time because I have an ***urgent***[3] ***matter***[4] to ***handle***[5].
I strongly suggest you go with an English ***interpreter***[6]. That would make things much easier in New York.

If you have any further questions, please feel free to contact me.

Yours sincerely,
Hugh Jackman

译文

亲爱的雷诺兹先生:

很抱歉，这次我不能陪同您一起去纽约参加展览会了，因为我有急事需要处理。
我强烈建议您找一位英语口译人员陪同您一起去，这样在纽约办事会容易得多。

如有任何其他问题，请随时联系我。

休·杰克曼 谨上

语法重点解析

1 解析重点1 I can't accompany you to New York for the exhibition this time.

当我们和别人约定一起参加商展，而后来由于临时有急事不得不告知对方自己无法前往时，该怎么说呢？I can't accompany you to New York for the exhibition this time（这次我不能陪同您一起去纽约参加展览会了）中的 accompany 这个词用在这个地方很恰当，比用 go together with sb 要简洁（Conciseness）许多。

2 解析重点2 That would make things much easier in New York.

如果对方不懂英文，而你又不能陪同协助他，这时可以建议他找一个英语口译人员陪同。提出的建议要能够解决他的实际问题，也就是能对他的事情有所帮助。That would make things much easier in New York（这样在纽约办事会容易得多），也可以说：It would be of great help to you.（这将对你有很大帮助。）

高频例句

1. **We can't give you *assistance*[7] for certain reasons.**
   因为某些理由，我们无法给您提供协助。
2. **We were unable to help you find the information.**
   我们无法帮您找资料。
3. **I am afraid I cannot help you there.**
   我恐怕无法帮助你。
4. **I'm sorry that I can't help you.**
   很抱歉，我无法帮助你。
5. **I'm sorry that I can't be of any assistance to you about that.**
   很抱歉，这件事我帮不上你的忙。
6. **I suggest you ask for help from other departments.**
   我建议您向其他部门寻求帮助。
7. **I am very regretful that I can't help you with this.**
   很遗憾，这方面我帮不上忙。
8. **Sorry, we can't supply *technical*[8] assistance to your company.**
   对不起，我们不能给贵公司提供技术支持。

必背关键单词 

1. ***accompany*** [əˈkʌmpənɪ] v. 随行；陪伴；伴随
2. ***exhibition*** [ˌeksɪˈbɪʃn] n. 展览；陈列
3. ***urgent*** [ˈɜːdʒənt] adj. 急迫的；紧急的
4. ***matter*** [ˈmætə(r)] n. 事情；问题
5. ***handle*** [ˈhændl] v. 处理；操作
6. ***interpreter*** [ɪnˈtɜːprɪtə(r)] n. 口译员；解释者
7. ***assistance*** [əˈsɪstəns] n. 协助；援助
8. ***technical*** [ˈteknɪkl] adj. 技术上的；技能的

10 无法介绍客户

这样写就对了

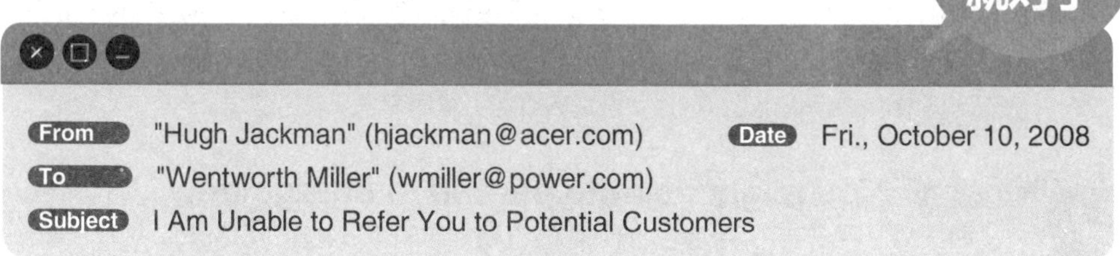

From "Hugh Jackman" (hjackman@acer.com) Date Fri., October 10, 2008
To "Wentworth Miller" (wmiller@power.com)
Subject I Am Unable to Refer You to Potential Customers

Dear Mr. Miller,

I am sorry that **I am unable to *refer*[1] you to potential customers for your products** because I have left the ***chamber***[2] of ***commerce***[3] for nearly five years. **You might ask the person in *charge*[4] now.**

If there's anything else that I can help you with, please let me know.

Yours sincerely,
Hugh Jackman

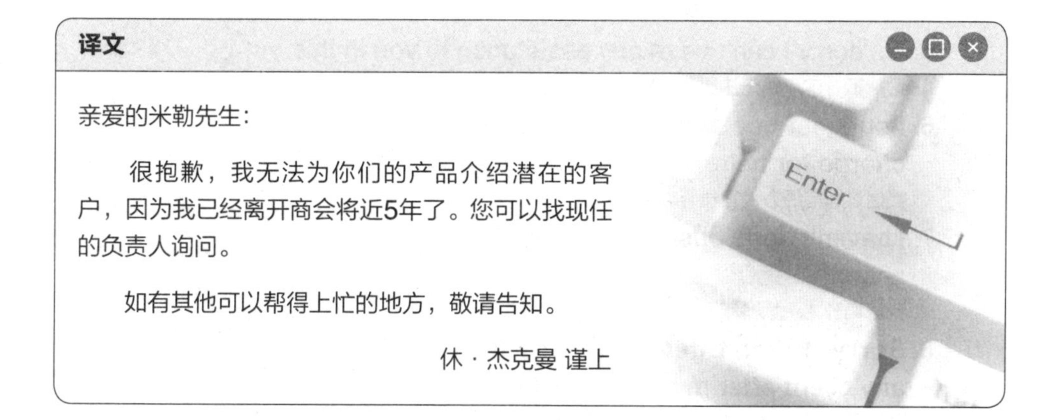

译文

亲爱的米勒先生：

很抱歉，我无法为你们的产品介绍潜在的客户，因为我已经离开商会将近5年了。您可以找现任的负责人询问。

如有其他可以帮得上忙的地方，敬请告知。

休·杰克曼 谨上

语法重点解析

1 解析重点1 I am unable to refer you to potential customers for your products.

帮别人介绍客户或是推荐客户时，英文该如何表达呢？除了我们最常用的 introduce sb to sb（将某人介绍给某人），我们还可以用另外一个表达方法，那就是上述邮件中出现的：...refer you to potential customers for your products（为你们的产品介绍潜在的客户），这里使用了 refer sb to sb，也是“帮某人介绍、推荐某人”的意思。

2 解析重点2 You might ask the person in charge now.

如果你已经离职了，还有人请求你的帮助，这时可以礼貌（Courtesy）地建议对方，让他去找目前的负责人询问。You might ask the person in charge now（你们可以找现在的负责人询问）中的 in charge 这个短语是“主管；负责”的意思。

1. **I couldn't refer you to some *clients*[5].**
   我无法为您介绍客户。
2. **You said that you would like me to introduce some customers.**
   您说贵公司想要我介绍一些客户。
3. **I am afraid I cannot help you with that.**
   恐怕我帮不上你那个忙。
4. **I have no clients to introduce to you.**
   我没有客户可以介绍给你。
5. **I'm sorry I can't be of any assistance to you in this.**
   很抱歉，这事我无法帮助您。
6. **I suggest you ask the *official*[6] in charge for help.**
   我建议您向经办的官员寻求帮助。
7. **I haven't done business for several years.**
   我已离开生意场多年了。
8. **Sorry, I didn't get in touch with any client after my *resignation*[7].**
   对不起，我辞职后就没有再联系任何客户了。

必背关键单词

1. ***refer*** [rɪˈfɜ:(r)] v. 参考；提及；介绍
2. ***chamber*** [ˈtʃeɪmbə(r)] n. 房间；卧室
3. ***commerce*** [ˈkɒmɜ:s] n. 商业；贸易
4. ***charge*** [tʃɑ:dʒ] n. 费用；职责
5. ***client*** [ˈklaɪənt] n. 客户
6. ***official*** [əˈfɪʃl] n. 官员
7. ***resignation*** [ˌrezɪɡˈneɪʃn] n. 辞职

11 婉谢邀请

这样写就对了

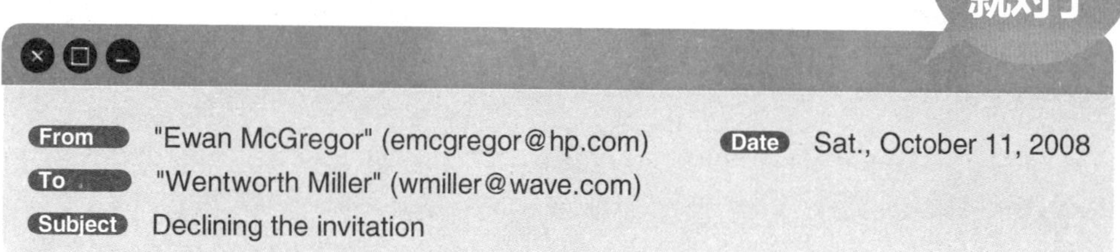

From "Ewan McGregor" (emcgregor@hp.com) Date Sat., October 11, 2008
To "Wentworth Miller" (wmiller@wave.com)
Subject Declining the invitation

Dear Mr. Miller,

I'm so glad that you ***invited***[1] me to attend the ***ceremony***[2] to be ***held***[3] on Monday, October 13 in Times ***Square***[4].

Unfortunately, I have another meeting to ***attend***[5], so **I have to decline with much regret your kind invitation. And I wish the *event*[6] a success.**

Yours sincerely,
Ewan McGregor

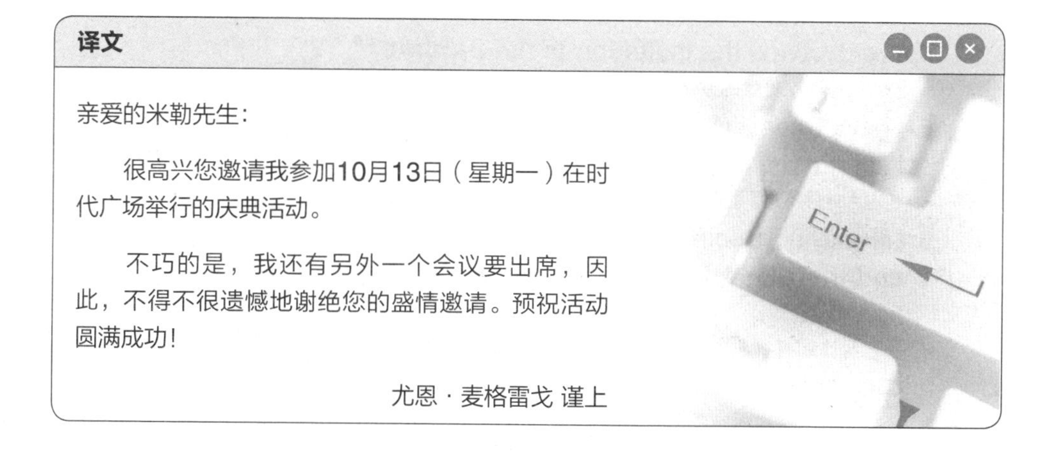

译文

亲爱的米勒先生：

很高兴您邀请我参加10月13日（星期一）在时代广场举行的庆典活动。

不巧的是，我还有另外一个会议要出席，因此，不得不很遗憾地谢绝您的盛情邀请。预祝活动圆满成功！

尤恩 · 麦格雷戈 谨上

语法重点解析

1 解析重点1 I have to decline with much regret your kind invitation.

一般我们在婉拒别人的邀请时，会说：

I very much regret that I have to decline your invitation.
（非常抱歉，我要婉谢您的邀请了。）

Unfortunately, I have to decline your invitation.（不巧的是，我不得不婉谢您的邀请。）

I have to decline with much regret your kind invitation.（我不得不很遗憾地婉谢您的好心邀请。）

2 解析重点2 And I wish the event a success.

虽然拒绝了别人的邀请，不能出席相关活动了，但是，我们还是要说一些祝福的话语，例如：And I wish the event a success（祝福活动圆满成功），这就显示了我们对他人的礼貌（Courtesy）和尊重。

1. There are some reasons causing me to decline the invitation.
因为一些原因，我婉谢了这次邀请。

2. I shall have to refuse your invitation because of a prior *engagement*[7].
我因有约在先，所以只好婉谢您的邀请。

3. We have to decline the invitation to your party.
我们不得不婉谢您的聚会邀请。

4. I've declined the invitation to the *banquet*[8].
我已婉谢这次宴会的邀请。

5. I have to decline the invitations for I am not feeling well.
我因为身体不适而婉谢邀请。

6. I already have something else planned and therefore have to decline the invitation.
我已经计划好了要去做别的事情，因此不得不婉谢邀请。

必背关键单词

1. ***invite*** [ɪnˈvaɪt] v. 邀请；招待
2. ***ceremony*** [ˈserəmənɪ] n. 庆典
3. ***hold*** [həʊld] v. 握住；拿着；持有；举办
4. ***square*** [skweə(r)] n. 正方形；广场
5. ***attend*** [əˈtend] v. 参加
6. ***event*** [ɪˈvent] n. 事件
7. ***engagement*** [ɪnˈɡeɪdʒmənt] n. 约会
8. ***banquet*** [ˈbæŋkwɪt] n. 宴会

12 无法变更日期

From "Paul Walker" (pwalker@global.com)　**Date** Sun., October 12, 2008
To "Topher Grace" (tgrace@hotmail.com)
Subject We Can't Change the Date

Dear Mr. Grace,

We regret very much that **we are unable to *postpone*[1] the *fixed*[2] *date*[3] of our *appointment*[4]**, because we have another client who is quite willing to cooperate with us.

An early reply would be appreciated as we wish to reach a *prompt*[5] decision.

Yours sincerely,
Paul Walker

译文

亲爱的格雷斯先生：

很遗憾，我们无法延后我们约定好的会面时间，因为已经有另外一名客户很想跟我们合作。

本公司希望能尽快下决定，故请您尽早回复。

保罗 · 沃克 谨上

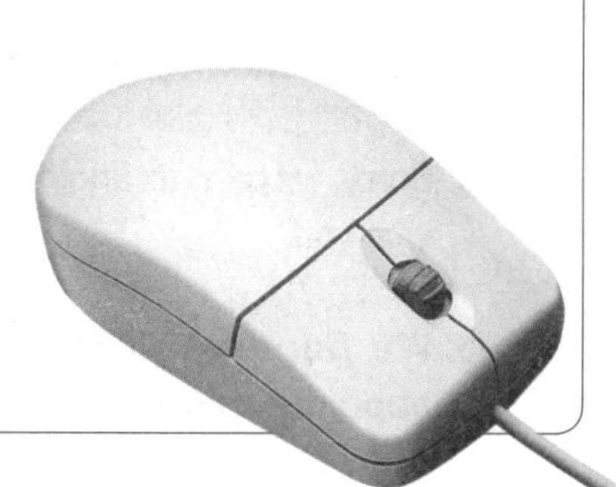

语法重点解析

1 解析重点1 We are unable to postpone the fixed date of our appointment.

如果我们无法满足对方更改日期的要求，我们可以说：We are unable to postpone the fixed date of our appointment（我们无法延后我们约定的时间），postpone 是“延后；延期”的意思。the fixed date（约好的日期）也可以改成 the set date 或 the target date。

2 解析重点2 An early reply would be appreciated as we wish to reach a prompt decision.

邮件中提到有别的客户很想跟我们合作，所以我们催促收件人尽快回复消息。我们可以说：An early reply would be appreciated（如能早日回复，将不胜感激）。而后面的 We wish to reach a prompt decision（本公司希望能尽快下决定）又说明了己方不同意更改日期的另外一个原因，同时，这也是在希望得到对方的理解，而能尽快回复。

1. **We hope everything would be ready by a *target*[6] date.**
   我们希望在预定日期之前将一切准备就绪。
2. **We can't accept your request to change the set time.**
   我们不能接受您更改约定日期的要求。
3. **I'm sorry, but another time would be *inconvenient*[7] for us.**
   很抱歉，但是我们不方便另约时间。
4. **We have told our manager the date of the appointment.**
   我们已经把这个会面时间告诉了经理。
5. **Sorry, we can't fix another appointment with you.**
   对不起，我们不能再另约时间与您见面了。
6. **We hope you can arrive at the appointed time.**
   我们希望您能在约定的时间到达。
7. **Please *remember*[8] to appear on the day appointed.**
   请记得在约定的日期出现。

必背关键单词

1. ***postpone*** [pəˈspəʊn] v. 延期；延后
2. ***fixed*** [fɪkst] adj. 固定的；不变的
3. ***date*** [deɪt] n. 日期；约会
4. ***appointment*** [əˈpɔɪntmənt] n. 指定；约定；指派；任命
5. ***prompt*** [prɒmpt] adj. 即时的
6. ***target*** [ˈtɑːgɪt] n. 目标；靶子
7. ***inconvenient*** [ˌɪnkənˈviːnɪənt] adj. 不方便的；打扰的
8. ***remember*** [rɪˈmembə(r)] v. 记得

13 拒绝延迟交货

From "Paul Walker" (pwalker@ftp.com)
Date Mon., October 13, 2008
To "Orlando Bloom" (obloom@nhk.com)
Subject We Can't Accept Any Delay in Delivery

Dear Mr. Bloom,

I am sorry that we can't accept your request for ***delay***[1] in delivery. **We are in urgent need of these *building*[2] materials**, or we can't ***complete***[3] the ***task***[4]. **If you can't deliver them on time, we shall have to cancel the order.**

Thank you for your understanding.

Yours sincerely,
Paul Walker

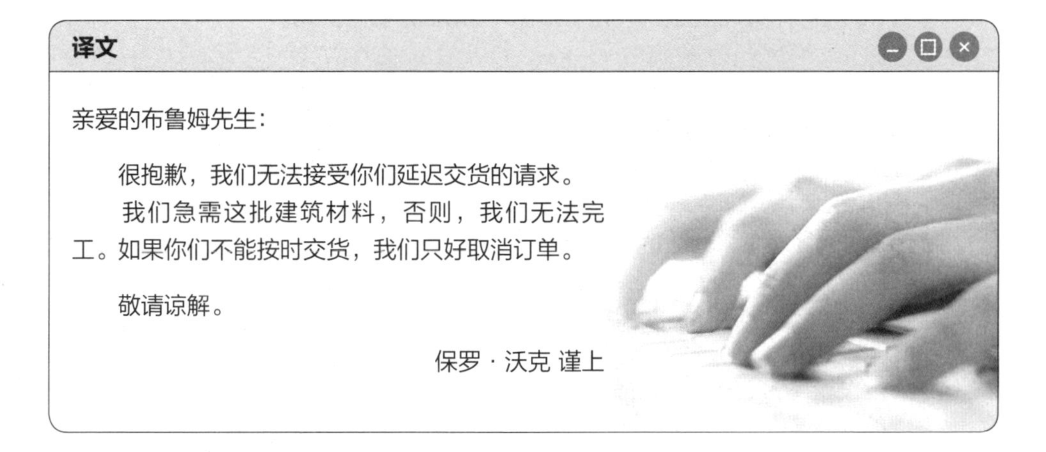

译文

亲爱的布鲁姆先生：

很抱歉，我们无法接受你们延迟交货的请求。

我们急需这批建筑材料，否则，我们无法完工。如果你们不能按时交货，我们只好取消订单。

敬请谅解。

保罗 · 沃克 谨上

语法重点解析

1

解析重点1 **We are in urgent need of these building materials.**

表达“急需”，除了用动词 need 或者 want，还可以使用短语的形式，例如：in great need of...，in dire need of... 和 in urgent need of...。请看下面的例句：

We are in urgent need of these building materials.（我们急需这批建筑材料。）

We are in want of these building materials.（我们急需这批建筑材料。）

2

解析重点2 **If you can't deliver them on time, we shall have to cancel the order.**

如果我们实在很需要这批货物，而对方又不能按时交货的话，我们只能取消订单，向别的厂商订购，这时我们就需要向对方表明自己的态度。If you can't deliver them on time, we shall have to cancel the order（如果你们不能按时交货，我们只好取消订单了），shall have to... 比起用 must 和 have to 在语气上要显得相对委婉一些。

1. **We are informed that you couldn't deliver the goods on time.**
   我们被告知你们无法按时交货。
2. **We can't accept your request to postpone the delivery.**
   我们不能接受您延迟交货的要求。
3. **The delay will cause great *inconvenience*[5] to us.**
   延迟会对我们造成很大的不便。
4. **We will be *forced*[6] to cancel this order.**
   我们将要被迫取消这次订单。
5. **We are so sorry we can't wait any longer.**
   很抱歉，我们不能再等了。
6. **If you still *insist*[7] on that, we will have to cancel the order.**
   如果您还是坚持的话，我们将不得不取消订单。
7. **We hope you can understand the current *situation*[8].**
   希望你们可以了解目前的情况。

必背关键单词

1. ***delay*** [dɪˈleɪ] n. 耽搁
2. ***building*** [ˈbɪldɪŋ] n. 建筑物
3. ***complete*** [kəmˈpliːt] v. 完成
4. ***task*** [tɑːsk] n. 任务
5. ***inconvenience*** [ˌɪnkənˈviːnɪəns] n. 不便；麻烦
6. ***force*** [fɔːs] v. 强迫；施压
7. ***insist*** [ɪnˈsɪst] v. 坚持；强调
8. ***situation*** [ˌsɪtʃʊˈeɪʃn] n. 情况；形势

14 无法接受临时取消订单

From "Paul Walker" (pwalker@delivery.com) Date Tues., October 14, 2008
To "Ben Affleck" (baffleck@goods.com)
Subject We Can't Accept the Cancellation

Dear Mr. Affleck,

According to our contract, **we can cancel the order only if you inform us three weeks *prior* [1] to the delivery date**. However, you just did this without ***notifying*** [2] us in advance.
Moreover, **we haven't got a *reasonable* [3] *explanation* [4] from your side yet**. You'll have to give us a ***satisfactory*** [5] ***answer*** [6] for the matter mentioned above.

Yours sincerely,
Paul Walker

译文

亲爱的阿弗莱克先生:

根据合约规定，我们能在发货日期前3周接受取消订单的要求。然而，贵公司未提前通知就取消订单了。

而且，我们至今还未从你方得到一个合理的解释。请贵公司就以上事件给我们一个满意的答复!

保罗 · 沃克 谨上

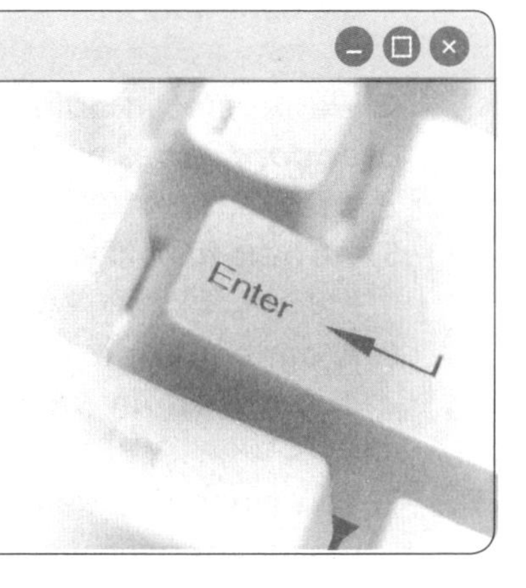

语法重点解析

1 解析重点1 **We can cancel the order only if you inform us three weeks prior to the delivery date.**

一般取消订单是有规定期限的，当对方超出规定的期限而取消订单，就是违反了合约约定，这一点是需要向对方说清楚的。We can cancel the order only if you inform us three weeks prior to the delivery date.（我们只能接受发货前3周的订单取消要求。）这里请注意一下时间的表达方法：three weeks prior to the delivery date（比发货日期提前3周的时间）。

2 解析重点2 **We haven't got a reasonable explanation from your side yet.**

当对方擅自取消订单，而且未做任何解释说明的时候，我们可以这样告诉对方：We haven't got a reasonable explanation from your side yet（我们至今尚未从你方得到一个合理的解释）。这里使用现在完成时，强调从过去直到现在都没有得到对方的解释。

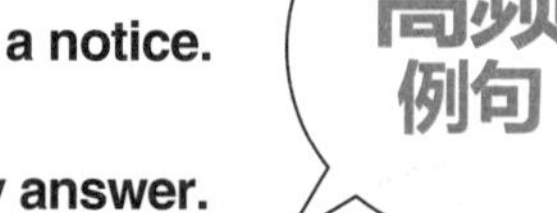

1. **We can't cancel an order without receiving a notice.**
   未收到通知，我们无法取消订单。
2. **You should be able to give us a satisfactory answer.**
   贵公司得给我们一个满意的答复。
3. **We must comply with the terms of the contract.**
   我们双方都必须按照合约条款行事。
4. **We persist in doing business in *accordance*[7] with the contract.**
   我们公司坚持按照合约行事。
5. **We have to hold you to the contract.**
   我们不得不要求你们按照合约行事。
6. **A cancellation notice must be *submitted*[8] in writing form.**
   取消通知必须以书面形式提交。
7. **You must give us a *legitimate*[9] reason for cancellation.**
   你们必须给我们一个合情合理的取消理由。

必背关键单词 

1. ***prior*** [ˈpraɪə(r)] adv. 在前；居先
2. ***notify*** [ˈnəʊtɪfaɪ] v. 通知；报告
3. ***reasonable*** [ˈriːznəbl] adj. 合理的
4. ***explanation*** [ˌekspləˈneɪʃn] n. 说明；解释
5. ***satisfactory*** [ˌsætɪsˈfæktərɪ] adj. 令人满意的
6. ***answer*** [ˈɑːnsə(r)] n. 回答；答案
7. ***accordance*** [əˈkɔːdns] n. 给予；一致；和谐
8. ***submit*** [səbˈmɪt] v. 屈服；提交
9. ***legitimate*** [lɪˈdʒɪtɪmət] adj. 合法的；合理的

15 无法履行合同

这样写就对了

From	"Paul Walker" (pwalker@delivery.com)	Date	Tues., October 14, 2008
To	"Ben Affleck" (baffleck@goods.com)		
Subject	We Can't Fulfill the Contract		

Dear Mr. Affleck,

We are so sorry that **we can't *continue*[1] to fulfill the contract on *account*[2] of the quality problem**.
Many customers complained that these computers didn't run well after the ***purchase***[3]. **Due to this, we want to *suspend*[4] the supply and hope you can give us a *solution*[5] to the problem immediately**.

Yours sincerely,
Paul Walker

译文

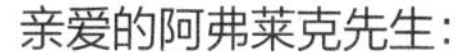

亲爱的阿弗莱克先生:

很抱歉，由于品质的问题，我们无法继续履行合同。

众多用户购买电脑之后，抱怨电脑使用情况不佳。鉴于此，我们想暂停你们的供货，并希望你们能立即给出一个解决方案。

保罗 · 沃克 谨上

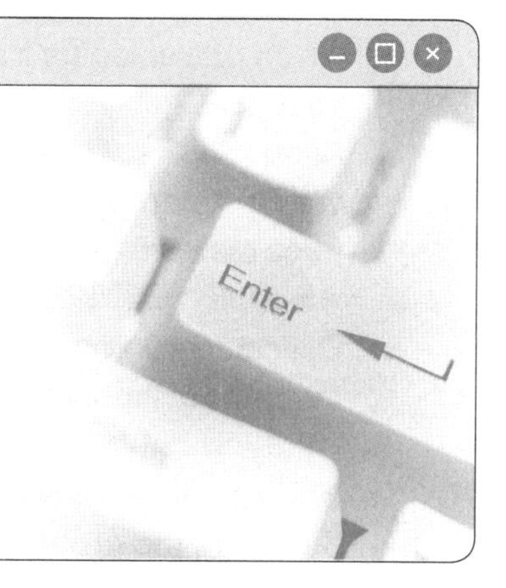

语法重点解析

1 解析重点1 We can't continue to fulfill the contract on account of the quality problem.

由于某些原因，而不能继续履行合同，我们可以说：We can't continue to fulfill the contract on account of the quality problem（由于品质的问题，我们无法继续履行合同），这里需要注意的是，我们给出的理由必须合情合理，否则将很难得到对方的理解，还会使问题上升到法律层面。

2 解析重点2 Due to this, we want to suspend the supply and hope you can give us a solution to the problem immediately.

由于产品出现品质问题而叫停供货，可以这样说：Due to this, we want to suspend the supply（鉴于此，我们想暂停供货），句子中的 due to... 有“由于……”的意思。同时，还别忘了要求对方给出相关解决方案：We hope you can give us a solution to the problem immediately.（希望你们能立即就该问题给我方一个解决方案。）

1. **We can't accept the delay in your execution of the contract.**
   我们不能接受贵公司在履行合同时发生的延误。
2. **We would like to cancel the order *by virtue of*[6] the inferior quality of the goods.**
   因为货物品质低劣，我们想取消订单。
3. **A *huge*[7] earthquake happened several days ago, which prevented us from fulfilling the contract.**
   几天前的大地震致使我方无法履行合同。
4. **We can't perform the contract due to force majeure.**
   由于不可抗力因素致使我方无法履行合约。
5. **Our company was incapable of fulfilling the terms of the contract.**
   我们公司已经失去了履行合约的能力。
6. **We want to *dissolve*[8] the contract due to all the problems.**
   由于所有问题，我们想跟贵公司解除合约。

必背关键单词

1. ***continue*** [kən'tɪnju:] v. 继续
2. ***account*** [ə'kaʊnt] n. 账目；理由
3. ***purchase*** ['pɜ:tʃəs] n. 购买
4. ***suspend*** [sə'spend] v. 悬挂；暂停
5. ***solution*** [sə'lu:ʃn] n. 溶解；解决
6. ***by virtue of*** [baɪ 'vɜ:tʃu: əv] phr. 由于；凭借
7. ***huge*** [hju:dʒ] adj. 巨大的
8. ***dissolve*** [dɪ'zɒlv] v. （议会等）解散；终止（商业协议）

Unit 13 道歉 Apology

01 发货失误的道歉

这样写就对了

From lcorporation@lcc.com Date August 11, 2008
To mrsimon@yahoo.com
Subject Apology for Error in Processing Order

Dear Mr. Simon,

Please ***accept***[1] our ***sincere***[2] ***apology***[3] for the ***error***[4] in the shipment of your order No.7563.
The correct items were shipped to you ***freight***[5] ***prepaid***[6] on August 10.Please let us know when you ***confirm***[7] the receipt.

We will make sure that such an error will not ***happen***[8] again.

Sincerely yours,
LC Corporation

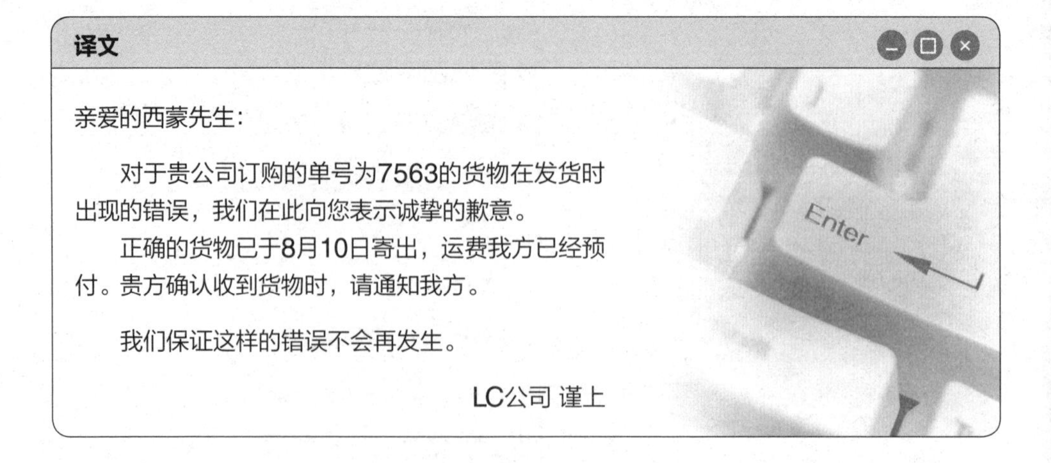

译文

亲爱的西蒙先生：

对于贵公司订购的单号为7563的货物在发货时出现的错误，我们在此向您表示诚挚的歉意。

正确的货物已于8月10日寄出，运费我方已经预付。贵方确认收到货物时，请通知我方。

我们保证这样的错误不会再发生。

LC公司 谨上

语法重点解析

1 解析重点1 freight prepaid

短语 freight prepaid 是“预付运费”的意思。“货到付款”则为 COD，即 cash on delivery 或 collect on delivery 的缩写，另外 payment on delivery 也是“货到付款”的意思。请看以下例句：

The bill of parcels should be marked as “freight prepaid”.（包裹单上应该注明“运费预付”字样。）

Cash on delivery is a rule of Buying and Selling.（一手交钱一手交货是买卖规则。）

2 解析重点2 make sure

短语 make sure 是“确信；确定；弄清楚”的意思。文中是指为了不再发生同样的错误而要“确认”的意思。也可以用 make certain, be sure, assure 等来表达相同的意思。请看以下例句：

Please make certain of the date of the meeting.（请确认开会的日期。）

No one can be sure that the weather will be fine.（谁也不能保证天气晴好。）

1. **I am so sorry for *delivering*[9] the wrong goods.**
   我对发错货感到非常抱歉。
2. **I do apologize for the error in the shipment.**
   对于出货时产生的错误，我感到非常抱歉。
3. **I make an apology for causing so many troubles.**
   造成这么多的麻烦，我感到很抱歉。
4. **Please accept my apology for my oversight.**
   请原谅我的疏忽。
5. **Please inform us if the goods arrive.**
   如果货物送达，请通知我们。
6. **I can assure you of the quality of our goods.**
   我可以向你保证我们货物的品质。
7. **I assure you it won't happen again.**
   我向你保证此类事情不会再发生。

必背关键单词

1. ***accept*** [əkˈsept] v. 接受
2. ***sincere*** [sɪnˈsɪə(r)] adj. 真诚的
3. ***apology*** [əˈpɒlədʒɪ] n. 道歉
4. ***error*** [ˈerə(r)] n. 错误
5. ***freight*** [freɪt] n. 运费；货运
6. ***prepaid*** [ˌpriːˈpeɪd] adj. 预付的；已付的
7. ***confirm*** [kənˈfɜːm] v. 确认；证实
8. ***happen*** [ˈhæpən] v. 发生
9. ***deliver*** [dɪˈlɪvə(r)] v. 递送；运送

02 瑕疵品的道歉

这样写就对了

From lfgoods@lfc.com
To browde@yahoo.com
Subject Apology for Faulty Products
Date August 11, 2008

Dear Mr. Brown,

We are very sorry to hear that you found **defective goods** in our shipment. We will ***certainly***[1] accept the return of these items and send you ***replacements***[2] at once. **Please accept our pure-hearted apologies for any *inconvenience*[3] this may have *caused*[4] you**. I ***assure***[5] you that I have ***instructed***[6] the quality ***control***[7] manager to make certain this does not happen again.

Sincerely yours,
LF Corporation

译文

亲爱的布朗先生：

对于运往贵公司的货物中出现了瑕疵品，我们感到十分抱歉。

我们当然会接受退货，新的货物也会立即寄出。对于可能给您带来的不便，请接受我们真诚的道歉。我向您保证，我已经指示质量控制经理保证不会再发生此类的事情。

LF公司 谨上

1

解析重点1 **defective goods**

defective goods 表示“瑕疵品”，也可以用 faulty products, faulty materials, factory second 等来表达。请看以下例句：

Factory second items can save you a lot of money if you're on a tight budget.（如果你的预算很紧，工厂的瑕疵品可以帮你省不少钱。）

They will replace the faulty goods or they will get no more orders from us.（他们得把这批瑕疵品换掉，不然以后别想收到我们的订单。）

2

解析重点2 **Please accept our pure-hearted apologies for any inconvenience this may have caused you.**

这句话的意思是“对于可能给您带来的不便，请接受我们真诚的道歉”。注意这里的“接受”用的是 accept 而不是 receive，因为前者指主观心理上的“接受”，而后者只是客观上的“收到”，并不一定是“接受”。pure-hearted 是“真诚的”，在此用来表示道歉的诚意，也可以使用 sincere, genuine, heartfelt 等单词表达“真诚的”。请看以下例句：

He has a genuine desire to help us.（他真心诚意地愿意帮助我们。）

Please send my heartfelt regards to your parents.（请代我向你的父母致上最真诚的问候。）

1. **I am so sorry for delivering the *defective*[8] goods.**
   我非常抱歉寄送了有瑕疵的货物。
2. **We will, of course, accept the return of these items.** 我们当然会接受这些货品的退货。
3. **I make an apology for all the trouble.**
   给您带来的一切麻烦，我感到很抱歉。
4. **Please accept my apology for any inconvenience my mistake has caused you.**
   由于我的失误给您造成不便，请接受我的道歉。
5. **I do apologize for the error in the color of the products.**
   对于货品颜色上的错误，我真的很抱歉。

必背关键单词 

1. ***certainly*** [ˈsɜːtnlɪ] adv. 当然地
2. ***replacement*** [rɪˈpleɪsmənt] n. 更换；替代品
3. ***inconvenience*** [ˌɪnkənˈviːnɪəns] n. 不便
4. ***cause*** [kɔːz] v. 引起；造成
5. ***assure*** [əˈʃʊə(r)] v. 保证；担保；使确认
6. ***instruct*** [ɪnˈstrʌkt] v. 指示；命令
7. ***control*** [kənˈtrəʊl] v. 控制
8. ***defective*** [dɪˈfektɪv] adj. 有瑕疵的

03 商品毁损的道歉

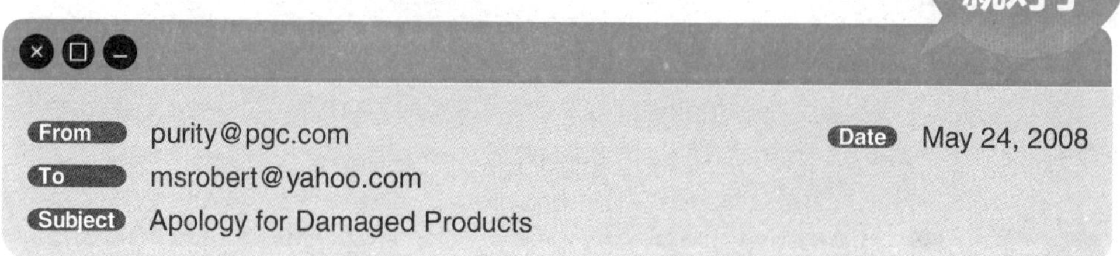

From purity@pgc.com
Date May 24, 2008
To msrobert@yahoo.com
Subject Apology for Damaged Products

Dear Ms. Robert,

We're ***extremely***[1] sorry to hear that the ***ornamental***[2] glass you ordered was broken during the ***transportation***[3].
We did instruct the forwarding company to ***handle***[4] your products very carefully, but something ***obviously***[5] went wrong in the ***container***[6].
Please wait while we negotiate with the forwarding company about how to settle the matter best.

Truly yours,
Purity Glass Corporation

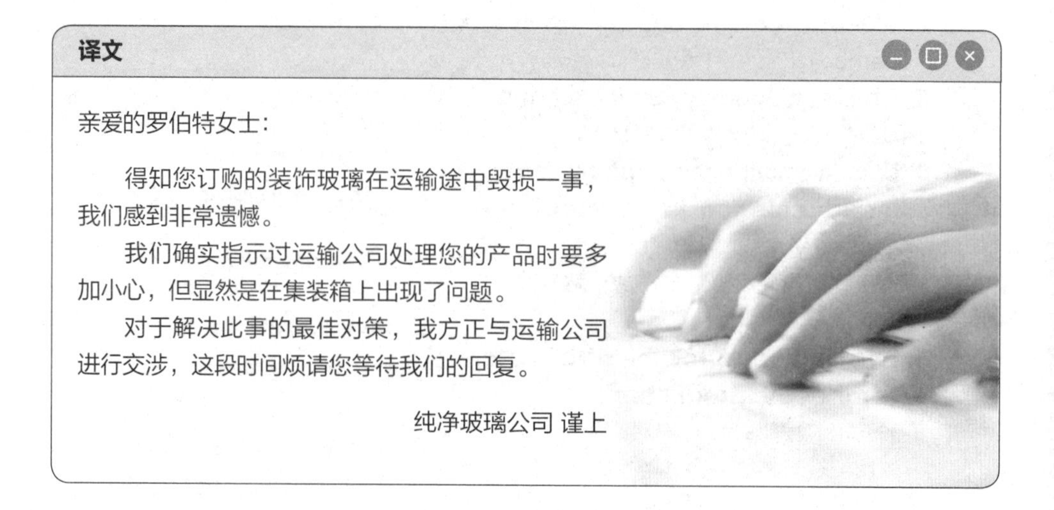

译文

亲爱的罗伯特女士:

得知您订购的装饰玻璃在运输途中毁损一事，我们感到非常遗憾。

我们确实指示过运输公司处理您的产品时要多加小心，但显然是在集装箱上出现了问题。

对于解决此事的最佳对策，我方正与运输公司进行交涉，这段时间烦请您等待我们的回复。

纯净玻璃公司 谨上

语法重点解析

1 解析重点1 We're extremely sorry to hear that...

We're extremely sorry to hear that... 这个句子的意思是"得知……，我们感到非常遗憾"，副词 extremely 意为"极度；非常"，表示程度之深。"听说 / 获悉 / 知道……，我们感到非常抱歉 / 遗憾"也可以用以下方式表达：We're terribly sorry to hear that... / We're extremely sorry to learn that... / We're extremely sorry to know that...

2 解析重点2 negotiate with

negotiate with sb 是"与某人交涉；与某人谈判"的意思，而字面相近的 negotiate about sth 则是"就某事进行交涉"的意思。请对照以下例句：

We've decided to negotiate with the employers about our wage claim.（我们决定就工资问题与雇主谈判。）

The two sides are negotiating about the contract.（双方正就合约问题进行交涉。）

1. **I am so sorry for *damaging*[7] the goods you ordered.**
   对于您订购的货物发生损坏的情况，我感到非常抱歉。
2. **We will certainly exchange your goods.**
   我们当然会为您更换货品。
3. **I make an apology for causing you unnecessary trouble.**
   给您造成不必要的麻烦，我感到很抱歉。
4. **Please accept my sincere apology for our carelessness during the shipment.**
   对于运输中的失误，请接受我真诚的道歉。
5. **I do apologize for damaging your products during transportation.**
   对于在运输中毁损了您的货品，我向您道歉。
6. **Please accept our pure-hearted apologies for impairing your goods.**
   对于此次损坏您的货物，请接受我真诚的道歉。

必背关键单词

1. ***extremely*** [ɪkˈstriːmlɪ] adv. 极度；非常
2. ***ornamental*** [ˌɔːnəˈmentl] adj. 装饰的
3. ***transportation*** [ˌtrænspɔːˈteɪʃn] n. 运输
4. ***handle*** [ˈhændl] v. 处理
5. ***obviously*** [ˈɒbvɪəslɪ] adv. 明显地；显而易见地
6. ***container*** [kənˈteɪnə(r)] n. 集装箱
7. ***damage*** [ˈdæmɪdʒ] v. 损坏；毁损

04 交货延迟的道歉

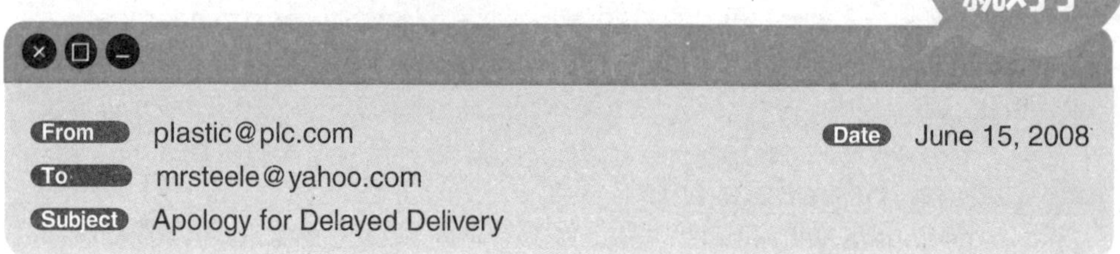

From plastic@plc.com Date June 15, 2008
To mrsteele@yahoo.com
Subject Apology for Delayed Delivery

Dear Mr. Steele,

Please accept our ***profound***[1] apologies for the ***late***[2] delivery of goods to your company.
The ***delay***[3] was due to a mix-up at our freight company, and we will make sure we work with those who can ***ensure***[4] that all delivery ***deadlines***[5] are met in the future.
We hope that you will *forgive*[6] us our *unintentional*[7] mistake and *continue*[8] to purchase items from us.

Truly yours,
PL Co.

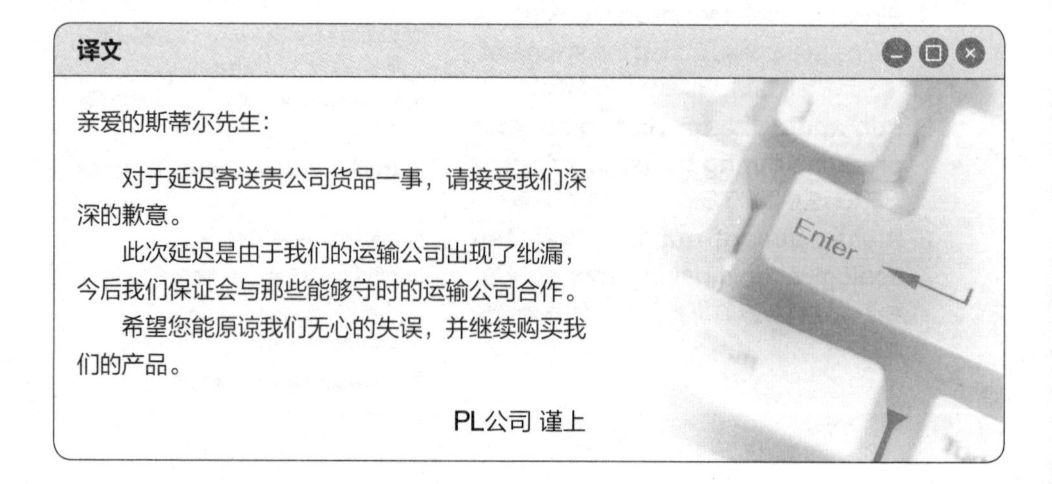

译文

亲爱的斯蒂尔先生：

对于延迟寄送贵公司货品一事，请接受我们深深的歉意。
此次延迟是由于我们的运输公司出现了纰漏，今后我们保证会与那些能够守时的运输公司合作。
希望您能原谅我们无心的失误，并继续购买我们的产品。

PL公司 谨上

1 解析重点1 **mix-up**

mix-up是"混乱；失误"的意思，类似的表达还可以用mistake, error, oversight, carelessness 等单词。请看以下例句：

There's been an awful mix-up over the dates!（日期问题混乱得不能再混乱了！）

Even an oversight in the design might result in heavy losses.（哪怕是设计中的一点点疏忽也可能导致重大的损失。）

2 解析重点2 **We hope that you will forgive us our unintentional mistake and continue to purchase items from us.**

这句话的意思是"希望您能原谅我们无心的失误，并继续购买我们的产品"。forgive 意为"原谅；谅解"；unintentional 意为"无心的；无意的"；purchase items from us 意为"从我们这里购买产品"，也可以理解为"继续与我们合作"或"继续支持我们"。所以这个句子也可以这样表达：We hope that you can understand the whole situation and continue to cooperate with us.

1. **I am so sorry for delaying the goods you ordered.**
   对于您订购的货物所发生的延误，我感到非常抱歉。
2. **We will certainly cooperate with a reliable *freight* [9] company.**
   我们日后一定会与可靠的货运公司合作。

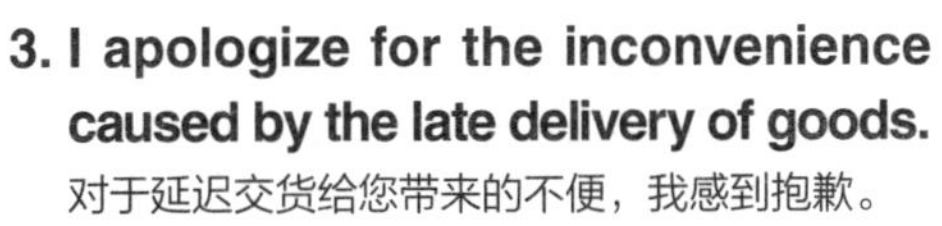

3. **I apologize for the inconvenience caused by the late delivery of goods.**
   对于延迟交货给您带来的不便，我感到抱歉。
4. **Please accept my sincere apology for our carelessness during the shipment.**
   对于我们运输中的疏忽，请接受我们真诚的道歉。
5. **I do apologize for the late delivery of your products.**
   对于延迟交货，我感到非常抱歉。

必背关键单词

1. ***profound*** [prəˈfaʊnd] adj. 深深的；深切的
2. ***late*** [leɪt] adj. 迟到的；晚的
3. ***delay*** [dɪˈleɪ] n. 延迟；耽搁
4. ***ensure*** [ɪnˈʃʊə(r)] v. 确定；保证；担保
5. ***deadline*** [ˈdedlaɪn] n. 最后期限；截止时间
6. ***forgive*** [fəˈɡɪv] v. 原谅；宽恕
7. ***unintentional*** [ˌʌnɪnˈtenʃənl] adj. 无心的；无意的
8. ***continue*** [kənˈtɪnjuː] v. 继续
9. ***freight*** [freɪt] n. 运输

05 货款滞纳的道歉

From financial@smg.com
Date April 24, 2008
To benjaminbarton@yahoo.com
Subject Apology for Late Payment

Dear Mr. Barton,

We are sorry for our ***tardiness***[1] in the payment of your invoice No.57908135. We were ***embarrassed***[2] to discover that your invoice was ***misplaced***[3] when we ***moved***[4] to another place. I have instructed our ***financial***[5] staff to ***transfer***[6] the full amount to your account at once.
We will make certain all payments are made on time in the ***future***[7]. Please accept our ***sincere***[8] apologies for any inconvenience this may have caused you.

Truly yours,
SMG Co.

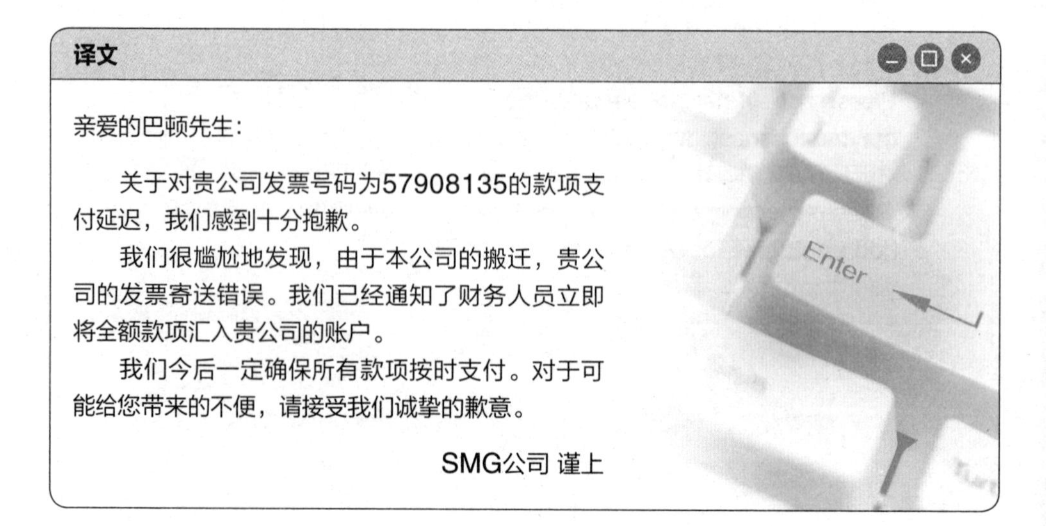

译文

亲爱的巴顿先生：

关于对贵公司发票号码为57908135的款项支付延迟，我们感到十分抱歉。

我们很尴尬地发现，由于本公司的搬迁，贵公司的发票寄送错误。我们已经通知了财务人员立即将全额款项汇入贵公司的账户。

我们今后一定确保所有款项按时支付。对于可能给您带来的不便，请接受我们诚挚的歉意。

SMG公司 谨上

语法重点解析

1 解析重点1 tardiness

tardiness 为名词，意为“缓慢；延迟”，它的形容词形式是 tardy“晚的；迟的”。对于货物的延迟还可以用 delay, lateness 等词。请看以下例句：

Steven was tardy this morning and alleged that his bus was late.（史蒂文今天早上迟到的说辞是公交车误点了。）

What's the cause of the delay?（是什么原因导致延误？）

2 解析重点2 We were embarrassed to discover that your invoice was misplaced when we moved to another place.

这句话的意思是“我们很尴尬地发现，由于本公司的搬迁，贵公司的发票放错了地方”。这个句子还可以这样表达：We felt most embarrassed to find that your invoice was misplaced when we moved to another place.

1. **We feel terribly sorry for the late *payment*[9].**
   对于延迟付款，我们感到非常抱歉。
2. **We're embarrassed to find that your invoice was misplaced.**
   非常尴尬的是我们发现贵公司的发票放错了地方。
3. **I do apologize for the inconvenience caused by the late payment.**
   对于延迟付款给您带来的不便，我感到很抱歉。
4. **Please accept my genuine apology for our oversight.**
   对于我们的失误，我真心诚意地向您道歉。
5. **I do apologize for delaying your payment.**
   对于延迟付款，我感到非常抱歉。
6. **Please accept our pure-hearted apologies for our tardiness in the payment.**
   对于此次延迟付款，请接受我诚挚的歉意。

必背关键单词

1. ***tardiness*** [ˈtɑ:dɪnəs] n. 延迟
2. ***embarrassed*** [ɪmˈbærəst] adj. 尴尬的；感到为难的
3. ***misplace*** [ˌmɪsˈpleɪs] v. 误置；把……放错地方
4. ***move*** [mu:v] v. 搬迁
5. ***financial*** [faɪˈnænʃl] adj. 金融的；财政的；财务的
6. ***transfer*** [trænsˈfɜ:(r)] v. 转移
7. ***future*** [ˈfju:tʃə(r)] n. 将来；未来
8. ***sincere*** [sɪnˈsɪə(r)] adj. 真诚的
9. ***payment*** [ˈpeɪmənt] n. 付款；款项

06 发票错误的道歉

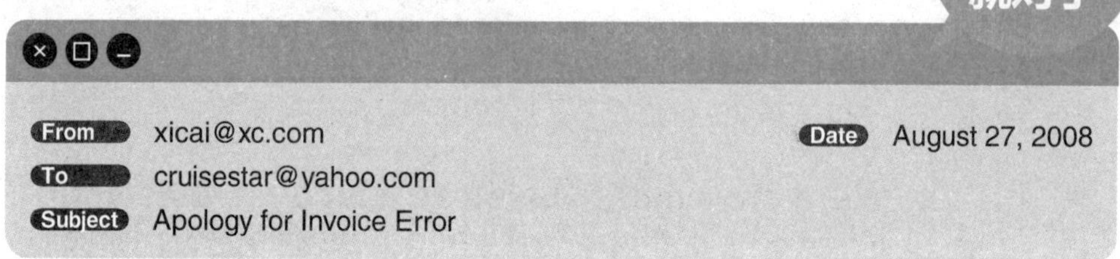

From xicai@xc.com
Date August 27, 2008
To cruisestar@yahoo.com
Subject Apology for Invoice Error

Dear Mr. Cruise,

I am ***extremely***[1] sorry to tell you that, as you ***pointed***[2] out, a mistake has been made on the ***invoice***[3] No. 7561852. I do apologize for any inconvenience it may cause.
We have ***remedied***[4] the ***situation***[5] by issuing a new invoice (No. 7561825) to ***revise***[6] the ***sum***[7] and are sending it by ***express***[8] delivery today.

We have taken ***measures***[9] to ensure that such an error does not happen again.

Sincerely yours,
XC Co.

译文

亲爱的克鲁斯先生：

非常抱歉地告诉您，正如您指出的，编号为7561852的发票出了错，我们为此可能给您带来的不便向您道歉。

我们已经重开了一张发票（编号为7561825）对此进行补救，修改了金额，今天会用快递将发票寄出。

我们已经采取了相关措施以确保此类错误不会再发生。

XC公司 谨上

语法重点解析

1 解析重点1 point out

短语 point out 意为“指出”，同样表达“指出；表明”的短语或单词还有 lay one's finger on, show, indicate 等。请看以下例句：

I can't quite lay my finger on what's wrong with the engine.（我无法确切地说出引擎的问题。）

He indicated his willingness with a nod of his head.（他点头表示愿意。）

2 解析重点2 remedy

remedy 在此做动词用，意为“纠正；补救”。也可以用 correct（改正）、fix（修正）、rectify（矫正）、redress（补救）等来表达此意。请看以下例句：

Your faults of pronunciation can be remedied.（你发音上的毛病是可以纠正的。）

I wish to correct my earlier misstatement.（我想更正我先前的不实之词。）

He do all that he possibly can to redress the wrong.（他尽了一切努力补救错误。）

1. We feel terribly sorry for the invoice error.
对于发票的错误我们感到非常抱歉。

2. The invoice was indeed issued in error.
发票的确开错了。

3. Please accept our sincere apology for the inconvenience that may have caused you.
对于可能给您带来的不便，请接受我们真诚的道歉。

4. We do apologize for issuing the wrong invoice.
对于开错发票一事，我们诚挚地向您道歉。

5. We will reopen the invoice for you at once.
我们会立刻帮您重开发票。

6. We have delivered the right invoice to you by express mail.
我们已经将正确的发票用快递寄给您了。

必背关键单词

1. ***extremely*** [ɪk'stri:mlɪ] adv. 极度；非常
2. ***point*** [pɔɪnt] v. 指出
3. ***invoice*** ['ɪnvɔɪs] n. 发票
4. ***remedy*** ['remədɪ] v. 纠正；补救
5. ***situation*** [ˌsɪtʃʊ'eɪʃn] n. 形势；局面；状况
6. ***revise*** [rɪ'vaɪz] v. 改正；修正
7. ***sum*** [sʌm] n. 金额；总数
8. ***express*** [ɪk'spres] n. 快递
9. ***measure*** ['meʒə(r)] n. 措施

07 汇款延迟的道歉

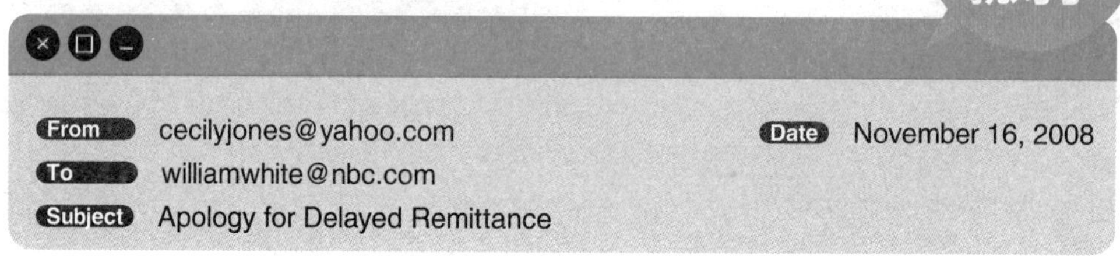

From cecilyjones@yahoo.com
Date November 16, 2008
To williamwhite@nbc.com
Subject Apology for Delayed Remittance

Dear Mr. White,

We are terribly sorry for the late ***remittance*** [1] this time. As the ***wines*** [2] are not yet sold, ***nor*** [3] are they likely to be for some time, we could not pay in time. Anyway, please accept our ***earnest*** [4] apology.
We can ensure that you will ***receive*** [5] the remittance on time in the future.

We ***appreciate*** [6] your ***understanding*** [7].

Yours faithfully,
Cecily Jones

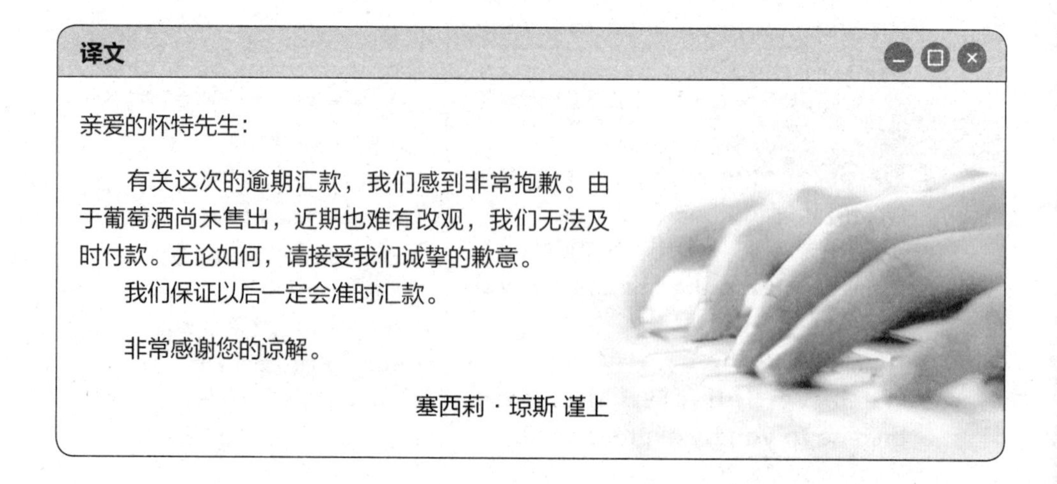

译文

亲爱的怀特先生：

有关这次的逾期汇款，我们感到非常抱歉。由于葡萄酒尚未售出，近期也难有改观，我们无法及时付款。无论如何，请接受我们诚挚的歉意。
我们保证以后一定会准时汇款。

非常感谢您的谅解。

塞西莉 · 琼斯 谨上

语法重点解析

1 解析重点1 **As the wines are not yet sold, nor are they likely to be for some time,...**

这句话的意思是“由于葡萄酒尚未售出，近期也难有改观”。连词 as 意为“由于；因为”，表示原因；nor 意为“也不；也没有”，一般会与 neither 连用，即 neither... nor...，意为“既不……也不……”，但是 nor 也可以单独使用。请看以下例句：

I never like fish, nor eat it.（我讨厌鱼，而且从不吃鱼。）

2 解析重点2 **Anyway,...**

短语 anyway 是“总之；不管怎样；无论如何”的意思。写信者虽然已解释了延迟汇款的原因，但还是应该道歉。表示无论是什么客观原因造成的汇款延迟，自己毕竟是有过错的。多加了这个短语可以表现出写信者道歉的诚意，对方也较易于接受。请看以下例句：

Anyway, I have hurt his self-respect.（不管怎样，我伤害了他的自尊心。）

1. **We feel terribly sorry for the late remittance.**
   对于延迟汇款，我们感到非常抱歉。
2. **It's because the goods have not been sold out yet.**
   是因为货物还没有卖完。
3. **Please accept our sincere *apology*[8] for any trouble it may cause.**
   因此可能给您造成的麻烦，请接受我们真诚的道歉。
4. **We are very sorry for remitting the *balance*[9] so late.**
   非常抱歉，我们这么晚才把余款汇过去。
5. **We will try our best to avoid late remittance from now on.**
   今后我们会尽最大的努力避免汇款延迟。
6. **Thank you very much for your understanding.**
   非常感谢您的理解。
7. **We are sure that such delay won't happen again.**
   我们保证这样的延迟不会再发生。

必背关键单词 

1. ***remittance*** [rɪˈmɪtns] n. 汇款
2. ***wine*** [waɪn] n. 葡萄酒
3. ***nor*** [nɔ:(r)] conj. 也不
4. ***earnest*** [ˈɜ:nɪst] adj. 诚挚的
5. ***receive*** [rɪˈsi:v] v. 收到
6. ***appreciate*** [əˈpri:ʃieɪt] v. 感谢；感激
7. ***understanding*** [ˌʌndəˈstændɪŋ] n. 理解；谅解
8. ***apology*** [əˈpɒlədʒɪ] n. 歉意
9. ***balance*** [ˈbæləns] n. 余额

08 延迟回复的道歉

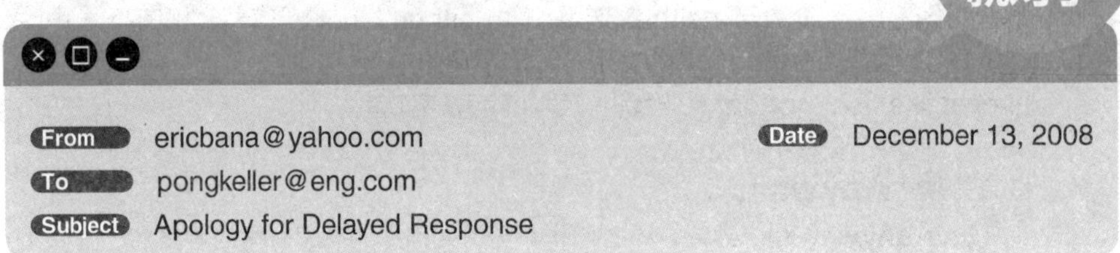

From ericbana@yahoo.com　　Date December 13, 2008

To pongkeller@eng.com

Subject Apology for Delayed Response

Dear Mr. Keller,

Please ***excuse***[1] me for **getting back to you** so late.
I *failed*[2] to *respond*[3] to your e-mails in time on account of *awfully*[4] busy work. I am very sorry for that.
I am very happy to accept your ***proposal***[5] and look forward to meeting with you when you come to New York and ***discussing***[6] about the ***related***[7] ***details***[8].

Best regards,
Eric Bana

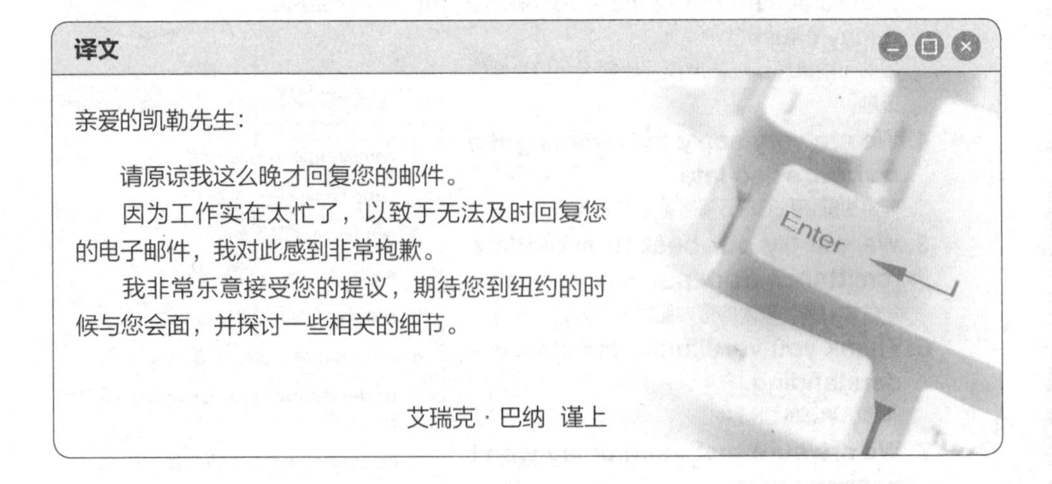

译文

亲爱的凯勒先生：

请原谅我这么晚才回复您的邮件。

因为工作实在太忙了，以致于无法及时回复您的电子邮件，我对此感到非常抱歉。

我非常乐意接受您的提议，期待您到纽约的时候与您会面，并探讨一些相关的细节。

艾瑞克 · 巴纳　谨上

语法重点解析

1 解析重点1 **get back to you**

get back to 意为“回复”。在英文书信写作中还可以用 respond, reply, write back 来表示“回复”的意思。请看以下例句：

How should we respond to this letter?（我们要怎么回复这封信呢？）

I'll write back to him tomorrow.（我明天会回信给他。）

2 解析重点2 **I failed to respond to your e-mails in time on account of awfully busy work.**

这句话的意思是“因为工作实在太忙了，以致于无法及时回复您的电子邮件”。短语 fail to do sth 表示“未能做某事”，相当于 not succeed in doing sth（未能成功地做某事）。短语 on account of 意为“由于”，表示原因，也可以替换为 owing to, because of 等。副词 awfully 意为“非常地；极端地”，用于修饰形容词 busy，表达繁忙程度非常严重。

1. **I feel terribly sorry for the late reply.**
   对于这么迟才回复您，我感到非常抱歉。
2. **Due to the busy work, I have little time to respond.**
   由于工作繁忙，我实在没时间回复。
3. **Please pardon me for not getting back to you earlier.**
   没能早点回复您的信件，敬请原谅。
4. **I am very sorry for replying so late.**
   这么晚才给您回信，我感到非常抱歉。
5. **I am pleased to accept your suggestion mentioned in your letter.**
   我非常乐意接受您在信中提到的建议。
6. **I look forward to meeting with you soon.**
   我期待尽快与您会面。
7. **I have been extremely busy and have fallen behind in my *correspondence*[9].**
   我实在太忙了，以致于无法回复。

必背关键单词

1. ***excuse*** [ɪkˈskjuːs] v. 原谅
2. ***fail*** [feɪl] v. 失败
3. ***respond*** [rɪˈspɒnd] v. 反应；答复
4. ***awfully*** [ˈɔːflɪ] adv. 非常地；极端地
5. ***proposal*** [prəˈpəʊzl] n. 提议
6. ***discuss*** [dɪˈskʌs] v. 讨论；商谈
7. ***related*** [rɪˈleɪtɪd] adj. 有关的
8. ***detail*** [ˈdiːteɪl] n. 细节；详情
9. ***correspondence*** [ˌkɒrəˈspɒndəns] n. 通信；信件

09 忘记取消订单的道歉

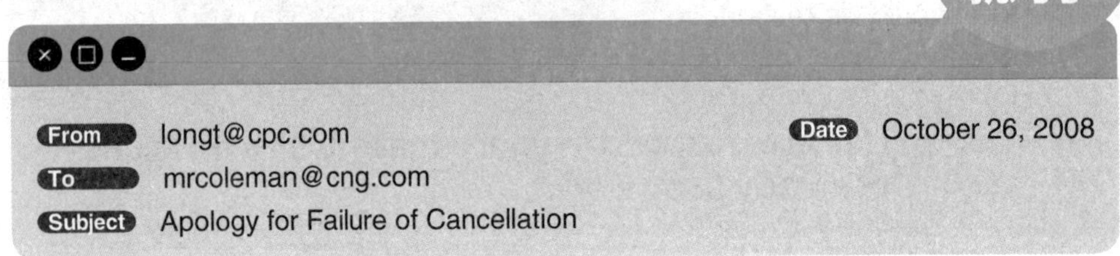

From longt@cpc.com

Date October 26, 2008

To mrcoleman@cng.com

Subject Apology for Failure of Cancellation

Dear Mr. Coleman,

We are very sorry for the ***failure***[1] to ***cancel***[2] your order No. 76541 in time. This was our ***oversight***[3] and we will accept the ***return***[4] of the goods sent to you.

We have ***instructed***[5] the ***department***[6] in charge to be more ***careful***[7] next time and make sure such an error won't happen again.

Please accept our apologies for any inconvenience it caused you.

Yours sincerely,

CP Co.

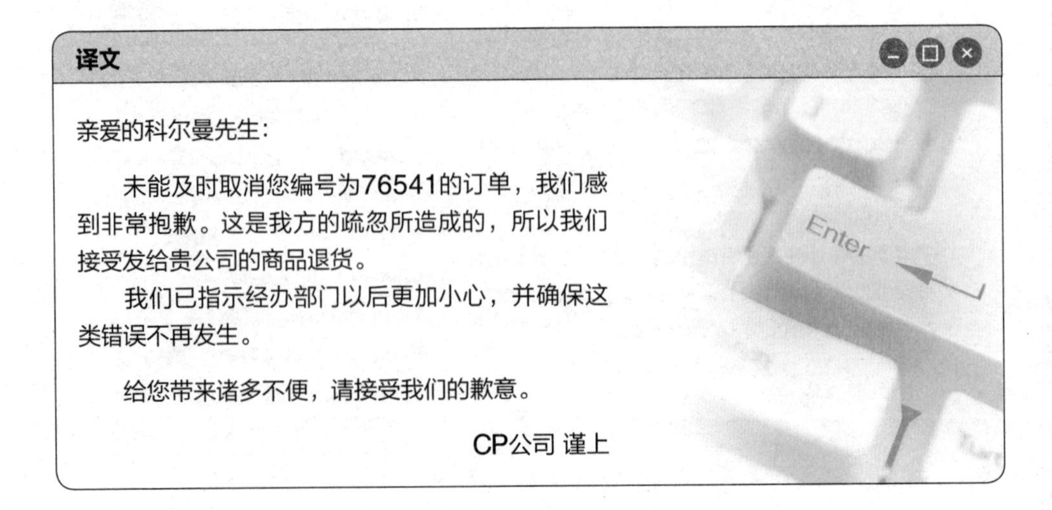

译文

亲爱的科尔曼先生：

未能及时取消您编号为76541的订单，我们感到非常抱歉。这是我方的疏忽所造成的，所以我们接受发给贵公司的商品退货。

我们已指示经办部门以后更加小心，并确保这类错误不再发生。

给您带来诸多不便，请接受我们的歉意。

CP公司 谨上

语法重点解析

1 解析重点1 failure

failure 在这里是指“疏忽；未履行；没做到”。关于 failure 这个单词的几种用法，请看以下例句：

Failure to follow customers' instructions can result in losing business.（未能遵守客户的指示会导致失去商机。）

If you enjoy being lazy, you are doomed to become a failure in this life.（如果你喜欢懒散，你这辈子注定一事无成。）

The movie theater was a failure and closed soon.（那家电影院营运不佳，不久就关门了。）

2 解析重点2 inconvenience

inconvenience 意为“不便；麻烦”。与它同词根的形容词 convenient 意为“方便的；合适的”；名词 convenience 意为“方便；便利”。与 inconvenience 意思相近的词还有 trouble，如 sorry to trouble you（抱歉给您添麻烦了）；sorry to have troubled you（麻烦您了）。请看以下例句：

Please accept our apologies for any inconvenience we have caused.（对于我们造成的不便，敬请原谅。）

He apologized to avoid trouble.（他通过道歉来避免麻烦。）

1. I feel terribly sorry for failing to cancel the order in time.

未能及时取消订单，我感到非常抱歉。

2. It is our fault for not cancelling the order.

未能取消订单是我们的过错。

3. Please forgive us for not cancelling your order.

未能帮您取消订单，敬请原谅。

4. We are *responsible*[8] for the error.

这次的失误是我们的责任。

5. We will take measures to avoid such an error.

我们会采取措施避免此类错误。

必背关键单词

1. ***failure*** [ˈfeɪljə(r)] n. 失败；不成功
2. ***cancel*** [ˈkænsl] v. 取消
3. ***oversight*** [ˈəʊvəsaɪt] n. 纰漏；疏忽
4. ***return*** [rɪˈtɜːn] n. 返还
5. ***instruct*** [ɪnˈstrʌkt] v. 指导；命令
6. ***department*** [dɪˈpɑːtmənt] n. 部门
7. ***careful*** [ˈkeəfl] adj. 仔细的；小心的
8. ***responsible*** [rɪˈspɒnsəbl] adj. 负有责任的；尽责的

10 商品目录错误的道歉

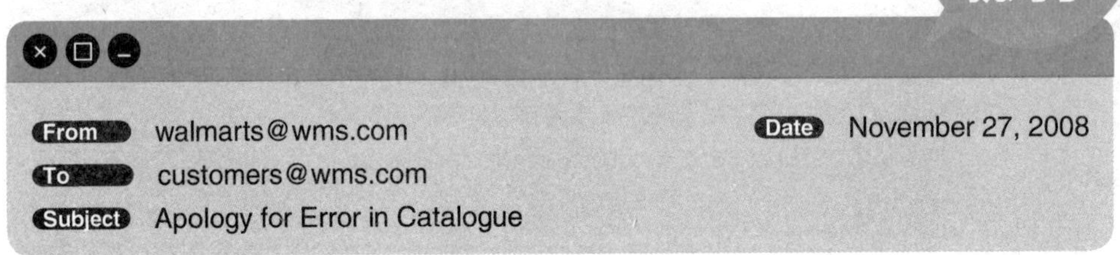
From walmarts@wms.com
Date November 27, 2008
To customers@wms.com
Subject Apology for Error in Catalogue

Dear Customers,

We are greatly sorry to inform you of an error ***found***[1] in the ***description***[2] of one of the ***items***[3] in our ***catalogue***[4].
Please note the *correction*[5] below:
(Page 51)***Cashmere***[6] ***Quilt***[7] (code No. 21-547-7)
This description should read:
Wool Comforter (code No. 21-547-7)

We apologize for this mistake and hope you will continue to be our *faithful*[8] customers.

Yours sincerely,
Wal-Mart Supermarket

译文

亲爱的顾客：

我们发现商品目录中有一件商品的描述有误，对此我们表示深深的歉意。
请注意以下更正部分：
误：第51页　　山羊绒棉被（编号21-547-7）
正：第51页　　羊毛围巾（编号21-547-7）

对于此次错误，我们深感抱歉，也希望您今后能继续做我们忠实的顾客。

沃尔玛超市 谨上

语法重点解析

1 解析重点1 **Please note the correction below.**

这句话的意思是“请注意以下更正部分”。note 意为“注意”，表达“注意”的单词或短语还有：pay attention to, notice, have an eye on 等。correction 为动词 correct 的名词形式，意为“修正；改正”。below 为副词，意为“在……下方”，相当于 as follows。

2 解析重点2 **We apologize for this mistake and hope you will continue to be our faithful customers.**

这句话的意思是“对于此次错误，我们深感抱歉，并希望您今后能继续做我们忠实的顾客”。短语 apologize for sth 意为“为……而道歉；因……而道歉”，相当于 make an apology for sth。faithful customers 意为“忠实的顾客”。

1. **I feel very sorry for the error in the catalogue.**
   对于商品目录中的错误，我感到很抱歉。
2. **A mistake has been found in the description of our product.**
   在我们的商品描述中发现了一个错误。
3. **Please accept our sincere apology.**
   请接受我们真诚的道歉。
4. **We are terribly sorry for the error found in the catalogue.**
   对于在商品目录中发现的错误，我们感到非常抱歉。
5. **We will do our best to *avoid*[9] such an error.**
   我们会尽力避免这样的错误。
6. **We have corrected the error at once as soon as we found it.**
   我们一发现错误就马上改过来了。

必背关键单词

1. ***found*** [faʊnd] v. 发现（find 的过去式和过去分词）
2. ***description*** [dɪˈskrɪpʃn] n. 描述；说明
3. ***item*** [ˈaɪtəm] n. 物品
4. ***catalogue*** [ˈkætəlɒg] n. 目录
5. ***correction*** [kəˈrekʃn] n. 改正
6. ***cashmere*** [ˈkæʃmɪə(r)] n. 山羊绒
7. ***quilt*** [kwɪlt] n. 被子
8. ***faithful*** [ˈfeɪθfl] adj. 忠实的；忠诚的
9. ***avoid*** [əˈvɔɪd] v. 避免

11 延迟开具收据的道歉

这样写就对了

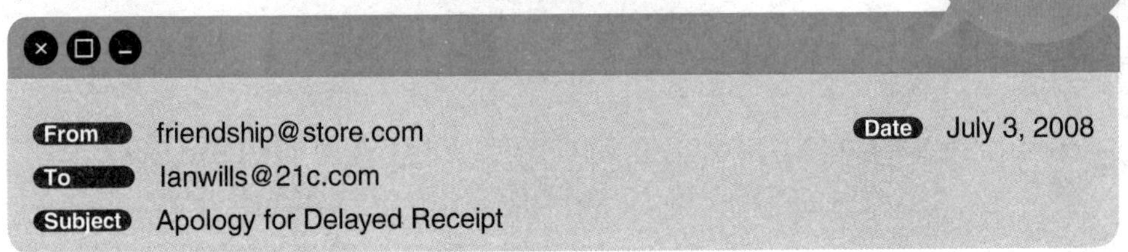
From friendship@store.com
To Ianwills@21c.com
Subject Apology for Delayed Receipt
Date July 3, 2008

Dear Mr. Wills,

I am very sorry for the ***delay***[1] in issuing the ***receipt***[2] for you. As you ***supposed***[3], I was ***overwhelmed***[4] by awfully busy work then so I forgot to issue the receipt.
I have sent you the receipt of your ***purchase***[5] from us by express ***mail***[6] today.
I do apologize for any inconvenience the delay may ***lead***[7] to you.

Thank you very much for your understanding.

With best regards,
Friendship Store

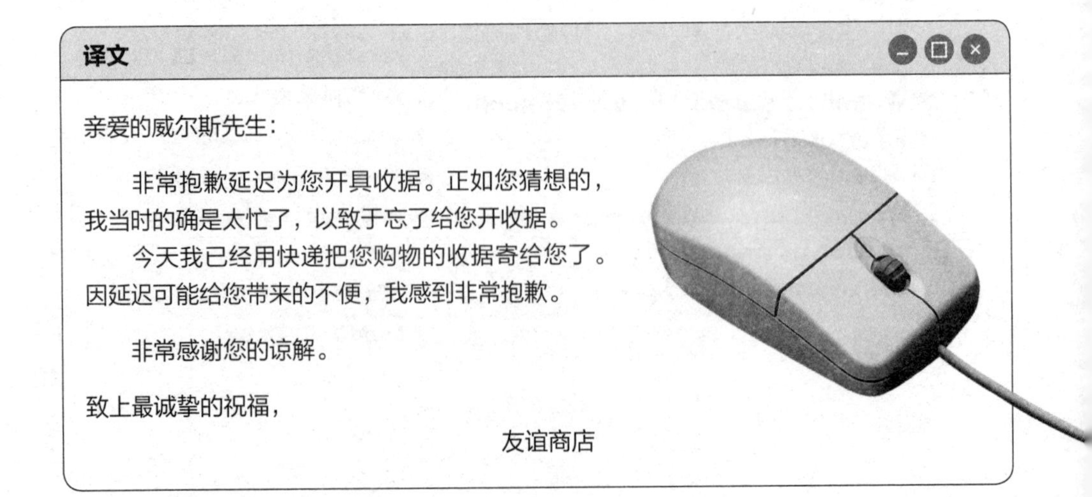
译文

亲爱的威尔斯先生：

非常抱歉延迟为您开具收据。正如您猜想的，我当时的确是太忙了，以致于忘了给您开收据。
今天我已经用快递把您购物的收据寄给您了。因延迟可能给您带来的不便，我感到非常抱歉。

非常感谢您的谅解。

致上最诚挚的祝福，
友谊商店

语法重点解析

1 解析重点1 overwhelm

overwhelm 有“战胜；覆盖；征服；压倒；使不知所措”的意思，在此语境中则可以理解为“被繁忙的工作所淹没；忙得不可开交”。这种表达非常有力度，书信中使用了这个词就会更具有信服力。请看以下例句：

He was overwhelmed by the death of his father.（他因父亲的去世而悲痛至极。）

2 解析重点2 lead to

短语 lead to 在此语境中的意思是“导致；引起”。表达同义的单词或短语还有：bring about（造成；导致）；result in（导致）；cause（造成；引起）等。为了避免重复使用，可以适当选择一些同义的单词或短语来表达。请看以下例句：

Such a mistake would perhaps lead to disastrous consequences.（这样的错误可能导致灾难性的后果。）

Gambling had brought about his ruin.（赌博毁了他。）

The reform resulted in tremendous change in our country.（改革使我们国家发生了巨大变化。）

1. ***Collect***[8] **your receipt, please.**
   请拿好您的收据。
2. **Please take care of your receipt.**
   请保存好您的收据。
3. **Please accept my *sincere*[9] apology.**
   请接受我真诚的道歉。
4. **I will make sure that such an error won't happen again.**
   我们保证此类失误不会再发生。
5. **Wait a minute, I'll make out your receipt at once.**
   请稍等，我立刻给您开收据。
6. **Thank you very much for your understanding.**
   非常感谢您的谅解。

必背关键单词

1. ***delay*** [dɪˈleɪ] n. 延迟
2. ***receipt*** [rɪˈsiːt] n. 收据
3. ***suppose*** [səˈpəʊz] v. 推测
4. ***overwhelm*** [ˌəʊvəˈwelm] v. 压倒；征服；使不知所措
5. ***purchase*** [ˈpɜːtʃəs] n. 购买；购买的物品
6. ***mail*** [meɪl] n. 邮政；邮递
7. ***lead*** [liːd] v. 指引
8. ***collect*** [kəˈlekt] v. 收集
9. ***sincere*** [sɪnˈsɪə(r)] adj. 真诚的

12 商品数量错误的道歉

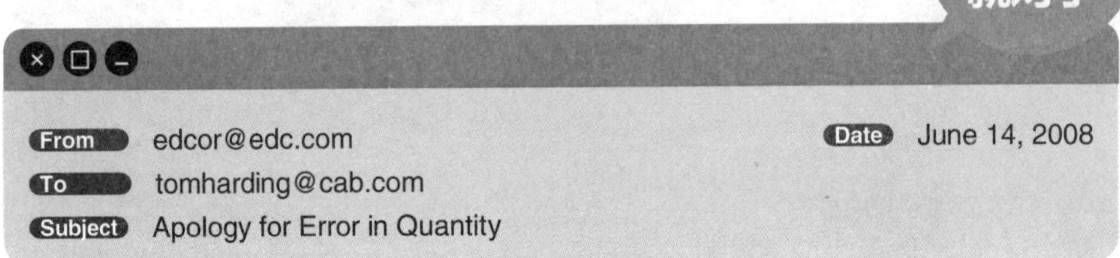

From edcor@edc.com
Date June 14, 2008
To tomharding@cab.com
Subject Apology for Error in Quantity

Dear Mr. Harding,

I am writing this letter ***specially***[1] to apologize to you for the error in ***quantity***[2] of the shoes we sent. We have sent you the rest of the shoes you ordered today. I have ***acknowledged***[3] that my ***fault***[4] has brought you great trouble. I hereby ***express***[5] my deep regret for this matter.
I ***promise***[6] that this mistake will not happen again. And I would ***appreciate***[7] it very much if you could give us a chance to show our ***sincerity***[8] on this matter.

Sincerely yours,
ED Co.

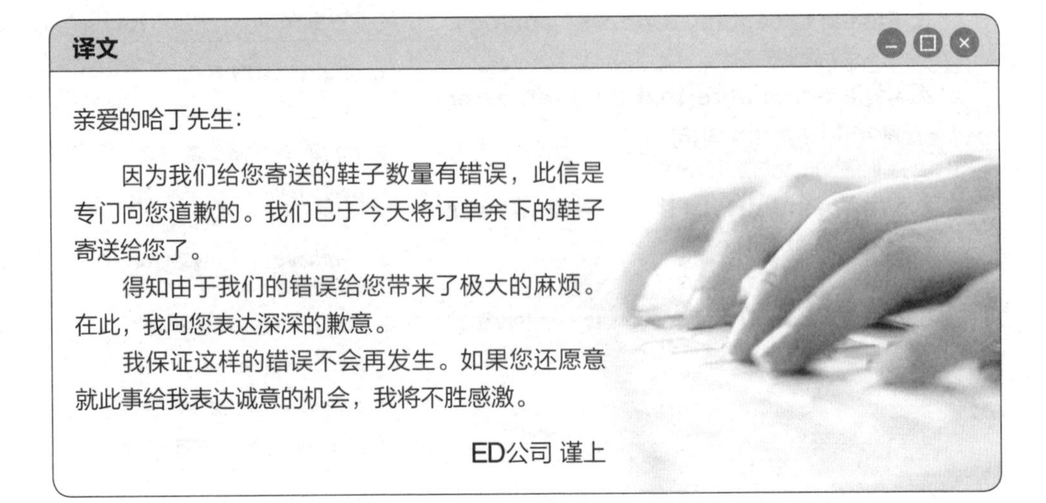

译文

亲爱的哈丁先生：

因为我们给您寄送的鞋子数量有错误，此信是专门向您道歉的。我们已于今天将订单余下的鞋子寄送给您了。

得知由于我们的错误给您带来了极大的麻烦。在此，我向您表达深深的歉意。

我保证这样的错误不会再发生。如果您还愿意就此事给我表达诚意的机会，我将不胜感激。

ED公司 谨上

1 解析重点1 specially

specially 和 especially 两个单词经常容易混淆。especially 意为“尤其；特别”，通常用来对前面所述的事件进一步地说明或补充；specially 意为“专门地；特地”，表示不是为了别的，而是为了某事，强调唯一目的。请对照以下例句：

He likes all subjects, especially English.（他喜欢所有的学科，尤其是英语。）

I specially made this cake for your birthday.（我为你的生日特别制作了这个蛋糕。）

2 解析重点2 I promise that this mistake will not happen again.

这句话的意思是“我保证这样的错误不会再发生”。promise 表示“承诺；允诺”，既可做动词，也可做名词。在本句中 promise 还可以替换成 assure, make sure, be sure, make certain 等。请看以下例句：

We are sure that such delay won't happen again.（我们保证这样的延迟不会再发生。）

I make certain that such an error won't happen again.（我保证这样的失误不会再发生。）

1. **I would like to express my apologies for the error in quantity.**
   对于商品数量上的错误，我感到非常的抱歉。
2. **We will send you the rest of your goods right away.**
   我们会马上把剩下的货物寄给您。
3. **Please accept our sincere apology for our carelessness.**
   对于我们的疏忽，请接受我们真诚的道歉。
4. **We will make certain that such an error won't happen again.**
   我们保证此类失误不会再发生。
5. **We would appreciate it if you could give us another chance.**
   如果您能再给我们一次机会，我们将不胜感激。

必背关键单词

1. ***specially*** [ˈspeʃəlɪ] adv. 特别地；专门地；格外地
2. ***quantity*** [ˈkwɒntətɪ] n. 数量
3. ***acknowledge*** [əkˈnɒlɪdʒ] v. 承认；告知收到
4. ***fault*** [fɔːlt] n. 错误；过失
5. ***express*** [ɪkˈspres] v. 表达
6. ***promise*** [ˈprɒmɪs] v. 承诺；允诺
7. ***appreciate*** [əˈpriːʃɪeɪt] v. 感激
8. ***sincerity*** [sɪnˈserətɪ] n. 真诚；诚心诚意

13 发货错误的道歉

这样写就对了

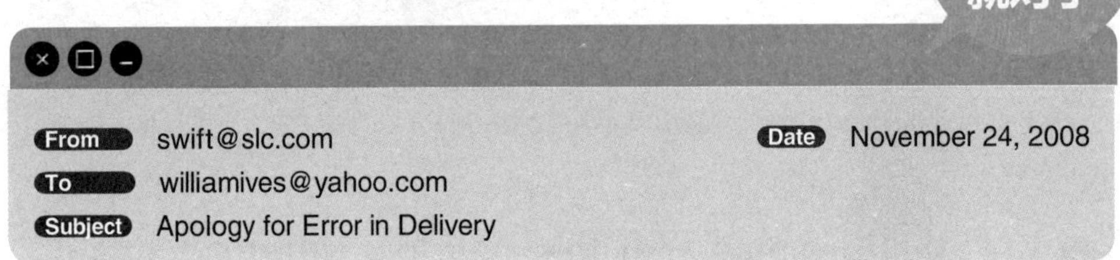

From swift@slc.com
To williamives@yahoo.com
Subject Apology for Error in Delivery
Date November 24, 2008

Dear Mr. Ives,

Please accept our ***genuine***[1] apologies for the error in ***delivery***[2]. We feel terribly sorry for delivering your goods to the ***wrong***[3] place.
We have sent your goods to the ***correct***[4] ***address***[5] today. We guess that it must have brought you great trouble because of our fault. **We hereby express our deep apologies for this error.**
I hope you could ***forgive***[6] us for what we did this time and **we can *guarantee*[7] that we will *avoid*[8] making such a mistake again**.

Sincerely yours,
Swift Logistics Company

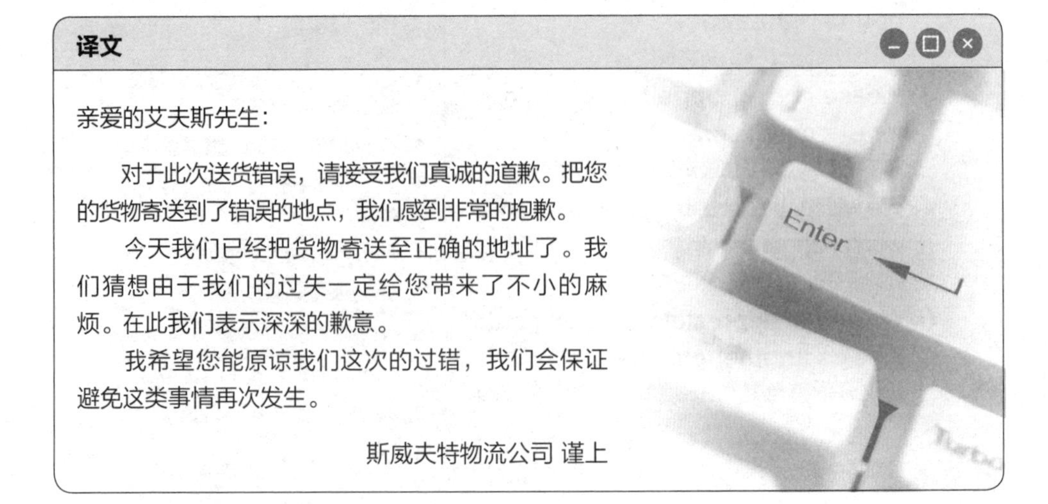

译文

亲爱的艾夫斯先生：

对于此次送货错误，请接受我们真诚的道歉。把您的货物寄送到了错误的地点，我们感到非常的抱歉。

今天我们已经把货物寄送至正确的地址了。我们猜想由于我们的过失一定给您带来了不小的麻烦。在此我们表示深深的歉意。

我希望您能原谅我们这次的过错，我们会保证避免这类事情再次发生。

斯威夫特物流公司 谨上

语法重点解析

1 解析重点1 We hereby express our deep apologies for this error.

这句话的意思是“在此我们表示深深的歉意”。hereby 为副词，意为“在此；特此；以此方式”，多用在正式的文件或声明中以示庄重。写信者在此使用这个词，说明是非常正式的道歉。请看以下例句：

We hereby revoke the agreement of January 1st 1982.（我们特此宣告1982年1月1日的协议无效。）

I hereby extend my hearty apologies to all of them.（在这里向所有人表示诚挚的歉意。）

2 解析重点2 We can guarantee that we will avoid making such a mistake again.

这句话的意思是“我们保证避免再次发生这类事情”。guarantee 有动词和名词两种词性，在此语境中则做动词意为“保证”。还可以用 assure, be sure, make sure 等单词或词组替换 guarantee。avoid doing sth 意为“避免做某事”。请看以下例句：

I will make guarantee to prove every statement I made.（我将保证证实我的每一项声明。）

We must try to avoid repeating these errors.（我们要避免重犯这些错误。）

高频例句

1. **I would like to express my apologies for the error in delivery.**
   对于发货错误，我感到非常的抱歉。
2. **We will send the goods to the right place at once.**
   我们会马上把货物送到正确的地方。
3. **Please accept our pure-hearted apology for our *fault*[9].**
   对于我们的过错，请接受我们诚恳的道歉。
4. **I will make certain that such an error won't happen again.**
   我保证此类失误不会再发生。
5. **We will send your goods to the correct address at once.**
   我们马上把贵公司的货物送至正确的地址。

必背关键单词

1. ***genuine*** [ˈdʒenjʊɪn] adj. 真诚的；真正的
2. ***delivery*** [dɪˈlɪvərɪ] n. 递送；发送
3. ***wrong*** [rɒŋ] adj. 错误的
4. ***correct*** [kəˈrekt] adj. 正确的
5. ***address*** [əˈdres] n. 地址
6. ***forgive*** [fəˈgɪv] v. 原谅；饶恕
7. ***guarantee*** [ˌgærənˈtiː] v. 保证；担保
8. ***avoid*** [əˈvɔɪd] v. 避免
9. ***fault*** [fɔːlt] n. 过错

14 汇款金额不足的道歉

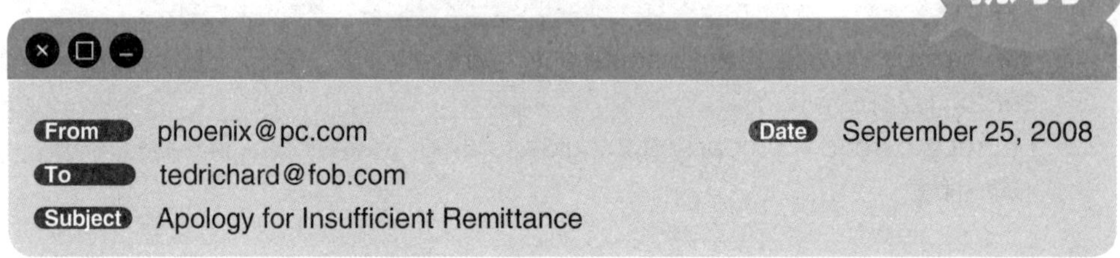

From phoenix@pc.com
Date September 25, 2008
To tedrichard@fob.com
Subject Apology for Insufficient Remittance

Dear Mr. Richard,

I am so sorry about the ***insufficient***[1] remittance I made. **I am afraid I *neglected*[2] to *include*[3] the *charges*[4] for *installation*[5] in the *total*[6] amount.**

I will send the balance by ***telegraphic***[7] transfer tomorrow, so you will receive it within three to five days.

Thank you very much for ***reminding***[8] **me of the error**.

Yours faithfully,
Phoenix Co.

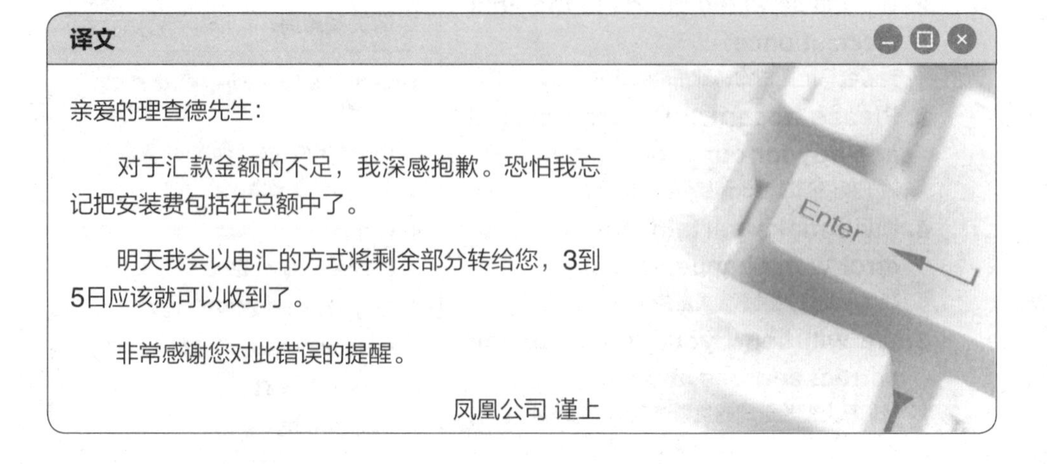

译文

亲爱的理查德先生：

对于汇款金额的不足，我深感抱歉。恐怕我忘记把安装费包括在总额中了。

明天我会以电汇的方式将剩余部分转给您，3到5日应该就可以收到了。

非常感谢您对此错误的提醒。

凤凰公司 谨上

语法重点解析

1 解析重点1 I am afraid I neglected to include the charges for installation in the total amount.

这句话的意思是“恐怕我忘记把安装费包括在总额中了”。

neglect 意为“疏忽；忽略”，在此句中还可以替换成 forgot。

charges for installation 意为“安装费”，表达“安装费”还可以说 installation fee, cost of installation, installation expenses 等。所以这句话还可以这样说：

I am afraid I forgot to include the installation fee in the total amount.

2 解析重点2 remind me of the error

短语 remind sb of sth 的意思是“就某事提醒某人；使某人想起某事”，还可以用 remind sb about sth 表示。请看以下例句：

You remind me of an old English lady.（你让我想起一位英国老妇人。）

I came to remind you about the meeting tomorrow.（我来是想提醒您明天的会议。）

1. **I would like to express my apologies for remitting insufficient funds.**
   对于汇款金额的不足，我感到非常抱歉。
2. **We will send the balance at once.**
   我们会马上把余额寄出。
3. **We will send the balance to you immediately.**
   我们会立刻把余额寄给您。
4. **Please accept our sincere apology for our negligence.**
   对于我们的疏忽，请接受我们真诚的道歉。
5. **We are sure such an error won't happen again.**
   我们保证此类过失不会再发生。
6. **Thank you very much for reminding me of the matter.**
   非常感谢您对此事的提醒。
7. **We *promise* [9] that the mistake will never occur again.**
   我们保证这样的错误绝对不会再发生。

必背关键单词

1. ***insufficient*** [ˌɪnsəˈfɪʃnt] adj. 不足的
2. ***neglect*** [nɪˈglekt] v. 疏忽；忽视
3. ***include*** [ɪnˈklu:d] v. 包含；包括
4. ***charge*** [tʃɑ:dʒ] n. 费用
5. ***installation*** [ˌɪnstəˈleɪʃn] n. 安装
6. ***total*** [ˈtəʊtl] adj. 全体的；总的
7. ***telegraphic*** [ˌtelɪˈgræfɪk] adj. 电报的；电信的
8. ***remind*** [rɪˈmaɪnd] v. 提醒
9. ***promise*** [ˈprɒmɪs] v. 保证；承诺

15 意外违反合同的道歉

这样写就对了

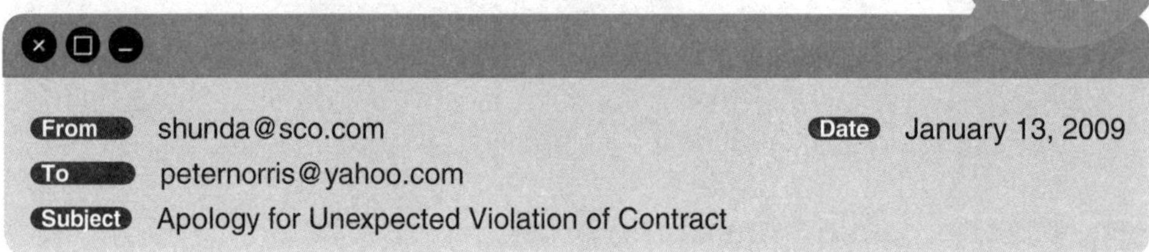
From shunda@sco.com
Date January 13, 2009
To peternorris@yahoo.com
Subject Apology for Unexpected Violation of Contract

Dear Mr. Norris,

We apologize for failure to deliver your goods at ***scheduled***[1] time in ***accordance***[2] with the ***contract***[3] we reached.

Unfortunately, owing to ***excessive***[4] demand last month, we were unable to fill all orders on time. However, we will ***compensate***[5] for all your ***economic***[6] ***loss***[7] due to the unexpected violation of contract.

Thanks very much for your understanding.

Sincerely yours,
Shunda Company

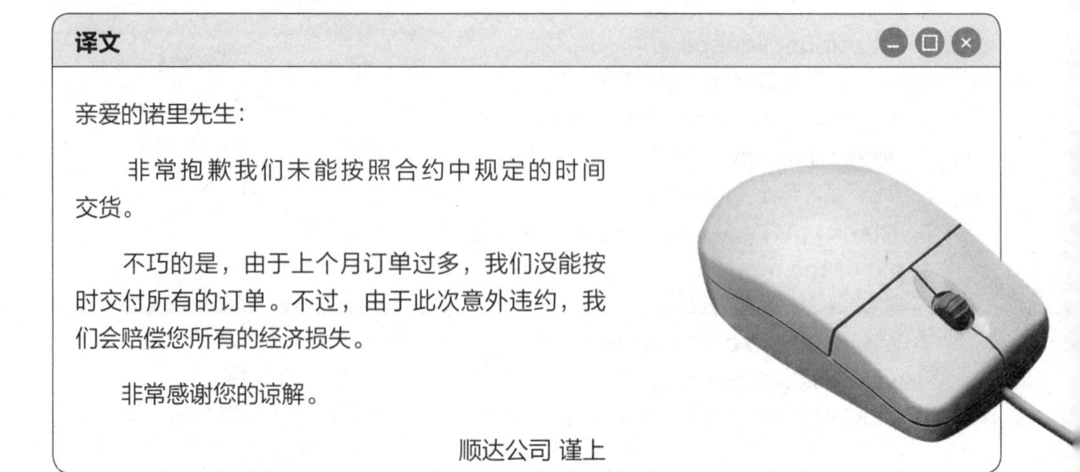
译文

亲爱的诺里先生：

非常抱歉我们未能按照合约中规定的时间交货。

不巧的是，由于上个月订单过多，我们没能按时交付所有的订单。不过，由于此次意外违约，我们会赔偿您所有的经济损失。

非常感谢您的谅解。

顺达公司 谨上

语法重点解析

1 解析重点1 in accordance with

短语 in accordance with 意为“与……一致；依照……”，一般在日常英语中使用频率不高，常用于非常正式的文件、声明或法律法规中，相当于短语 according to（按照；根据）。类似的表达还有：on the basis of（以……为基础；根据），in line with（跟……一致；符合），be based on（以……为基础；根据）等。请看以下例句：

Everything has been done in accordance with the rules.（所有这一切均是依据规定执行的。）

Our trade is conducted on the basis of equality.（我们是在平等的基础上进行贸易的。）

2 解析重点2 the unexpected violation of contract

the unexpected violation of contract 意思是“此次意外违约”。unexpected 做形容词意为“出于意料的；想不到的；意外的”。写信者用这个词是想表明其违反合约并非故意，而是有客观原因的。violation 为动词 violate 的名词形式，一般表示违反合同、公约、法律、规则等。常用的短语是 in violation of。请看以下例句：

You are in violation of tax regulations.（你触犯了税法。）

高频例句

1. I would like to apologize for the lengthy delay in shipping your order.

对于订单发货耽误这么长时间，我向您致以歉意。

2. We are doing everything we can to ensure that your order is shipped without further delay.

我们正在尽一切努力确保你方订货及时装运。

3. Please accept our sincere apology for our *violation*[8] of the contract.

对于我方违约一事，请接受我们真诚的歉意。

4. We will make certain that you will receive your shipment by next Tuesday at the latest.

我们能保证最迟下个星期二您就会收到货物。

必背关键单词

1. ***scheduled*** [ˈʃedjuːld] adj. 预定的
2. ***accordance*** [əˈkɔːdns] n. 符合；一致
3. ***contract*** [ˈkɒntrækt] n. 合约
4. ***excessive*** [ɪkˈsesɪv] adj. 过多的
5. ***compensate*** [ˈkɒmpenseɪt] v. 补偿
6. ***economic*** [ˌiːkəˈnɒmɪk] adj. 经济的
7. ***loss*** [lɒs] n. 损失
8. ***violation*** [ˌvaɪəˈleɪʃn] n. 违反

Unit 14 恭贺 Congratulations

01 恭贺添丁

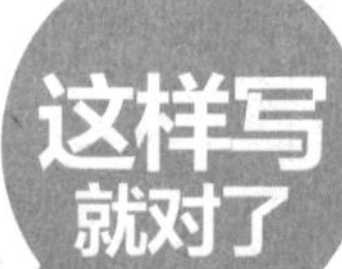

From tomclark@yahoo.com
Date August 11, 2008
To brownde@yahoo.com
Subject Congratulations on Your Baby Boy!

Dear Mr. Brown,

Please accept our ***wholehearted***[1] congratulations on the ***safe***[2] ***arrival***[3] of your **baby boy**. We are sure you'll make a wonderful father.
We believe that the mother and the baby are both very ***healthy***[4]. We can hardly wait to see the baby.
We *pray*[5] that the baby will grow up to be a good citizen of the country and bring ***honor***[6] and ***glory***[7] to the family and ***motherland***[8].

Affectionately yours,
Mr. and Mrs. Tom Clark

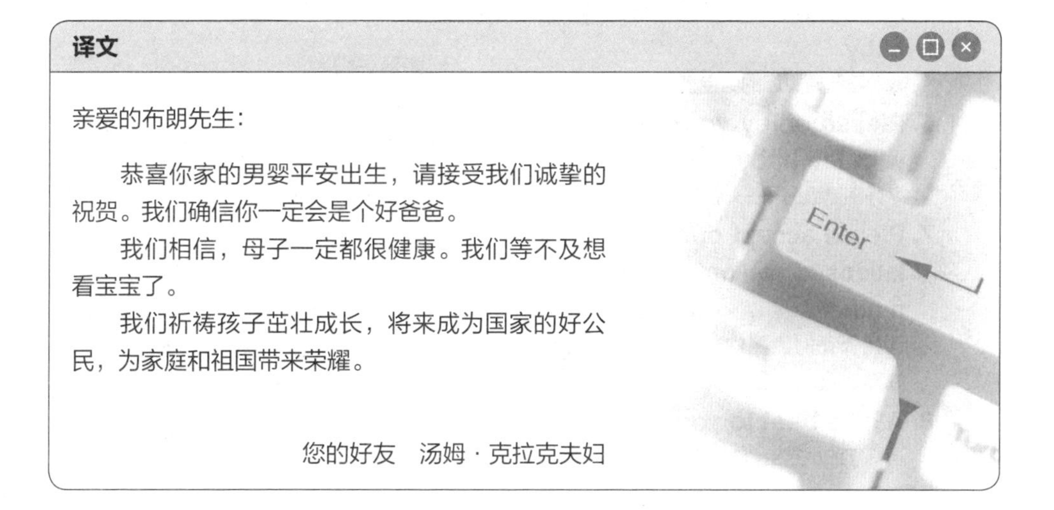
译文

亲爱的布朗先生：

恭喜你家的男婴平安出生，请接受我们诚挚的祝贺。我们确信你一定会是个好爸爸。
我们相信，母子一定都很健康。我们等不及想看宝宝了。
我们祈祷孩子茁壮成长，将来成为国家的好公民，为家庭和祖国带来荣耀。

您的好友　汤姆·克拉克夫妇

语法重点解析

1 解析重点1 baby boy

一般外国人恭贺对方生小孩时，会把孩子的性别写出来，男孩是 baby boy，女孩则为 baby girl。

2 解析重点2 We pray that...

句型 We pray that... 意为“我们祈祷……”，pray 为“祈祷；请求；祈求”的意思。这个句型表达了写信者对新生儿衷心的祝福，也可以换成 We wish that... / We hope that... 请看下面的例句：

I pray that fate may preserve you from all harm.（我祈祷，愿命运保佑你一切平安。）

We hope that you would be happy with Smith.（我们衷心地祝福你和史密斯幸福美满。）

1. **How wonderful to hear that you had a baby girl!**
   我听说您有了个宝贝女儿，这真是太好了！
2. **We are also very pleased to hear of your great news.**
   听到您这个好消息，我们也非常开心。
3. **I am sure that you will be a good mother.**
   我相信您会是一个好妈妈。
4. **Please take care of yourself after labor.**
   产后请照顾好你自己。
5. **I can hardly wait to see your new baby.**
   我等不及要看您的宝宝了。
6. **Please enjoy your every moment with the baby.**
   好好享受跟孩子在一起的每一刻。
7. **Please accept our warm congratulations to you on becoming a young mother.**
   请接受我们热烈的祝贺，祝贺您成为年轻的妈妈。
8. **All the best to your little baby.**
   祝福您的小宝宝幸福平安。

必背关键单词 

1. ***wholehearted*** [ˌhəʊlˈhɑːtɪd] adj. 全心全意的；真挚的
2. ***safe*** [seɪf] adj. 安全的
3. ***arrival*** [əˈraɪvl] n. 抵达；到达
4. ***healthy*** [ˈhelθɪ] adj. 健康的
5. ***pray*** [preɪ] v. 祈祷；请求
6. ***honor*** [ˈɒnə(r)] n. 荣誉；敬意
7. ***glory*** [ˈglɔːrɪ] n. 光荣；壮丽
8. ***motherland*** [ˈmʌðəlænd] n. 祖国

02 恭贺生日

这样写就对了

From lisa@hotmail.com　Date February 26, 2008
To felicity@yahoo.com
Subject Happy Birthday!

Dear Felicity,

You may not like to be ***reminded***[1] that you are a year older today. But that is not going to ***prevent***[2] me from saying "Happy Birthday" to you!
If my ***memory***[3] does not ***fail***[4] me, it is your 26th birthday today. Please accept my best wishes for this ***occasion***[5]. I hope that it is going to be a very happy day and that there will be many happy ***returns***[6].
I'm sending you a ***bouquet***[7] of flowers and a birthday card, which I hope you will enjoy.

Yours affectionately,
Lisa Green

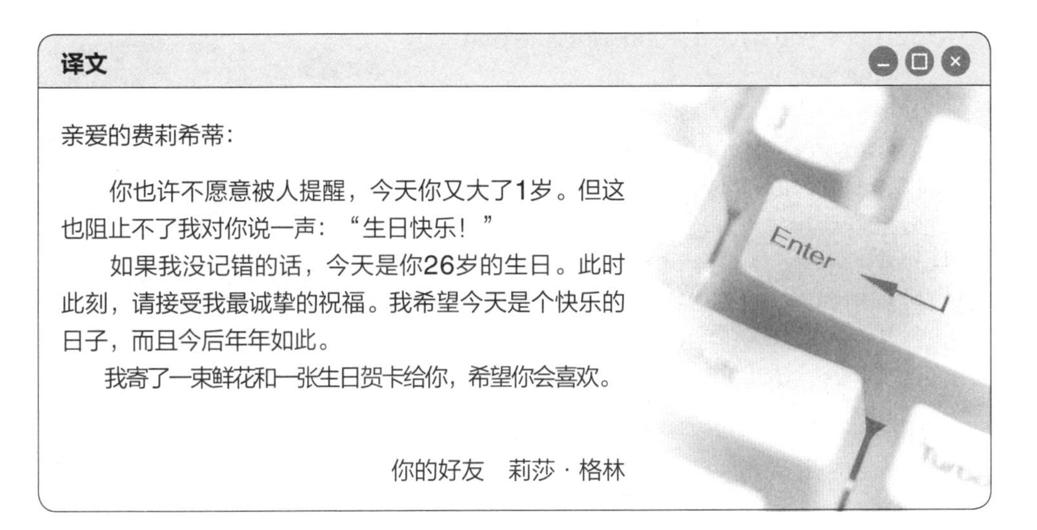

译文

亲爱的费莉希蒂:

你也许不愿意被人提醒，今天你又大了1岁。但这也阻止不了我对你说一声：“生日快乐！”

如果我没记错的话，今天是你26岁的生日。此时此刻，请接受我最诚挚的祝福。我希望今天是个快乐的日子，而且今后年年如此。

我寄了一束鲜花和一张生日贺卡给你，希望你会喜欢。

你的好友　莉莎·格林

语法重点解析

1 解析重点1 **But that is not going to prevent me from saying "Happy Birthday" to you!**

这句话的意思是"但这也阻止不了我对你说一声'生日快乐'"这是一种非常俏皮的表达方式，多用于亲密的朋友之间。

短语 prevent sb from doing sth 意为"阻止某人做某事"，相当于 stop sb from doing sth。请看下面的句子：

They did not prevent him from expressing his views.（他们没有阻止他发表自己的观点。）

Nothing can stop me from carrying out my plan.（没有什么能阻止我执行我的计划。）

2 解析重点2 **If my memory does not fail me,...**

这句话的意思是"如果我没记错的话"，相当于 If I do not remember in error / by error / mistakenly。fail 除了有"失败"的意思，还有"忘记""未做成""使失望"的意思。如 words fail me 的意思是"说不出话来；不能用语言表达出来"。请看下面的句子：

My books are friends that never fail me.（我的书是从不让我失望的朋友。）

1. **Please *accept*[8] my sincere blessing on your birthday.**
   请接受我真诚的生日祝福。
2. **Congratulations on your 24th birthday.**
   祝你24岁生日快乐。
3. **I hope you have a wonderful time with your family.**
   愿您和家人一起度过美好时光。
4. **I am sorry I cannot be there with you to celebrate this special day.**
   不能和你一起庆祝这个特别的日子，我感到很遗憾。
5. **I am sure my thoughts are with you.**
   我确信我的心与你同在。

必背关键单词 

1. ***remind*** [rɪˈmaɪnd] v. 提醒；使想起
2. ***prevent*** [prɪˈvent] v. 阻止；预防
3. ***memory*** [ˈmeməri] n. 记忆；回忆
4. ***fail*** [feɪl] v. 失败；忘记
5. ***occasion*** [əˈkeɪʒn] n. 时刻；时机；特殊场合
6. ***return*** [rɪˈtɜːn] n. 返回；归还
7. ***bouquet*** [bʊˈkeɪ] n. 花束
8. ***accept*** [əkˈsept] v. 接受；承兑

03 恭贺 金榜题名

From steven@163.com　Date July 21, 2008
To wilson@yahoo.com
Subject Congratulations on Admission to University

Dear Wilson,

I was ***extremely***[1] happy to hear the news of your ***admission***[2] to Yale University last week and I am writing to send my wholehearted congratulations to you.
As an old saying goes, "No ***pain***[3], no ***gain***[4]", your success today is closely ***related***[5] with you ***persistent***[6] hard work during the past three years. I hope you will make greater ***achievement***[7] in the university.

Wishing you success and I hope we can keep in close contact.

Sincerely yours,
Steven

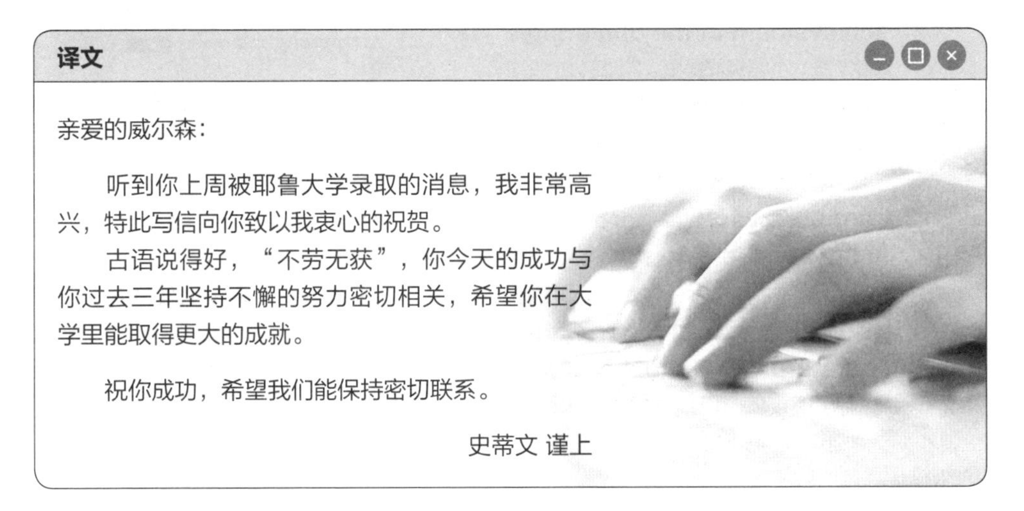
译文

亲爱的威尔森：

听到你上周被耶鲁大学录取的消息，我非常高兴，特此写信向你致以我衷心的祝贺。
古语说得好，“不劳无获”，你今天的成功与你过去三年坚持不懈的努力密切相关，希望你在大学里能取得更大的成就。

祝你成功，希望我们能保持密切联系。

史蒂文 谨上

语法重点解析

1 解析重点1 **no pain, no gain**

“不劳则无获”是一句谚语，也可以翻译成“一分耕耘，一分收获”。类似于Every drop of sweat counts.

2 解析重点2 **...is closely related with...**

短语 be closely related with 的意思是“与……密切相关”。类似的表达还有：be bound up with（与……有密切关系），have something to do with（和……有点关系），be relevant to（与……有关）。它们的区别是相关程度略有差异。请看下面的句子：

Economic progress is closely bound up with educational development.（经济的进步与教育的发展密切相关。）

He had something to do with the British Embassy.（他和英国大使馆有些关系。）

1. **Congratulations on your admission to MIT!**
   祝贺你考上麻省理工学院！
2. **I just heard that you were accepted to the University of Cambridge.**
   我刚刚听说你被剑桥大学录取了。
3. **I hope you will make greater achievement in the university.**
   希望你在大学里能取得更大的成就。
4. **Please *cherish*[8] every minute in the university.**
   请珍惜在大学里的每一分钟。
5. **I am sure you will be outstanding in the college.**
   我相信你在大学里会非常优秀。
6. **Your success is bound up with your hard work in the past.**
   你的成功与你过去的努力是密切相关的。

必背关键单词 

1. ***extremely*** [ɪkˈstriːmlɪ] adv. 极其；非常
2. ***admission*** [ədˈmɪʃn] n. 准许进入；承认，坦白
3. ***pain*** [peɪn] n. 痛苦
4. ***gain*** [geɪn] n. 收益；利润
5. ***relate*** [rɪˈleɪt] v. 联系
6. ***persistent*** [pəˈsɪstənt] adj. 持续的；坚持不懈的
7. ***achievement*** [əˈtʃiːvmənt] n. 成就；成绩
8. ***cherish*** [ˈtʃerɪʃ] v. 珍惜；爱护

04 恭贺获奖

这样写就对了

From terry@hotmail.com　　Date May 25, 2008
To martin@yahoo.com
Subject Congratulations on Winning the Award

Dear Martin,

I couldn't be happier to learn that you have got the ***scholarship*** [1] to Yale University and I think no one could have been more ***deserving*** [2] than you. It's great that your dream to study ***abroad*** [3] has finally come true. Congratulations!
Your hard work and your ***industriousness*** [4] ***combined*** [5] with your gift ***indicate*** [6] a ***splendid*** [7] future. I wish you all the best in your study.

Sincerely yours,
Terry

译文

亲爱的马丁：

　　欣闻你顺利获得耶鲁大学的奖学金，我非常高兴，我觉得没有人比你更有资格了。你出国留学的梦想终于实现了，祝贺你！
　　你的努力、你的勤奋再加上你的天赋，预示着你会拥有辉煌的前程。预祝你学业顺利。

特里 谨上

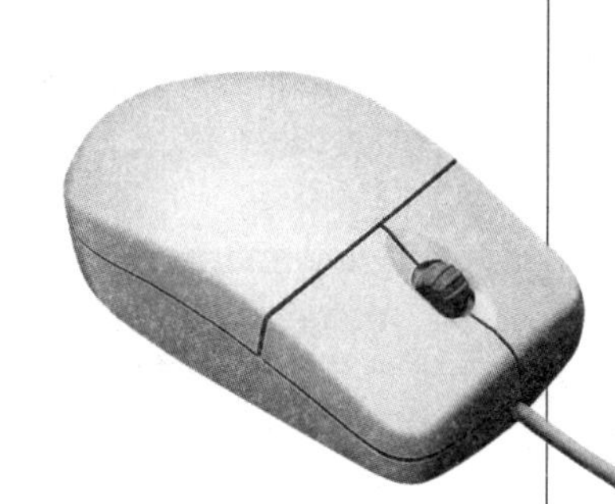

语法重点解析

1 解析重点1 I think no one could have been more deserving than you.

这句话的意思是“我觉得没有人比你更有资格了”。这实际上是一个省略了 that 的宾语从句。当 think 等动词后接的宾语从句为含有 not 的否定句时，该否定词应移至主句；但有些否定词，如 no, never, hardly, few, little, seldom 等，则不必转移。与 think 类似用法的动词还有 believe, suppose, imagine, expect 等。请看下面的句子：

I don't think it will rain tomorrow.（我觉得明天不会下雨。）

I believe my brother has never been late for school.（我相信哥哥上学从来没迟到过。）

2 解析重点2 combine with

短语 combine with 的意思是“与……结合”。表达相同意思的短语还有：be combined with, in combination with 等。请看下面的句子：

We think it is important that theory shall be combined with practice.（我们认为理论联系实际是重要的。）

He carried on the business in combination with his friends.（他与朋友们合伙做生意。）

高频例句

1. **I am so happy to hear that you have been awarded a scholarship to Harvard.**
   我真高兴听说你获得了哈佛大学的奖学金。
2. **Congratulations on winning the big prize!**
   祝贺获得大奖！
3. **You are so lucky.** 你真是个幸运儿。
4. **I'm almost *jealous*[8] of your good fortune.**
   我都要嫉妒你的好运了。
5. **He won the Oscar Award for the best actor.**
   他获得了奥斯卡最佳男演员奖。
6. **The novel earned him a literary award.**
   这篇小说为他赢得了文学奖。

必背关键单词  

1. ***scholarship*** [ˈskɒləʃɪp] n. 奖学金
2. ***deserve*** [dɪˈzɜːv] v. 值得；应得
3. ***abroad*** [əˈbrɔːd] adv. 到国外；在国外
4. ***industriousness*** [ɪnˈdʌstrɪəsnɪs] n. 勤劳；勤奋
5. ***combine*** [kəmˈbaɪn] v. 联合；结合
6. ***indicate*** [ˈɪndɪkeɪt] v. 象征；暗示
7. ***splendid*** [ˈsplendɪd] adj. 壮观的；极好的
8. ***jealous*** [ˈdʒeləs] adj. 嫉妒的

05 恭贺升迁

From thomas@hotmail.com
Date June 27, 2008
To jackgreen@yahoo.com
Subject Congratulations on Your Promotion

Dear Mr. Green,

I am very delighted to have the ***confirmation*** [1] today of your promotion to the office of the Vice-minister of Foreign Affairs. It is an ***outstanding*** [2] achievement in a very ***competitive*** [3] field, and I should like to offer my warmest congratulations.

May I also take this opportunity to thank you for the help which you have always so ***readily*** [4] **given in all** ***circumstances*** [5]. I'm looking forward to working with you in your new ***responsibilities*** [6] as in your earlier ones for the best interests of our two countries.

My colleagues join me in wishing you every happiness and success in the important tasks that lie before you.

Sincerely yours,
Thomas Ray

译文  

亲爱的格林先生：

今天你晋升为外交部副部长的消息得到了证实，我非常高兴。这是在这一充满激烈竞争的领域取得的显著成就，请接受我最热烈的祝贺。

我也想借此机会对您曾经给予我的种种帮助表示感谢。我期待着在您就任新职后同您一如既往地进行有利于我们两国的合作。

我的同事们与我一同祝您幸福，并祝您在重要的岗位上收获成功！

汤马斯 · 雷 谨上

语法重点解析

1 解析重点1 **May I also take this opportunity to thank you for the help which you have always so readily given in all circumstances.**

这句话的意思是“我也想借此机会对您曾经给予我的种种帮助表示感谢”。这个句子里面 help 为先行词，关系代词 which 引导定语从句。May I also take this opportunity... 则是一个非常礼貌客气的句型，意思是“我能借此机会……”，常用于非常正式的场合。请看下面的句子:

May I take this opportunity to welcome all of you and to wish the Congress every success.（我谨借此机会欢迎各位代表，并祝大会取得圆满成功。）

2 解析重点2 **My colleagues join me in wishing you...**

这句话的意思是“我的同事与我一同祝您……”。英文书信的祝福语经常会遇到这样的句型，用来表示多个人一块向某人送祝福。短语 join in 意为“参加；加入”。请看下面的句子:

My wife joins me in congratulating you on your promotion!（我的妻子和我一同祝贺你升职！）

1. **Congratulations on your *promotion*[7]!**
   祝贺你升职！
2. **You definitely deserve it.**
   这绝对是你应该得到的。
3. **I wish you every success in your new position.**
   祝你在新的岗位上大获全胜。
4. **Please accept my warmest congratulations.**
   请接受我最热烈的祝贺。
5. **I was extremely pleased to learn of your promotion.**
   听说你高升了，我非常高兴。

必背关键单词

1. ***confirmation*** [ˌkɒnfəˈmeɪʃn] n. 证实；确认
2. ***outstanding*** [aʊtˈstændɪŋ] adj. 杰出的；地位显著的
3. ***competitive*** [kəmˈpetətɪv] adj. 竞争的；有竞争力的
4. ***readily*** [ˈredɪlɪ] adv. 乐意地；欣然地
5. ***circumstance*** [ˈsɜːkəmstəns] n. 条件；环境
6. ***responsibility*** [rɪˌspɒnsəˈbɪlətɪ] n. 责任；职责
7. ***promotion*** [prəˈməʊʃn] n. 晋升；推销

06 恭贺新婚

这样写就对了

From thomas@hotmail.com　Date August 18, 2008
To catherinej@yahoo.com
Subject Congratulations on Your Marriage

Dear Catherine,

I am ***thrilled***[1] and ***delighted***[2] to receive the ***announcement***[3] of your ***marriage***[4] last night.
My wife joins me hereby in expressing our sincere congratulations and send our best wishes for every happiness that life can ***bring***[5].

We wish you a wonderful ***honeymoon***[6]!

Affectionately[7] yours,
Thomas Ray

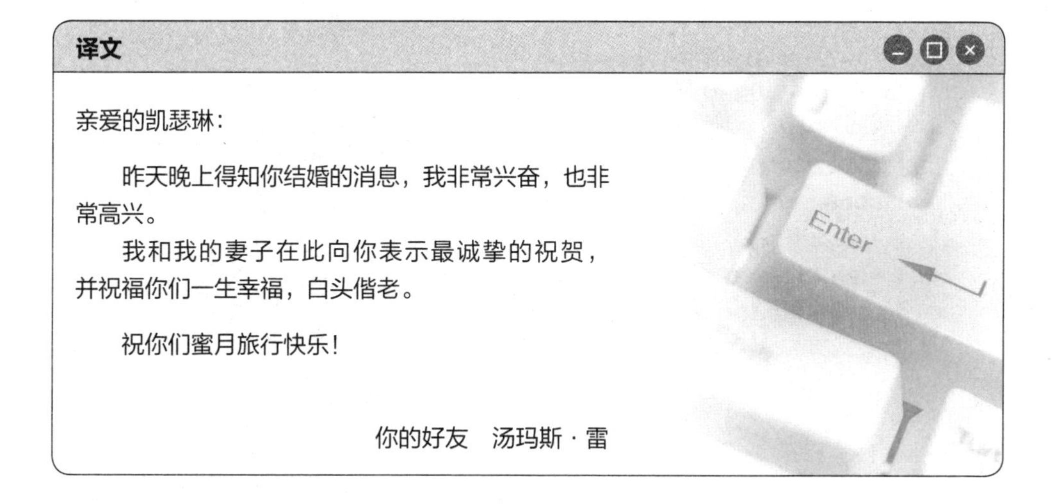

译文

亲爱的凯瑟琳:

昨天晚上得知你结婚的消息，我非常兴奋，也非常高兴。
我和我的妻子在此向你表示最诚挚的祝贺，并祝福你们一生幸福，白头偕老。

祝你们蜜月旅行快乐!

你的好友　汤玛斯 · 雷

语法重点解析

1 解析重点1 **thrilled**

thrill 这个词做动词意为“使兴奋；使激动”，做名词意为“兴奋；一阵强烈的感觉”。thrilled 则是形容词，表示“非常兴奋的”，相当于 excited（兴奋的；激动的）。所以在英文书信写作中表达“兴奋；激动；高兴”等意时，除了用 happy, pleased, glad, delighted 之外，还可以用 thrilled 或者 excited。请看下面的句子：

He was thrilled at the good news.（那个好消息使他兴奋极了。）

Are you excited about graduating from high school?（你高中毕业了，会不会感到很兴奋呢？）

2 解析重点2 **We wish you a wonderful honeymoon!**

这句话的意思是“祝你们蜜月旅行快乐”。honeymoon 意为“蜜月”，是个合成词，honey 本意为“蜂蜜；甜蜜”，moon 为“月亮”，合在一起翻译成“蜜月”。类似的合成词还有 bookmark（书签），football（足球）等。

1. **Congratulations on your wedding!**
   祝贺你们新婚快乐！
2. **You two are meant for each other.**
   你们俩是天生的一对。
3. **Please accept my wholehearted congratulations on your marriage!**
   请接受我对你们婚姻最衷心的祝福！
4. **I am looking forward to attending your wedding in May.**
   我期盼着5月参加你们的婚礼。
5. **Please accept my warmest congratulations.**
   请接受我最热烈的祝贺。
6. **I wish the two of you a happy and healthy life together.**
   我祝你们幸福健康、白头偕老。

必背关键单词 

1. ***thrilled*** [θrɪld] adj. 非常兴奋的；极为激动的
2. ***delighted*** [dɪˈlaɪtɪd] adj. 高兴的；欣喜的
3. ***announcement*** [əˈnaʊnsmənt] n. 公告；宣告
4. ***marriage*** [ˈmærɪdʒ] n. 结婚；婚姻
5. ***bring*** [brɪŋ] v. 带来；造成
6. ***honeymoon*** [ˈhʌnimuːn] n. 蜜月
7. ***affectionately*** [əˈfekʃənətli] adv. 热情地；挚爱地

07 恭贺乔迁

这样写就对了

From jim@hotmail.com　Date September 18, 2008
To lisa@yahoo.com
Subject Congratulations on Moving to a New House

Dear Lisa,

It's so great that you ***finally***[1] have your own house with a beautiful ***view***[2] !
And thanks for telling me your new home ***address***[3].
Congratulations on your ***move***[4] to the new house next week and it would be my great ***honor***[5] to visit your new home ***someday***[6].
I have sent a bouquet of flowers to you and hope you will like it. Best wishes to you and all your family members!

Sincerely yours,
Jim

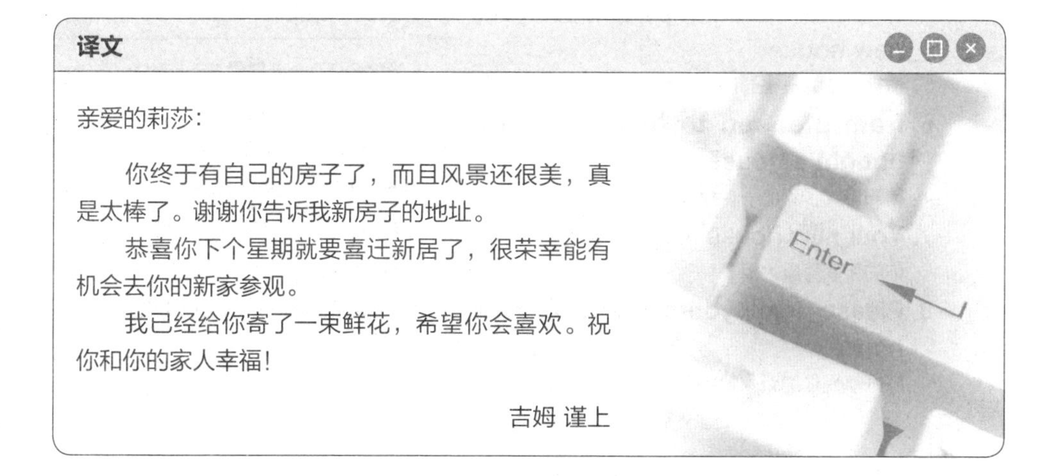

译文

亲爱的莉莎：

你终于有自己的房子了，而且风景还很美，真是太棒了。谢谢你告诉我新房子的地址。

恭喜你下个星期就要喜迁新居了，很荣幸能有机会去你的新家参观。

我已经给你寄了一束鲜花，希望你会喜欢。祝你和你的家人幸福！

吉姆 谨上

语法重点解析

1 解析重点1 **It's so great that...**

It's so great that... 这个句型的意思是"……真是太好了"。在这个句型中 it 为形式主语。为了防止句子头重脚轻，通常把形式主语 it 放在主语位置，而把真正主语搁置于后。所以此书信中"It's so great that you finally have your own house"这句话真正的主语是 you finally have your own house。请看下面的句子：

It is a pity that we won't be able to go to the south to spend our summer vacation.（不能去南方过暑假真是太可惜了。）

2 解析重点2 **someday**

someday 是"有朝一日，将来有一天"的意思，常与 one day 互换使用；而 some time 则是"在未来的某时；一段时间"的意思，小心不要搞混了。

1. Thank you for telling me your new home address.
谢谢你告诉我你的新家地址。

2. You finally have your own house in Shanghai.
你终于在上海有了自己的房子。

3. Congratulations on a *well-situated*[7] house.
恭喜你有了地段这么好的房子。

4. We hope to visit your new house someday.
我们希望有朝一日去参观你的新家。

5. Please accept one of my paintings as an ornamental picture for your new house.
请接受我的一幅画作为您新家的装饰画吧。

6. I am pleased to hear that you bought a house.
得知你买了套房子，我很高兴。

7. Your new house is fairly wonderful!
你的新家简直太棒了！

8. Please enjoy yourself in your new house.
好好享受你的新家吧。

必背关键单词 

1. ***finally*** [ˈfaɪnəlɪ] adv. 终于；最终
2. ***view*** [vjuː] n. 景色；视野
3. ***address*** [əˈdres] n. 地址；演说
4. ***move*** [muːv] n. 移动；搬家
5. ***honor*** [ˈɒnə(r)] n. 荣誉；敬意
6. ***someday*** [ˈsʌmdeɪ] adv. 有一天
7. ***well-situated*** [welˈsɪtʃʊeɪtɪd] adj. 地理位置佳的

08 恭贺生意兴隆

这样写就对了

From edmond@hotmail.com
To marco@yahoo.com
Subject Congratulations on Your Prosperous Business
Date October 10, 2008

Dear Marco,

I learned from today's newspaper that you have set up your own ***private***[1] company, which I think ***resulted***[2] from many years of your hard work and ***experience***[3] in ***managing***[4] foreign trade.
Please accept my warmest congratulations. I do hope your company enjoys a ***smooth***[5] ***development***[6] and you yourself will find the happiness and good luck in this new ***venture***[7].

Wishing you a flourishing business!

Yours faithfully,
Edmond

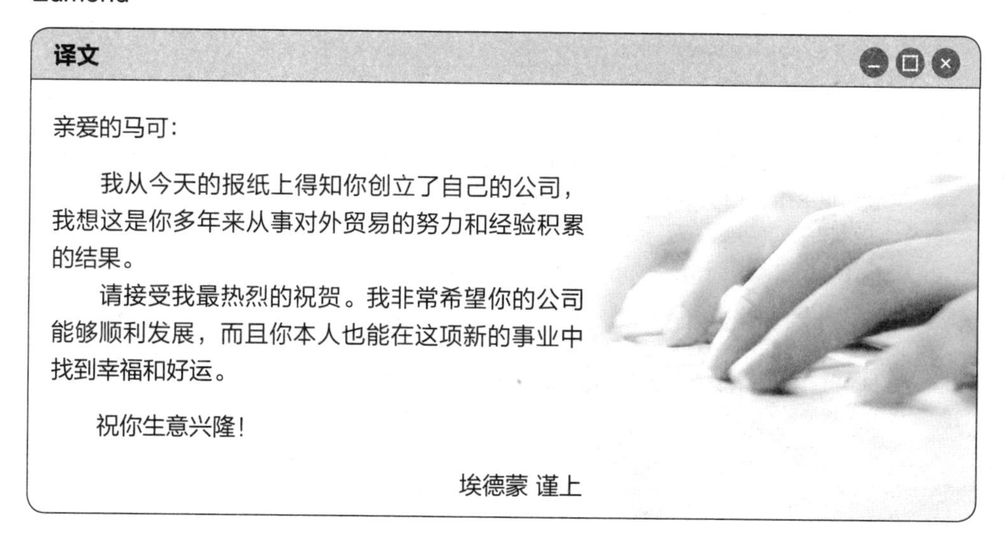

译文

亲爱的马可:

我从今天的报纸上得知你创立了自己的公司,我想这是你多年来从事对外贸易的努力和经验积累的结果。

请接受我最热烈的祝贺。我非常希望你的公司能够顺利发展,而且你本人也能在这项新的事业中找到幸福和好运。

祝你生意兴隆!

埃德蒙 谨上

语法重点解析

1 解析重点1 set up

短语 set up 有“建立；创立；安排；产生”的意思。在此语境中则是“创立”的意思，类似的词还有：establish（建立；成立），found（创办；成立；建立），build（建立；建造）等。表达“开公司”还可以用 run a company 或者 open a company 等短语。请看下面的句子：

They founded the company themselves.（他们自己创办了这家公司。）

I hope that I could open a company like you one day.（希望有一天我能像你一样开一家公司。）

2 解析重点2 result from

短语 result from... 的意思是“产生于……；由……引起；是……的结果”，是个表示原因的短语。类似表达还有：on account of, due to, owing to, because of, by reason of, in respect that 等。所以在要表达原因的时候可以适当选择不同的表达方式，以增加句子的多样性。请看下面的句子：

Nothing had resulted from our efforts.（我们的努力没有任何结果。）

The meeting was cancelled by reason of his illness.（由于他生病，所以会议取消了。）

高频例句

1. **May you succeed in business!**
   祝您生意兴隆！
2. **With best wishes for your prosperity.**
   祝愿你生意兴隆。
3. **The shopkeeper had laid his hopes on a revival of trade.**
   店主把希望寄托在贸易的复苏上。
4. **Smith's new store opened last week and it is going great guns.**
   史密斯的新商店上周开业了，而且生意兴隆。
5. **I'm sure you can build up a prosperous business.**
   我相信你一定可以开创一番成功的事业。
6. **He is running a prosperous business.**
   他的事业经营得很成功。

必背关键单词 

1. ***private*** [ˈpraɪvət] adj. 私人的；私营的
2. ***result*** [rɪˈzʌlt] v. 导致；产生
3. ***experience*** [ɪkˈspɪərɪəns] n. 经验；经历
4. ***manage*** [ˈmænɪdʒ] v. 经营；管理
5. ***smooth*** [smuːð] adj. 平稳的；顺利的
6. ***development*** [dɪˈveləpmənt] n. 发展；成长
7. ***venture*** [ˈventʃə(r)] n. 冒险；冒险事业

09 恭贺病愈

这样写就对了

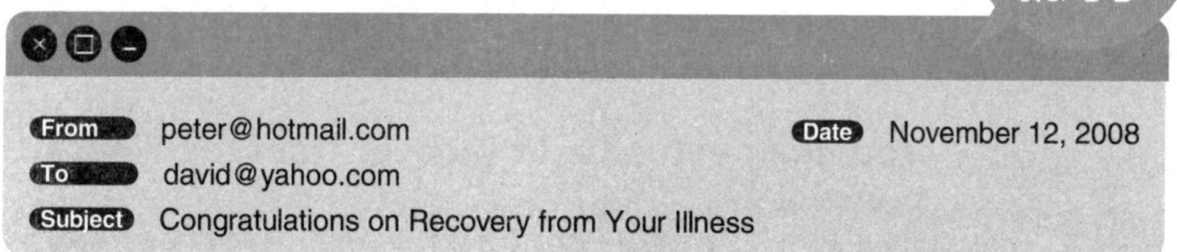

From peter@hotmail.com
Date November 12, 2008
To david@yahoo.com
Subject Congratulations on Recovery from Your Illness

Dear David,

To my relief, I heard that you have been discharged from hospital yesterday. I believe it is worth a ***celebrating*** [1] party to ***welcome*** [2] you back.
However, since you just left hospital and still need time to ***recover*** [3] from the ***illness*** [4] ***completely*** [5], please stay at home and have a good ***rest*** [6]. Take care of yourself and I will come to see you ASAP.

I hope you can get well soon.

Sincerely yours,
Peter

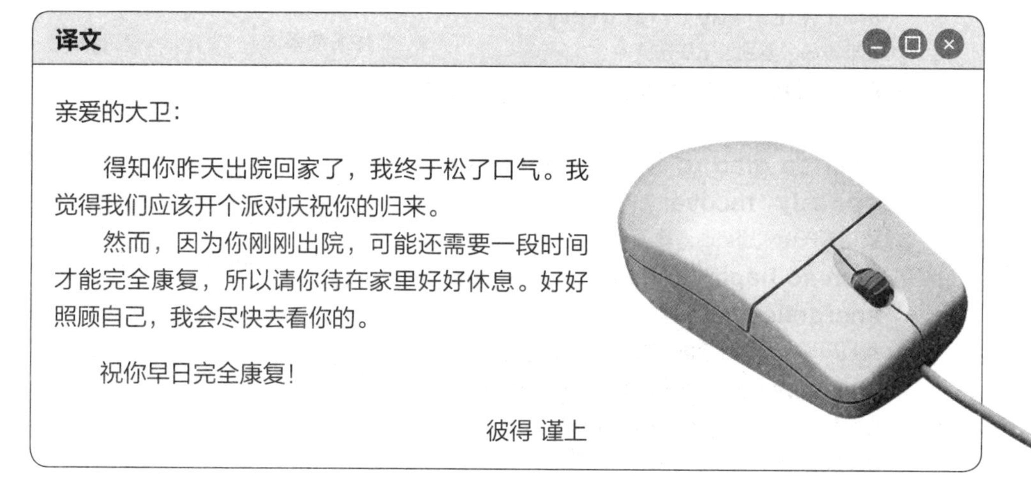
译文

亲爱的大卫：

得知你昨天出院回家了，我终于松了口气。我觉得我们应该开个派对庆祝你的归来。

然而，因为你刚刚出院，可能还需要一段时间才能完全康复，所以请你待在家里好好休息。好好照顾自己，我会尽快去看你的。

祝你早日完全康复！

彼得 谨上

语法重点解析

1 解析重点1 To my relief...

短语 to one's relief 意为“令某人感到放心的是……；使某人松了一口气；使某人安心”的意思。relief 为名词，意为“缓解；宽慰”。由于得知对方刚刚康复，写信者用这个短语则表达出了对于对方病愈的欣慰之情。通过这个短语，就能看出写信者和收信人之间关系亲密。

2 解析重点2 You still need time to recover from the illness completely.

这句话的意思是“你可能还需要一段时间才能完全康复”。这里面包含一个短语：need time to do sth，意为“需要时间做某事”。另一种相同含义的句型如下：

It takes three hours for you to go there.（你到那里去要花3个小时的时间。）

It took six months for her to prepare for the important examination.（她花了6个月的时间准备这场重要的考试。）

1. **Congratulations on your fast recovery.**
   恭喜你这么快就痊愈了。
2. **I heard that you have left hospital yesterday.**
   我听说你昨天已经出院了。
3. **I hope you can get well soon.**
   我希望您可以早日康复。
4. **I am so pleased to see that you are on the way to recovery.**
   我非常高兴看到您正在恢复中。
5. **Please accept my sincere regards.**
   请接受我真诚的问候。
6. **I am so glad to hear about your *speedy*[7] recovery.**
   我很高兴得知您快速恢复了。
7. **I am so happy to see that you are energetic again.**
   我真高兴又看到你充满活力的样子。
8. **I think we should throw a celebratory party to welcome you back.**
   我觉得应该开个派对庆祝你的归来。

必背关键单词 

1. ***celebrate*** ['selɪbreɪt] v. 庆祝
2. ***welcome*** ['welkəm] v. 欢迎；迎接
3. ***recover*** [rɪ'kʌvə(r)] v. 恢复；重新获得
4. ***illness*** ['ɪlnəs] n. 疾病
5. ***completely*** [kəm'pli:tlɪ] adv. 完整地；完全地
6. ***rest*** [rest] n. 休息
7. ***speedy*** ['spi:dɪ] adj. 快速的

10 恭贺梦想成真

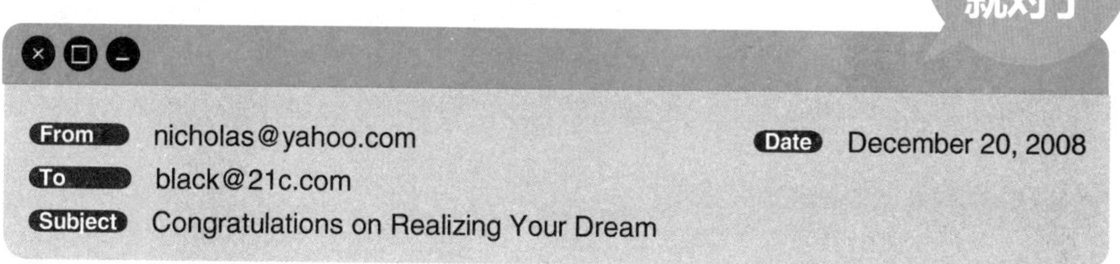
From nicholas@yahoo.com　Date December 20, 2008
To black@21c.com
Subject Congratulations on Realizing Your Dream

Dear Mr. Black,

I'm happy to learn that you have got the ***opportunity***[1] to ***travel***[2] abroad and I'm writing to express my hearty congratulations to you.
As far as I know, traveling abroad has been one of your ***lifelong***[3] dreams. Finally, you have won the ***chance***[4] by yourself ***through***[5] years of hard work and you absolutely ***deserve***[6] it. Congratulations on the ***realization***[7] of your dream again!

Wishing you a ***fabulous***[8] journey abroad!

Sincerely yours,
Nicholas

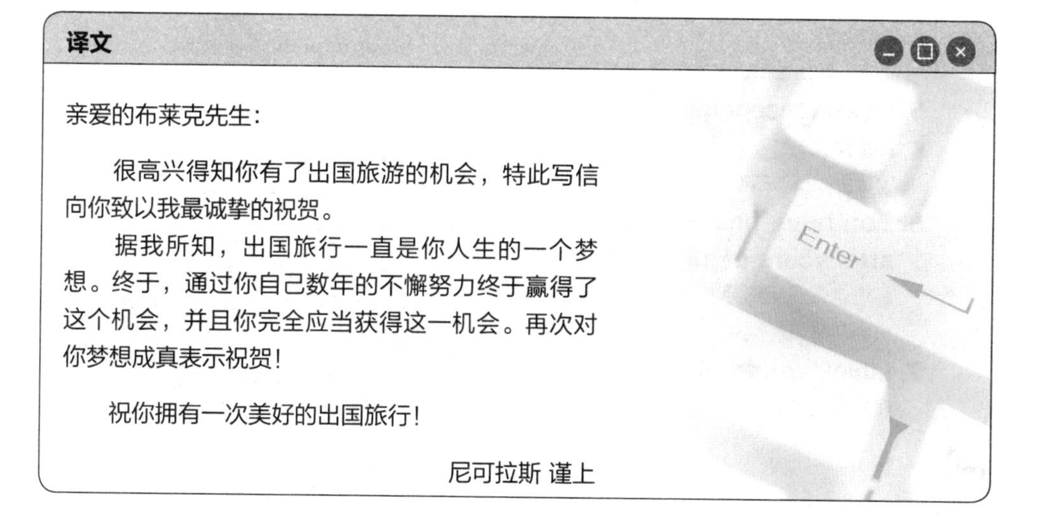
译文

亲爱的布莱克先生：

很高兴得知你有了出国旅游的机会，特此写信向你致以我最诚挚的祝贺。

据我所知，出国旅行一直是你人生的一个梦想。终于，通过你自己数年的不懈努力终于赢得了这个机会，并且你完全应当获得这一机会。再次对你梦想成真表示祝贺！

祝你拥有一次美好的出国旅行！

尼可拉斯 谨上

语法重点解析

1 解析重点1 As far as I know...

短语 as far as I know 是“据我所知”的意思。表达“据我所知”的短语还有：for what I can tell, to the best of my knowledge 等。请看下面的句子：

To the best of my knowledge, this famous singer loves spicy food.（据我所知，这位著名的歌唱家喜欢辛辣的食物。）

He might be in the library as far as I know.（据我所知，他可能在图书馆。）

2 解析重点2 deserve

deserve 意为“值得；应受；应得”，它是个中性词，既可以用于褒义的句子，也可以用于贬义的句子。请对照以下的句子：

He deserves a reward for his efforts.（他的努力值得奖赏。）

He deserved to be punished.（他应受惩罚。）

First deserve and then desire.（先做到受之无愧，再邀功请赏。）

1. **Congratulations on the realization of your dream.**
   祝贺你梦想成真。
2. **I am so happy to hear that your dream has come true.**
   我很高兴听说你的梦想实现了。
3. **You have always dreamed of a rapid rise to fame.**
   你一直都梦想一举成名。
4. **To be an excellent doctor has always been your dream.**
   成为一名出色的医生一直是你的梦想。
5. **Please accept my sincere congratulations.**
   请接受我真诚的祝贺。
6. **You have finally won the chance after years of struggle.**
   经过这么多年的努力奋斗，你终于赢得了这次机会。
7. **I think you totally deserve it.**
   我觉得这完全是你应得的。

必背关键单词 

1. ***opportunity*** [ˌɒpəˈtjuːnətɪ] n. 机会；时机
2. ***travel*** [ˈtrævl] v. 旅行
3. ***lifelong*** [ˈlaɪflɒŋ] adj. 持续一生的
4. ***chance*** [tʃɑːns] n. 机会
5. ***through*** [θruː] prep. 由于；因为
6. ***deserve*** [dɪˈzɜːv] v. 值得；应得
7. ***realization*** [ˌriːəlaɪˈzeɪʃn] n. 实现；认识
8. ***fabulous*** [ˈfæbjələs] adj. 极美好的

Unit 15 慰问吊唁 Consolations and Condolences

01 生病慰问

From "John Walker" (jwalker@excellent.com)
Date Sat., November 1, 2008
To "Ben Jackson" (bjackson@cool.com)
Subject Get Well Soon!

Dear Mr. Jackson,

I have ***missed***[1] you so much and you have been on my mind ever since you went to the ***hospital***[2].
I hope that by the time this note reaches you, you will be ***feeling***[3] a great deal better. I am sure that now it will not be long before you are entirely and completely yourself again. Everyone at the ***office***[4] hopes you can be back soon.

Yours sincerely,
John Walker

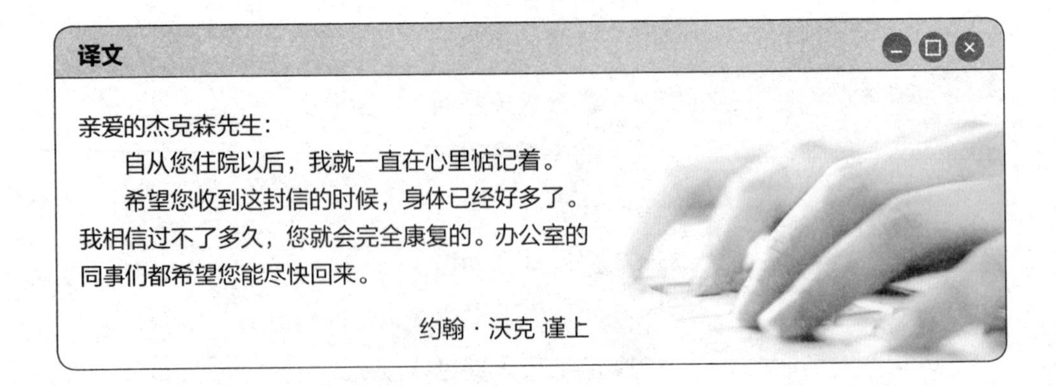

译文

亲爱的杰克森先生:

自从您住院以后，我就一直在心里惦记着。

希望您收到这封信的时候，身体已经好多了。我相信过不了多久，您就会完全康复的。办公室的同事们都希望您能尽快回来。

约翰 · 沃克 谨上

1 解析重点1 **You have been on my mind ever since you went to the hospital.**

表达对病人的挂念，除了用 miss 这个单词以外，还可以用 on my mind，这样虽然比用一个单词要繁琐一些，但是却能更好地表达那种挂念之情。

2 解析重点2 **It will not be long before you are entirely and completely yourself again.**

同事生病住院了，我们祝福他能早日康复，及早出院。可以用 it will not be long before... 这个句型来表达：It will not be long before you are entirely and completely yourself again.（过不了多久，你就会完全康复的。）

1. **I was very much *upset*[5] about the news of your illness.**
   听到你生病的消息，我很不安。
2. **We hope you will come back as soon as possible.**
   我们希望你能尽快回来。
3. **I was so sorry to learn of your *illness*[6].**
   听说你生病了，我深感遗憾。
4. **We all hope you can get well soon.**
   我们都希望你能早日康复。
5. **Everyone at the office misses you so much.**
   办公室所有的人都很想念你。
6. **We are anxious to know whether you feel any better now.**
   我们很想知道你现在是否感觉好一点了。
7. **Please don't worry about your work and have a good *rest*[7].**
   请不要担心你的工作，好好休息一下吧！
8. **I am very glad to hear you are making such a rapid *recovery*[8].**
   听说你这么快就康复了我很高兴。

必背关键单词 

1. ***miss*** [mɪs] v. 想念；怀念
2. ***hospital*** [ˈhɒspɪtl] n. 医院
3. ***feel*** [fiːl] v. 感觉；觉得
4. ***office*** [ˈɒfɪs] n. 办公室
5. ***upset*** [ʌpˈset] adj. 沮丧的；心烦意乱的
6. ***illness*** [ˈɪlnəs] n. 疾病
7. ***rest*** [rest] n. 休息
8. ***recovery*** [rɪˈkʌvərɪ] n. 恢复

02 意外事故慰问

这样写就对了

From "Simon Smith" (si_smith@gmail.com)
Date Sun., November 2, 2008
To "Jack Chen" (chen_j111@hotmail.com)
Subject Get Well Soon!

Dear Mr. Chen,

I was so sorry to ***hear*** [1] of the news that you had a car ***accident*** [2] and was taken to the ***local*** [3] hospital yesterday.
Luckily, you're not ***hurt*** [4] very badly. How are you feeling now? Everybody here sends his best wishes to you and wishes you a ***quick*** [5] recovery.

Yours sincerely,
Simon Smith

译文

亲爱的陈先生：

听说您昨天出了车祸并且被送到当地的医院治疗，我感到很遗憾。

幸运的是，您的伤并不是很严重。不知道您现在状况如何？这里的每个人都祝福您能早日康复。

西蒙 · 史密斯 谨上

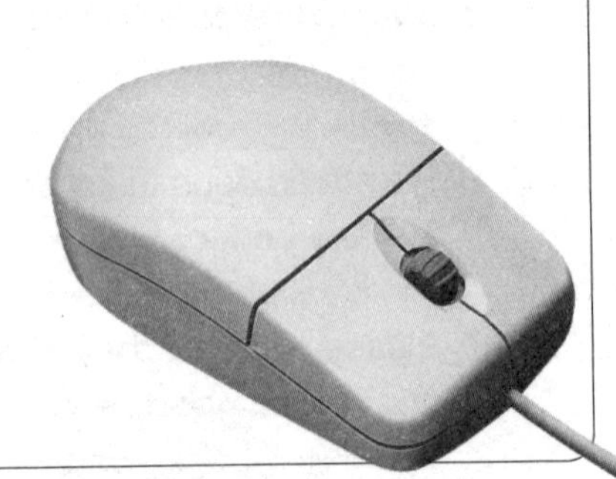

语法重点解析

1 解析重点1 **I was so sorry to hear of the news that you had a car accident.**

听到对方出车祸的消息后，我们为了表达自己对受伤者的关心，可以这样说：It's sad to hear of the news that you have a car accident（听说你出了车祸，我很难过），这里的 hear 有“听说”的意思。

2 解析重点2 **How are you feeling now?**

对方遭遇事故而住院治疗时，我们一般会询问对方现在的情况如何。How are you feeling now?（不知你现在状况如何？）这既是出于礼貌的一种询问，也包含着我们对受伤者的关怀。

1. **I heard that you were hit by a bus, but not seriously hurt.**
   听说你被公交车撞到了，不过伤得不重。
2. **I heard that you *encountered*[6] a car accident.**
   我听说你出了车祸。
3. **The *runaway*[7] of the traffic accident is wanted by the police.**
   警方正在通缉交通事故的逃逸者。
4. **Your right *knee*[8] was hurt in the traffic accident.**
   你的右膝在这次交通事故中受伤了。
5. **You were badly injured in the traffic accident.**
   在这起交通事故中，你伤得非常严重。
6. **I was more than upset by the traffic accident.**
   我为这次的交通事故感到难过.
7. **Let us hope that you will recover for only a very short time.**
   我们希望你能很快康复。
8. **We hope that you will soon be out and about again.**
   我们希望你很快就能重新下床走动。

必背关键单词 

1. ***hear*** [hɪə(r)] v. 听说；得知
2. ***accident*** [ˈæksɪdənt] n. 事故；偶发事件；机遇
3. ***local*** [ˈləʊkl] adj. 当地的
4. ***hurt*** [hɜːt] adj. 受伤的
5. ***quick*** [kwɪk] adj. 快的
6. ***encounter*** [ɪnˈkaʊntə(r)] v. 遭遇
7. ***runaway*** [ˈrʌnəweɪ] n. 逃跑者
8. ***knee*** [niː] n. 膝部；膝盖

03 遭逢地震慰问

From "Colin Will" (cw1113@hot.com) Date Mon., November 3, 2008
To "Hugh Jackson" (hj1981@wind.com)
Subject My Deep Sympathies

Dear Mr. Jackson,

I was dreadfully sorry to hear of the ***earthquake*** [1] which ***destroyed*** [2] your ***beautiful*** [3] house in New York.
Please accept my deepest *sympathies* [4] for you and your *family* [5]. If there is anything I can do to help, please feel free to call me.

Yours sincerely,
Colin Will

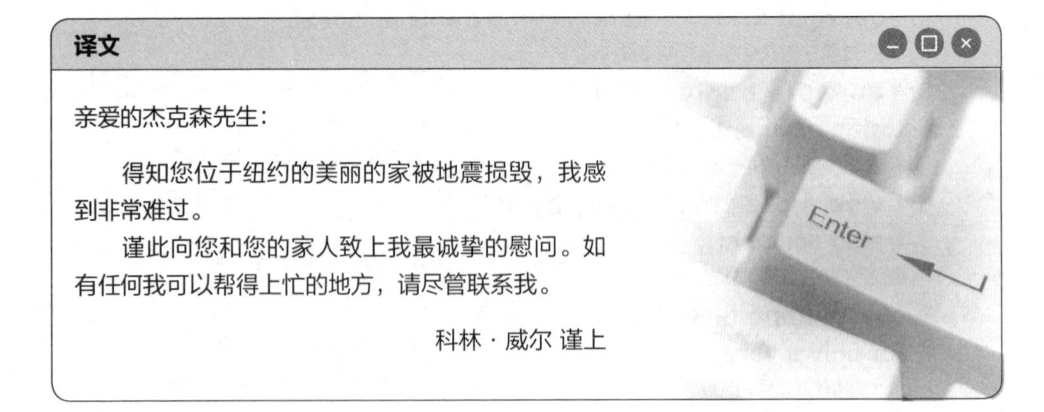

译文

亲爱的杰克森先生：

得知您位于纽约的美丽的家被地震损毁，我感到非常难过。

谨此向您和您的家人致上我最诚挚的慰问。如有任何我可以帮得上忙的地方，请尽管联系我。

科林 · 威尔 谨上

语法重点解析

1 解析重点1 Please accept my deepest sympathies for you and your family.

对方家中遭遇地震导致房屋损毁，我们在向他们表达慰问的时候，应该这样说：Please accept my deepest sympathies for you and your family（谨此向你和你的家人致上我最诚挚的慰问），sympathy 在这里是“同情”之意，而 deepest 进一步加强了语气，表现出最诚挚的慰问。

2 解析重点2 If there is anything I can do to help, please feel free to call me.

向受到类似地震灾害的朋友们表达了自己的慰问之后，我们还要尽量给他们提供一些比较实际的帮助。因此我们可以这样说：If there is anything I can do to help, please feel free to call me.（如有任何我能帮得上忙的地方，请尽管联系我。）

高频例句

1. **A section of the city where you lived was *rubbed*[6] out in the earthquake.**
   你所居住的那座城市的一部分在地震中遭到了毁灭。
2. **This earthquake *shocked*[7] all of us.**
   这次地震使我们感到相当震惊。
3. **The earthquake overturned your houses.**
   地震把你的房屋摧毁了。
4. **An earthquake took place last week.**
   上周发生了一场地震。
5. **We hope your whole family is well.**
   希望你们全家平安。
6. **I hope there is no one injured in this earthquake.**
   希望在这次地震中没有人受伤。
7. **I was most relieved to learn that none of you *suffered*[8] serious injury.**
   令我欣慰的是，你们并未受到重大的伤害。
8. **The earthquake claimed hundreds of lives.**
   此次地震已造成数百人死亡。

1. ***earthquake*** [ˈɜːθkweɪk] n. 地震
2. ***destroy*** [dɪˈstrɔɪ] v. 损毁；毁坏
3. ***beautiful*** [ˈbjuːtɪfl] adj. 美丽的
4. ***sympathy*** [ˈsɪmpəθɪ] n. 同情
5. ***family*** [ˈfæməlɪ] n. 家庭
6. ***rub*** [rʌb] v. 摩擦
7. ***shock*** [ʃɒk] v. 使震惊
8. ***suffer*** [ˈsʌfə(r)] v. 受苦；遭受

04 遭逢火灾慰问

From "Alex Bloom" (abloom@connect.com)　　Date Tues., November 4, 2008
To "Timmy Miller" (tmiller@link.com)
Subject My Deep Sympathies

Dear Mr. Miller,

I heard that a *fire*[1] *broke*[2] out at *midnight*[3] last night and *ruined*[4] your house.
I am anxious to know whether all of your family is all right. I hope no one's injured in it. Please accept my sympathy and do let me know if there is anything I can help you.

Yours sincerely,
Alex Bloom

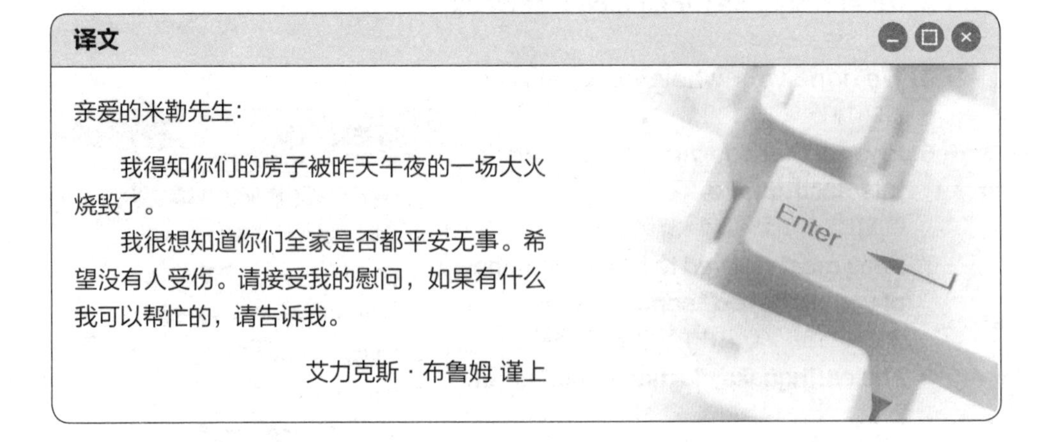

译文

亲爱的米勒先生：

我得知你们的房子被昨天午夜的一场大火烧毁了。

我很想知道你们全家是否都平安无事。希望没有人受伤。请接受我的慰问，如果有什么我可以帮忙的，请告诉我。

艾力克斯 · 布鲁姆 谨上

语法重点解析

1 解析重点1 I heard that a fire broke out at midnight last night and ruined your house.

一般当他人遭受灾害时，我们都会先提到对方的遭遇，以引出我们的话题。I heard that a fire broke out at midnight last night and ruined your house（我得知你们的房子在昨天午夜的一场大火中被烧毁了），这句话基本上把对方的大致遭遇又重新说了一遍，以便引出后面的慰问。

2 解析重点2 I am anxious to know whether all of your family is all right.

对方遭受了火灾，我们一定想了解对方是不是全家都平安，以及是否需要自己的帮忙等，因此我们可以说：I am anxious to know whether all of your family is all right（我很想知道你们全家是否都平安），I am anxious to know... 是“很想知道某事”的意思。

1. **Your house was damaged by the fire.**
   你们的房子在火灾中被烧毁了。
2. **You had to live at a *hotel*[5] for several weeks.**
   你们不得不在一家旅馆住上几个星期。
3. **Maybe you can live with us for a *while*[6].**
   也许你们可以过来跟我们住一段时间。
4. **If you don't mind, I think we can help you find a house.**
   如果你们不介意的话，我们可以帮你们找个房子。
5. **I am writing to tell you how deeply *distressed*[7] I am.**
   我写信是想告诉你们，我对此感到很难过。
6. **I hope everything will go well soon.**
   我希望一切能很快好起来。
7. **I was most relieved to learn that you are not seriously hurt.**
   令我欣慰的是，你受伤的并不严重。
8. **I am anxious to know about the present condition.**
   我急于知道目前的情况。

必背关键单词

1. *fire* [ˈfaɪə(r)] n. 火
2. *break* [breɪk] v. 打破
3. *midnight* [ˈmɪdnaɪt] n. 午夜
4. *ruin* [ˈruːɪn] v. 毁灭
5. *hotel* [həʊˈtel] n. 旅馆
6. *while* [waɪl] n. 一段时间
7. *distressed* [dɪˈstrest] adj. 痛苦的；忧伤的

05 遭逢水灾慰问

Dear Mr. Cruise,

I am extremely sad to hear that your house was ***washed***[1] away in the ***flood***[2].
I hope all of you are in ***safety***[3] and will have a new house in the near future. And I also hope that your life will return to ***normal***[4] soon. If I can be of any help, please let me know.

Yours sincerely,
Aaron Kidd

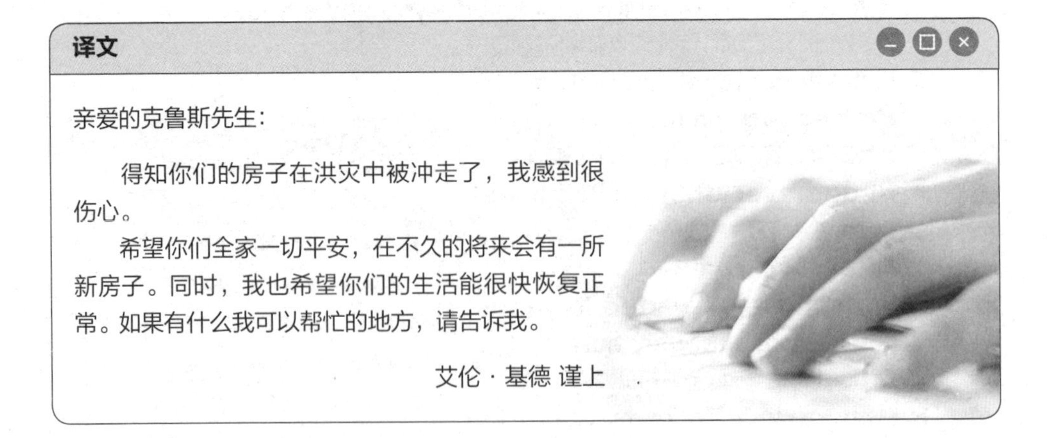

译文

亲爱的克鲁斯先生：

得知你们的房子在洪灾中被冲走了，我感到很伤心。

希望你们全家一切平安，在不久的将来会有一所新房子。同时，我也希望你们的生活能很快恢复正常。如果有什么我可以帮忙的地方，请告诉我。

艾伦 · 基德 谨上

语法重点解析

1

解析重点1 **I am extremely sad to hear that your house was washed away in the flood.**

这句话也说明了对方遭受水灾的情况。I am extremely sad to hear that your house was washed away in the flood（得知你们的房子在洪灾中被冲走了，我感到很伤心），在句子中，wash away 的意思是“（被水）冲走”。I am extremely sad to hear that... 后面接的是一个从句，表达的是从句的内容让我感到很伤心难过的意思。

2

解析重点2 **I also hope that your life will return to normal soon.**

遭受灾害之后，人们在心理上或多或少会留有灾害的阴影，会有心绪不宁的情况。那么，我们要抚慰一下他们受伤的心。I also hope that your life will return to normal soon.（我也希望你们的生活能很快恢复正常。）

高频例句

1. **The damages *created*[5] by the huge floods are exceptionally serious.**
   这场严重的洪灾所造成的损失非比寻常。
2. **I hope you can *reconstruct*[6] your home very soon.**
   我希望你们能很快重建家园。
3. **It's said that nobody got hurt in the *catastrophe*[7].**
   据说在这次大灾难中，没有人受伤。
4. **When the river burst its bank, the field is inundated.**
   河岸决堤后，田地遭洪水淹没。
5. **A lot of villages were *absorbed*[8] into the flood.**
   很多村庄被洪水吞没了。
6. **Numerous people were suffering from the flood.**
   很多人正饱受洪灾之苦。
7. **The cost of the flood damage is impossible to quantify.**
   这次洪灾的损失是无法估量的。

必背关键单词 

1. ***wash*** [wɒʃ] v. 冲刷；洗
2. ***flood*** [flʌd] n. 洪水
3. ***safety*** [ˈseɪftɪ] n. 安全
4. ***normal*** [ˈnɔːml] adj. 标准的；正常的
5. ***create*** [krɪˈeɪt] v. 引起；产生
6. ***reconstruct*** [ˌriːkənˈstrʌkt] v. 重建；改组
7. ***catastrophe*** [kəˈtæstrəfɪ] n. 大灾难
8. ***absorb*** [əbˈsɔːb] v. 吸收

06 讣文

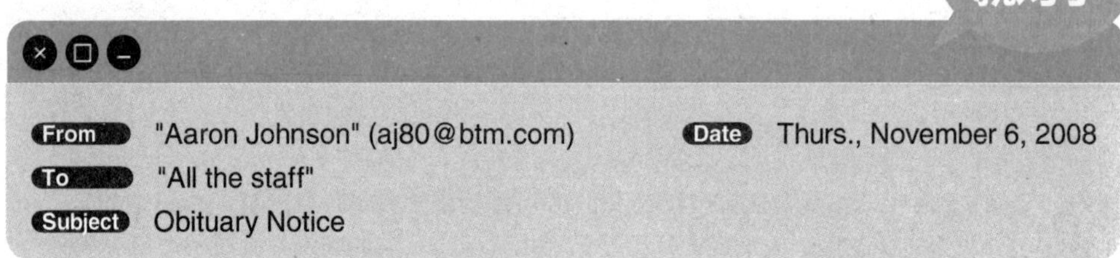
From "Aaron Johnson" (aj80@btm.com)
Date Thurs., November 6, 2008
To "All the staff"
Subject Obituary Notice

Dear Madam or Sir,

Chris Evans, one of the computer engineers of our company, died of a ***heart*** [1] ***attack*** [2] on November 5, 2008. He was born in New York and just turned sixty when he died.
He was an excellent employee and ***contributed*** [3] a lot to our company. Let's show our ***respect*** [4] for the dead and our ***condolences*** [5] to his family.

Yours sincerely,
Aaron Johnson

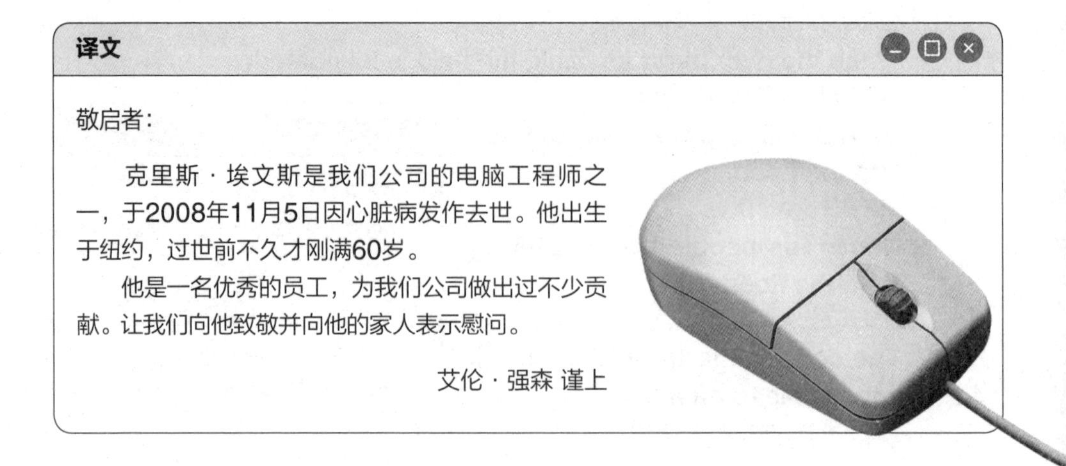
译文

敬启者：

克里斯·埃文斯是我们公司的电脑工程师之一，于2008年11月5日因心脏病发作去世。他出生于纽约，过世前不久才刚满60岁。

他是一名优秀的员工，为我们公司做出过不少贡献。让我们向他致敬并向他的家人表示慰问。

艾伦·强森 谨上

语法重点解析

1 解析重点1 Chris Evans, one of the computer engineers of our company, died of a heart attack on November 5, 2008.

发布讣文的时候，要说明去世者的出生地、担任职位、死亡时间和死亡原因等内容。例如邮件中就有这样的描述：Chris Evans, one of the computer engineers of our company, died of a heart attack on November 5, 2008.（克里斯·埃文斯是我们公司的电脑工程师之一，于2008年11月5日因心脏病发作去世。）

2 解析重点2 Let's show our respect for the dead and our condolence to his family.

对于去世的人我们要表示敬意，而对于他的那些亲人们，我们则要劝说他们节哀顺变，不要太悲伤。这时，我们通常说：Let's show our respect for the dead and our condolence to his family.（让我们向死者致敬并向他的家人表示我们的慰问。）

1. He died, and was survived by his wife and three children.
他去世了，抛下了他的妻子和三个孩子。

2. I heard of his death and felt deep regret.
我听到他去世的消息，感到万分悲痛。

3. He did a great job in our company.
他曾在公司表现得相当优秀。

4. We feel so sorry that we lost such a good man.
我们为失去这样一个好人而感到无比惋惜。

5. We were so *distressed*[6] to *announce*[7] the news about his death.
我们十分悲痛地宣布了他去世的消息。

6. He will be remembered by all of us.
我们永远都不会忘记他。

7. Let us show our deepest *pity*[8] to his family.
让我们向他的家人表示最深切的同情。

8. He departed this life at the age of sixty.
他于60岁时去世。

必背关键单词

1. ***heart*** [hɑːt] n. 心
2. ***attack*** [əˈtæk] n. 攻击
3. ***contribute*** [kənˈtrɪbjuːt] v. 捐献；捐助；贡献出
4. ***respect*** [rɪˈspekt] n. 尊重
5. ***condolence*** [kənˈdəʊləns] n. 哀悼；吊唁；慰问
6. ***distressed*** [dɪˈstrest] adj. 痛苦的；忧伤的
7. ***announce*** [əˈnaʊns] v. 宣告
8. ***pity*** [ˈpɪtɪ] n. 同情

07 吊唁 同事逝世

From "Ronan King" (rk_see@sadness.com) Date Fri., November 7, 2008
To eroberts@mail.com
Subject My Condolences

Dear Mrs. Roberts,

The news of Mike's ***death***[1] ***shocked***[2] me. Please accept my deep sympathy. I have known Mike for many years, and we have ***enjoyed***[3] working together all the time. I must say he was an ***intelligent***[4], kind-hearted and ***just***[5] man. I will miss him a great deal.

Yours sincerely,
Ronan King

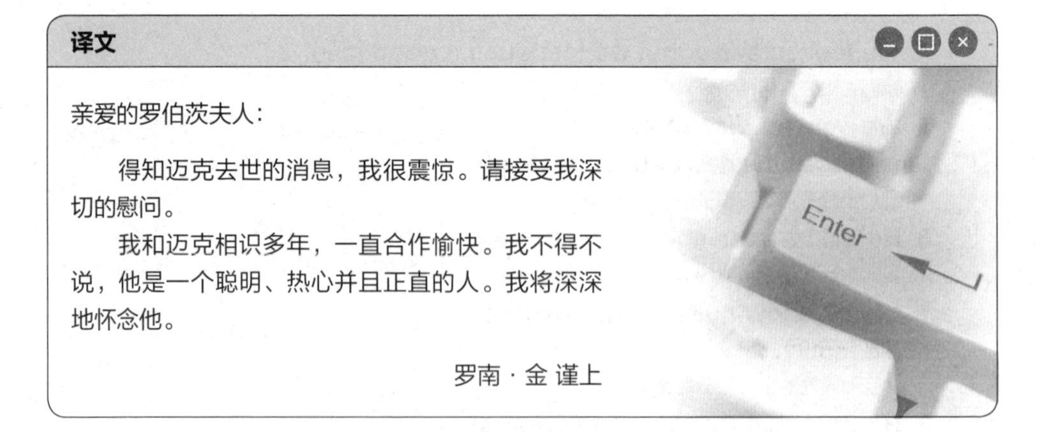

译文

亲爱的罗伯茨夫人：

得知迈克去世的消息，我很震惊。请接受我深切的慰问。

我和迈克相识多年，一直合作愉快。我不得不说，他是一个聪明、热心并且正直的人。我将深深地怀念他。

罗南·金 谨上

1 解析重点1 **The news of Mike's death shocked me.**

表示听到对方去世的消息很震惊时，可以这样表达：The news of Mike's death shocked me.（得知迈克去世的消息，我很震惊。）

2 解析重点1 **I must say he was an intelligent, kind-hearted and just man.**

一般我们会在吊唁函中提及自己对去世的人的一些评价。I must say he was an intelligent, kind-hearted and just man.（我不得不说，他是一个聪明、热心并且正直的人。）这既是回顾自己与其相处的情谊，也向他的家人表达了安慰之情。

1. **My deepest condolences to you. May you have strength to bear this great affliction.**
   我致上最深切的慰问，并希望您能节哀。
2. **The news of his death was a great shock.**
   他去世的消息令人震惊。
3. **Please accept my sincere and deep sympathy.**
   请接受我真诚并深切的慰问。
4. **We have been *co-workers* [6] for several years.**
   我们是多年的老同事。
5. **I have enjoyed working with him.**
   我们相处愉快。
6. **We got along well with each other in the company.**
   我们在公司相处很融洽。
7. **I will really miss him very much.**
   我会很怀念他。
8. **I must say he was one of my best *friends* [7].**
   他可以说是我最好的朋友之一。

必背关键单词

1. ***death*** [deθ] n. 死亡
2. ***shock*** [ʃɒk] v. 震惊
3. ***enjoy*** [ɪnˈdʒɔɪ] v. 享受；欣赏
4. ***intelligent*** [ɪnˈtelɪdʒənt] adj. 有智慧（才智）的
5. ***just*** [dʒʌst] adj. 正直的；公平的
6. ***co-worker*** [ˈkəʊˌwɜːkə] n. 同事
7. ***friend*** [frend] n. 朋友

08 吊唁 领导逝世

From "Michael Smith" (msmith@seed.com)
Date Sat., November 8, 2008
To "Adam Torres" (adamt@ppl.com)
Subject My Condolences

Dear Mr. Torres,

We are so sad to *read*[1] the announcement of Mr. Walker's death in the *Daily News*. I am writing to express our deepest regret.
He was a good *leader*[2] who *led*[3] us from *victory*[4] to victory. He will be long remembered by all of us. Please extend our deepest condolences to his family.

Yours sincerely,
Michael Smith

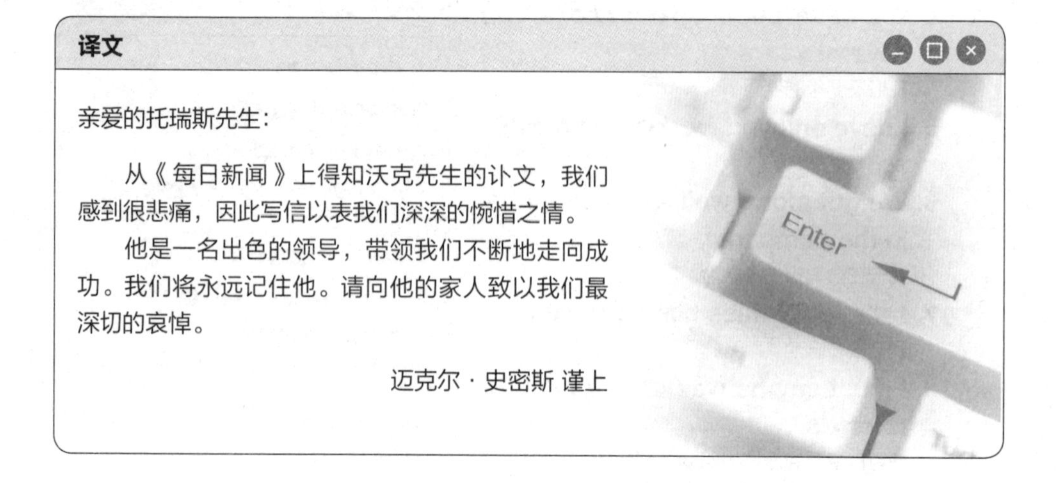

译文

亲爱的托瑞斯先生:

从《每日新闻》上得知沃克先生的讣文，我们感到很悲痛，因此写信以表我们深深的惋惜之情。

他是一名出色的领导，带领我们不断地走向成功。我们将永远记住他。请向他的家人致以我们最深切的哀悼。

迈克尔 · 史密斯 谨上

语法重点解析

1 解析重点1 We are so sad to read the announcement of Mr. Walker's death in the *Daily News*.

有时候我们会从报纸上得知某人去世的消息。We are so sad to read the announcement of Mr. Walker's death in the *Daily News*（从《每日新闻》上得知沃克先生的讣文，我们感到很悲痛），read the announcement in the... 的意思是"在……上看到告示"。

2 解析重点2 He was a good leader who led us from victory to victory.

通常在领导的吊唁文中，我们还要提及他的功绩和贡献，以表示我们对他的由衷感谢和尊敬。He was a good leader who led us from victory to victory.（他是一名出色的领导，带领我们不断地走向成功。）

高频例句

1. **He *excelled*[5] in all his work and made great achievements.**
   他在所有的工作上都表现得很出色，取得了巨大的成就。
2. **She contributed her time and energy to work.**
   她把时间和精力贡献给了工作。
3. **He has made an important *contribution*[6] to the company's success.**
   他对公司的成功做出了重要的贡献。

4. **He *spared*[7] no *efforts*[8] to develop our company.**
   他不遗余力地促进我们公司的发展。

5. **He did his best to promote the development of the company.**
   他竭尽全力促进公司发展。
6. **His death was a real grief to us.**
   他的去世实在令我们痛心。
7. **He was very modest about his great deeds.**
   他从不夸耀自己的功绩。
8. **Men pass away, but their deeds abide.**
   人会死去，但他们的功绩会永存。

必背关键单词

1. ***read*** [ri:d] v. 阅读；朗读

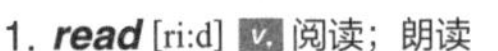

2. ***leader*** [ˈli:də(r)] n. 领袖；领导
3. ***lead*** [li:d] v. 领导
4. ***victory*** [ˈvɪktərɪ] n. 胜利
5. ***excel*** [ɪkˈsel] v. 胜过；优于
6. ***contribution*** [ˌkɒntrɪˈbju:ʃn] n. 贡献；捐献
7. ***spare*** [speə(r)] v. 节省；腾出
8. ***effort*** [ˈefət] n. 努力

09 吊唁 亲人逝世

这样写就对了

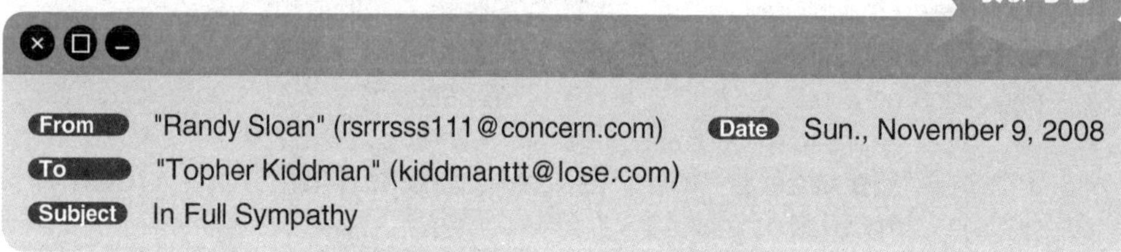

From "Randy Sloan" (rsrrrsss111@concern.com) Date Sun., November 9, 2008
To "Topher Kiddman" (kiddmanttt@lose.com)
Subject In Full Sympathy

Dear Topher,

I am most grieved to hear of the loss of your mother, my beloved aunt, and *hasten*[1] to offer my deepest sympathy.
You would find *comfort*[2] in the fact that you lived happily with your mother these years. She was also gratified by your ***filial***[3] ***piety***[4] and ***accomplishments***[5]. I hope that would ***ease***[6] your grief.

Please convey my deepest sympathy to all your family.

Yours sincerely,
Randy Sloan

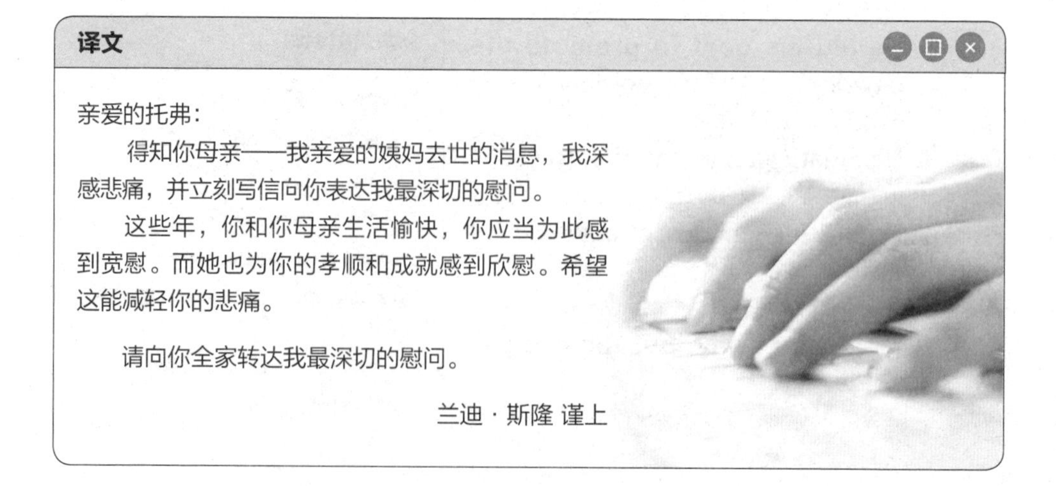

译文

亲爱的托弗：

得知你母亲——我亲爱的姨妈去世的消息，我深感悲痛，并立刻写信向你表达我最深切的慰问。

这些年，你和你母亲生活愉快，你应当为此感到宽慰。而她也为你的孝顺和成就感到欣慰。希望这能减轻你的悲痛。

请向你全家转达我最深切的慰问。

兰迪 · 斯隆 谨上

语法重点解析

1 解析重点1 **I am most grieved to hear of the loss of your mother, my beloved aunt, and hasten to offer my deepest sympathy.**

当对方的亲人不幸去世，我们要表达自己听到消息时的悲痛可以这样说：I am most grieved to hear of the loss of your mother（得知你母亲去世的消息，我深感悲痛）。同时，还要表达自己对他们家人的深切慰问：hasten to offer my deepest sympathy（立刻写信向你表示我最深切的慰问）。

2 解析重点2 **You would find comfort in the fact that you lived happily with your mother these years.**

对方因为亲人去世而悲痛欲绝，这时候，我们要试着去开导对方，让他节哀顺变，不要太悲伤。You would find comfort in the fact that you lived happily with your mother these years（这些年，你和母亲生活愉快，你应当为此感到宽慰），这句话就是在试图说服对方，他母亲在的时候，他们生活得很开心，这比什么都好，从而让对方放宽心胸，不要太过伤痛。

1. **The news of your father's death was a terrible shock to us.** 你父亲去世的噩耗使我们感到非常震惊。
2. **I am sorry to hear the news about your *deceased*[7] mother.**
   听到你母亲去世的消息，我很难过。
3. **I am sorry to hear that your brother has passed away.**
   听到你兄弟去世的消息，我很难过。
4. **I am deeply grieved to hear that your father has passed away.**
   听到你父亲去世的消息，我很难过。
5. **Everyone who has known him must have felt a great loss.**
   他的去世对于所有认识他的人来说，都是巨大的损失。
6. **That should help *soften*[8] your sorrow and grief a little.**
   这应该有助于减轻一点你的悲伤和痛苦。

必背关键单词 

1. ***hasten*** [ˈheɪsn] v. 加速
2. ***comfort*** [ˈkʌmfət] n. 安慰；慰问
3. ***filial*** [ˈfɪlɪəl] adj. 子女的；孝顺的
4. ***piety*** [ˈpaɪətɪ] n. 虔诚；虔敬
5. ***accomplishment*** [əˈkʌmplɪʃmənt] n. 达成；成就
6. ***ease*** [iːz] v. 减轻；缓和
7. ***deceased*** [dɪˈsiːst] adj. 已故的
8. ***soften*** [ˈsɒfn] v. （使）变轻柔；（使）变温和

10 答复唁电

这样写就对了

From "Timmy Cruise" (tcruise@uni.com) Date Mon., November 10, 2008
To "Paul Willman" (pw1247@atm.com)
Subject Thank You for Your Condolences!

Dear Mr. Willman,

Please accept my heart-felt thanks for your sympathy.
There is no greater ***solace***[1] than the ***knowledge***[2] that our friends are there and feel the same with us.
I will ***pull***[3] myself together and take ***care***[4] of our family ***affairs***[5].

Yours sincerely,
Timmy Cruise

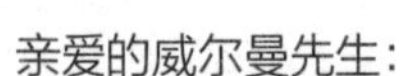

译文

亲爱的威尔曼先生：

衷心感谢您的慰问。

没有什么比知道我们的朋友在那里和我们有同样的感受更大的安慰了。

我会振作起来，并且好好处理家里的事情。

提米 · 克鲁斯 谨上

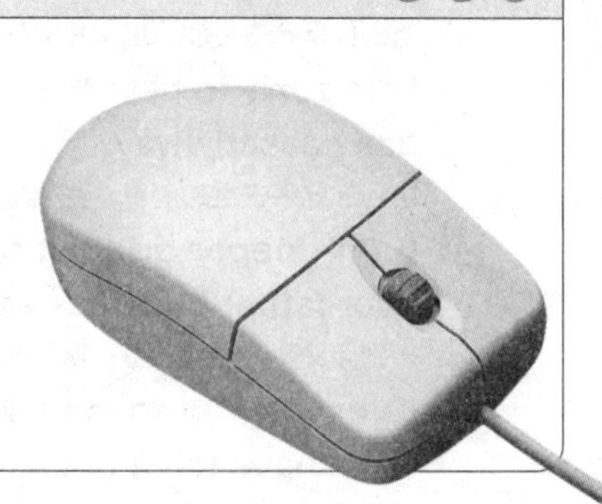

语法重点解析

1 解析重点1 There is no greater solace than...

当对方在自己亲人去世的时候，发来唁电表示慰问，我们也要回复对方的关心，向对方表示感谢。There is no greater solace than the knowledge that our friends are there and feel the same with us（得知我们的朋友正与我们一起哀悼，这对我们来说是最大的安慰），这里需要注意一个句型：There is no greater solace than...（没有比……更大的安慰了），句子中用的是 no + 比较级 + than 的结构。

2 解析重点2 I will pull myself together and take care of our family affairs.

当对方发唁电劝慰我们一番之后，我们也要让对方放心，说一些让双方都很宽慰的话。I will pull myself together and take care of our family affairs（我会振作起来，好好打理家务），句子中的 pull oneself together 是“振作起来”的意思。take care of... 是“照顾；处理”的意思。例如“处理事情”就可以说：take care of the matter。

1. **My *gratitude* [6] goes to our friends who wrote to show their support.**
   我感谢所有写信给我们的朋友，谢谢他们的支持。
2. **Your kind *expression* [7] of sympathy is deeply appreciated.**
   谢谢你的同情，万分感激。
3. **I know well what you must be feeling.**
   你此刻的心情，我能体会。
4. **Time will heal all the sorrows.**
   时间会治愈悲伤。
5. **Your expression of love and *concern* [8] was very much appreciated.**
   十分感谢您的关爱。
6. **Your friendship and support mean a lot to us.**
   您的友谊和支持对我们来说很重要。
7. **I want to thank you most sincerely for your kind words of sympathy.**
   我衷心感谢你的慰问。

必背关键单词

1. ***solace*** [ˈsɒləs] n. 安慰；慰藉
2. ***knowledge*** [ˈnɒlɪdʒ] n. 了解；理解
3. ***pull*** [pʊl] v. 拉；拖
4. ***care*** [keə(r)] n. 小心；照料；忧虑
5. ***affair*** [əˈfeə(r)] n. 事件
6. ***gratitude*** [ˈɡrætɪtjuːd] n. 感激；感谢
7. ***expression*** [ɪkˈspreʃn] n. 表达
8. ***concern*** [kənˈsɜːn] n. 担心；挂念；关怀

Part 3

英文E-mail

词汇篇

01 公司 部门名称

Accounting Department	财务部
Advertising Department	广告部
Branch Office	分公司
Business Office	营业部
Export Department	出口部
General Affairs Department	总务部
Head Office	总公司
Human Resources Department	人力资源部
Import Department	进口部
International Department	国际部
Management Department	管理部
Market Department	市场部
Personnel Department	人事部
Planning Department	企划部
Product Development Department	产品开发部
Public Relations Department	公关部
Real Estate Development Department	地产开发部
Research and Development Department	研发部
Sales Department	销售部
Sales Promotion Department	销售推广部
Secretarial Pool	秘书室

02 公司职位名称

English Instructor / Teacher	英语教师
Export Sales Manager	外销部经理
Financial Controller	财务主任
Financial Reporter	财务报告人
F.X. (Foreign Exchange) Clerk	外汇部职员
F.X. Settlement Clerk	外汇部结算员
Fund Manager	基金经理
General Manager / President	总经理
General Manager Assistant	总经理助理
General Manager's Secretary	总经理秘书
Hardware Engineer	硬件工程师
Import Liaison Staff	进口部联络员
Import Manager	进口部经理
Insurance Actuary	保险精算师
International Sales Staff	国际销售员
Interpreter	口译员
Legal Adviser	法律顾问
Line Supervisor	生产线主管
Maintenance Engineer	维修工程师
Management Consultant	管理顾问
Manager	经理
Manager for Public Relations	公关部经理
Manufacturing Engineer	制造工程师
Manufacturing Worker	生产工人
Market Analyst	市场分析员
Market Development Manager	市场开发部经理
Marketing Manager	市场部经理
Marketing Staff	营销人员
Marketing Assistant	营销助理

Marketing Executive	销售主管
Marketing Representative	销售代表
Mechanical Engineer	机械工程师
Mining Engineer	采矿工程师
Naval Architect	造船工程师
Office Assistant	办公室助理
Office Clerk	办公室职员
Operational Manager	业务经理
Operator	操作员；话务员
Package Designer	包装设计师
Passenger Reservation Staff	乘客票位预订员
Personnel Clerk	人事部职员
Personnel Manager	人事部经理
Plant / Factory Manager	工厂经理
Postal Clerk	邮政人员
Private Secretary	私人秘书
Product Manager	产品经理
Production Engineer	生产工程师
Programmer	程序设计师
Promotional Manager	推销部经理
Proof-reader	校对员
Purchasing Agent	采购（进货）员
Quality Control Engineer	质量管理工程师
Real Estate Staff	房地产职员
Recruitment Coordinator	招聘协调员
Regional Manger	区域经理
Research & Development Engineer	研发工程师
Restaurant Manager	饭店经理
Sales and Planning Staff	销售计划员
Sales Assistant	销售助理
Sales Clerk	店员；售货员
Sales Coordinator	销售协调员
Sales Engineer	销售工程师
Sales Executive	销售主管

Sales Manager	销售部经理
Salesperson	销售员
Seller Representative	销售代表
Sales Supervisor	销售监管员
Secretarial Assistant	秘书助理
Secretary	秘书
Securities Custody Clerk	保安人员
Security Officer	安全人员
Senior Accountant	高级会计
Senior Consultant / Adviser	高级顾问
Senior Secretary	高级秘书
Service Manager	客服部经理
Simultaneous Interpreter	同声传译员
Software Engineer	软件工程师
Supervisor	监管员
Systems Adviser	系统顾问
Systems Engineer	系统工程师
Systems Operator	系统操作员
Teacher	教师
Technical Editor	技术编辑
Technical Translator	技术翻译
Technical Worker	技术工人
Telecommunication Executive	电信主管
Tourist Guide	导游
Trade Finance Executive	贸易财务主管
Trainee Manager	培训部经理
Translation Checker	翻译核对员
Translator	翻译员
Trust Banking Executive	信托银行高级职员
Typist	打字员
Word Processing Operator	文字处理操作员

03 学校 科系及课程名称

1 学校科系

Department of Accounting	会计系
Department of Applied Mathematics	应用数学系
Department of Archaeology	考古学系
Department of Architecture	建筑系
Department of Art	美术系
Department of Astronomy	天文系
Department of Automation Engineering	自动化工程系
Department of Finance	金融系
Department of Biology	生物学系
Department of Botany	植物学系
Department of Business Administration	工商管理系
Department of Chemistry	化学系
Department of Chinese	中文系
Department of Civil Engineering	土木工程系
Department of Communication Engineering	通信工程系
Department of Computer Science	计算机科学系
Department of Dance	舞蹈系
Department of Diplomacy	外交系
Department of Economics	经济系
Department of Education	教育系
Department of Electronic Engineering	电子工程系
Department of English	英语系
Department of Environmental Engineering	环境工程系
Department of Food Engineering	食品工程系
Department of Food Science	食品科学系
Department of Foreign Languages	外语系

Department of Human Resources Management	人力资源管理系
Department of Industrial Management	工业管理系
Department of International Politics	国际政治系
Department of International Trade	国际贸易系
Department of Journalism	新闻系
Department of Law	法律系
Department of Library Management	图书管理系
Department of Literature	文学系
Department of Mathematics	数学系
Department of Mechanical Engineering	机械工程系
Department of Medicine	医学系
Department of Music	音乐系
Department of Philosophy	哲学系
Department of Physical Education	体育系
Department of Physics	物理系
Department of Tourism Management	旅游管理系

2 学校课程

Accounting and Finance	会计财务学
Accounting	会计学
Aesthetics	美学
Applied Mathematics	应用数学
Art Theory	艺术理论
Bioengineering	生物工程学
Civil Engineering	土木工程
Computer Application and Maintenance	计算机应用与维修
Constitutional Law and Administrative Law	宪法与行政法学
Criminal Jurisprudence	刑法学
Dance	舞蹈
Economics	经济学
Electronic Commerce	电子商务学
Engineering	工程学

English	英语
Ethics	伦理学
Film	电影艺术
Finance	金融学
Financial Management	财务管理
Fine Arts	美术
History of Economic Thought	经济思想史
History of Economics	经济史
History	历史学
Industrial Economics	工业经济学
International Trade	国际贸易
Jurisprudence	法理学
Labor Economics	劳动经济学
Legal History	法律史
Logic	逻辑学
Logistic Management	物流管理
Management Science	管理学
Marketing	市场营销
Music	音乐
Philosophy of Marxism	马克思主义哲学
Political Economy	政治经济学
Radio and Television Art	广播电视艺术学
Science of Economic Law	经济法学
Science of Law	法学
Science of Procedure Laws	诉讼法学
Science of Religion	宗教学
Software Engineering	软件工程
Statistics	统计学
Theater and Chinese Traditional Opera	戏剧戏曲学
Western Economics	西方经济学
World Economics	世界经济学

04 电脑使用相关词汇

access	读取
activate	激活
back	上一步
browser	浏览器
clear	清除
click	点击
close	关闭
code	编码
column	栏
command	命令
copy	复制
cut	剪切
data	数据；资料
database	资料库
delete	删除
document	文件
double click	双击
edit	编辑
e-mail	电子邮件
exception	异常；例外状况
execute	执行
exit	退出
file	文件
find	查找
finish	结束
folder	文件夹
font	字体

form	格式
full screen	全屏幕
function	函数
graphics	图片
homepage	主页
host	主机
icon	图标
image	图片
insert	插入
interface	界面
Internet	互联网
Kbytes	千字节
LAN	局域网
log off	注销
log in	登录
manual	手册
menu	菜单
next	下一步
OS (Operation System)	操作系统
online	线上
password	密码
paragraph	段落
paste	粘贴
print preview	打印预览
print	打印
program	程序
replace	替换
restart	重新启动
save	存储
scale	缩放比例
select	选择
search engine	搜索引擎
setup	安装
short cut	快捷方式
size	大小
status bar	状态栏
symbol	符号
table	表
text	文本
tool bar	工具栏
uninstall	卸载
update	更新
user	用户
virus	病毒
WAN	广域网
webpage	网页
website	网站
WWW (World Wide Web)	万维网
zoom in	放大
zoom out	缩小

05 国际贸易相关词汇

A/W = All Water
全水路运输（主要指由美国西岸中转至东岸或内陆点的货物的运输方式）

BAF = Bunker Adjustment Factor
燃油附加费

B/L = Bill of Lading
提单

C.A.D = Cash Against Documents
交单付现

CAF = Cost and Freight
成本加运费（指定目的港）

CC = Charges Collect = Freight Collect
运费到付：表示货物运送费用未付，应由买方在目的地 / 港支付

CFR(C&F, CNF) = Cost and Freight
成本加运费 / 到岸价格（指定目的港）

CFS = Container Freight Station
集装箱货运站

CIF = Cost, Insurance & Freight
成本、保险费加运费 / 到岸价格（指定目的港）

CIP = Carriage and Insurance Paid to
运费、保险费付至（指定目的地）

C/O = Certificate of Origin
原产地证书

C.O.D = Cash on Delivery
货到付现

CPT = Carriage Paid to
运费付至（指定目的地）

Cut Off = Closing Date = Cut off Date
结关日 / 截关日

C.W.O = Cash with Order
随订单付现

CY = Container Yard
集装箱堆场

D/A = Documents against Acceptance
承兑交单

DAF = Delivered at Frontier
边境交货

D/D = Demand Draft
即期汇票

D.D.C. = Destination Delivery Charge
目的地 / 港交货费用

DDP = Delivered Duty Paid
完税后交货

DDU = Delivered Duty Unpaid
未完税交货

DES = Delivered Ex Ship
目的港船上交货

DEQ = Delivered Ex Quay
目的港码头交货

D/O = Delivery Order
交货单

DOC = Document Charges
文件费

D/P = Documents against Payment
付款交单

EPS = Equipment Position Surcharges
设备位置附加费

ETA = Estimated Time of Arrival
预计到达时间

ETD = Estimated Time of Delivery
预计交货时间

FOB = Free on Board
船上交货价 / 离岸价格

GRI = General Rate Increase
正常费率增加

L/C = Letter of Credit = Commercial Letter of Credit
信用证 / 商业信用证

LCL = Less than Container Load
拼箱货

M/T = Mail Transfer
信汇

NVOCC = Non-Vesse Operating Common Carrier
无船承运人

O/A = On Account
赊账 / 记账（交易）；分期付款

O/B = On Board Date
装船日期

O/F = Ocean Freight
海运运费

ORC = Origin Receiving Charges
本地收货费用

PCS = Port Congestion Surcharge
港口拥挤附加费

P.O.D = Pay on Delivery
货到付款

PP = Prepaid = Charges Prepaid = Freight Prepaid (= carriage prepaid = carriage paid)
运费预付：表示货物运送费用已由卖方在起运地（港）付清

PSS = Peak Season Surcharges
旺季附加费

S/C = Sales Contract / Sales Confirmation
销售合同 / 销售确认书

S/O = Shipping Order
装运单；发货单

T/T = T.T. = Telegraphic Transfer
电汇

Part 4

英文E-mail 超值附赠

商用书信必抄200惯用句

因为商用书信有许多惯用表达，和我们平常写给朋友、家人的信件并不相同。如要委婉地表达要求、不满和拒绝，或想要真诚地表示谢意、歉意，都有一些固定的句型可以套用。套上外国人惯用的句型，不但能精准地表达意思，更能让国外客户惊呼“你的英文好地道！”以下为大家整理出了200句商用书信惯用必抄句，找到你需要的情境，大胆地用吧！

委婉表达希望对方快点回复的心情

1. Your prompt reply will be very much appreciated.

如果您能尽快回复，我将感激不尽。

prompt adj. 迅速的；及时的　　appreciate v. 感激

2. I look forward to hearing your opinions on this matter.

我很期待听到您对这件事的看法。

look forward to  期待　　opinion n. 意见

3. Feel free to communicate with me through e-mail any time.

欢迎随时通过e-mail与我联系。

feel free to... phr. 随意……；想要……就请自便
communicate v. 沟通

4. I'm anxiously awaiting your response.

我急切地等待您的回复。

anxiously adv. 焦急地；担忧地　　await v. 等待

5. I'm expecting a swift reply.

我希望你能尽快回复。

expect v. 预期 swift adj. 快速的

6. I would appreciate it if you could reply as soon as possible.

如果你能尽快回复，我会非常感激。

as soon as possible phr. 尽快

7. Please send me any feedback you have.

如果你有任何意见，请发给我。

feedback n. 反应；回馈

8. I look forward to your response.

我期待你的回复。

response n. 回答

9. I'm sorry, but this really can't wait.

我很抱歉，但这真的不能等了。

10. It's vital that I hear from you soon.

我要尽快收到你的回复，这至关重要。

vital adj. 至关重要的

11. If you have any objections, please let me know ASAP.

如果你反对的话，请尽快让我知道。

objection n. 反对；异议

12. I'm looking forward to receiving your acceptance of my offer.

我期待你能接受我提出的报价。

acceptance *n.* 接受　　offer *n.* 报价；提议

13. I'll really appreciate it if you can take care of this as soon as possible.

如果你能尽快处理，我会非常感激。

take care of sth *phr.* 照顾；处理某事

14. Do you mind replying as soon as you can? Time is of the utmost importance.

你介意尽快回复吗？时间非常紧急。

utmost *adj.* 最大的；最大限度的

15. Whenever you get a chance, please inform me on your decision.

请你一有机会就赶快告诉我你的决定。

inform *v.* 通知　　decision *n.* 决定

16. Please immediately inform me if you have any concerns.

如果你有什么疑虑，请尽快通知我。

immediately *adv.* 立即地　　concern *n.* 担心；顾虑

17. Please don't hesitate to contact me as soon as possible.

请不要犹豫，尽快跟我联系。

hesitate *v.* 犹豫　　contact *v.* 与某人联系

18. Awaiting your quick response.

期待你的快速回复。

19. I understand that you must be busy, but this is urgent.

我了解你一定很忙，但这非常紧急。

urgent adj. 紧急的

20. I'm sorry for bugging you for a reply, but we need to get this done as soon as possible.

我很抱歉打扰你请求回复，但我们真的得尽快完成这件事。

bug v. 打扰；烦扰

21. I'm sure you can understand that this is a most urgent matter.

我相信你能够了解这件事非常紧急。

22. I need to hear from you as soon as possible.

我必须尽快得到你的回复。

23. Please take a few minutes to let me know about your decision.

请花几分钟时间告诉我你的决定。

24. If we hear from you soon, we can start getting to work right away.

如果你尽快回复我们，我们就能马上开始工作。

get to work phr. 开始工作

25. Please send us an update at your earliest convenience.

您方便的话，请尽早发给我们最新信息。

update n. 更新　　convenience n. 便利

对自己的话可能会伤害到对方而表达歉意，或委婉表达避免伤害对方

1. I'm sorry if this sounds harsh.

如果这听起来很刺耳，我很抱歉。

harsh adj. 刺耳的；严厉的

2. I hope you don't feel offended.

我希望你不觉得被冒犯了。

offended adj. 被冒犯到的

3. I apologize, but I have to admit that I'm a bit disappointed.

我很抱歉，但我得承认我有点失望。

apologize v. 道歉　　admit v. 承认

4. I have to come out and tell you this. I'm very sorry.

我很抱歉，我不得不告诉你这件事。

5. Please understand that I don't mean to imply any dissatisfaction on my part.

请理解，我并不是要暗示我对你不满意。

imply v. 暗示；暗指　　dissatisfaction n. 不满

on sb's part phr. 以某人的立场而言

6. I respectfully disagree with your view.

恕我不同意你的看法。

respectfully adv. 恭敬地　　disagree v. 不同意

view n. 看法；视角

7. Please don't take offense.

请别生气。

take offense phr. 生气

8. Please forgive me for being frank with you.

请恕我直言。

forgive v. 原谅　　frank adj. 坦白的；直率的

9. I'm sorry, but I have to make it clear that I'm not very satisfied.

我很抱歉，但我得说清楚，我不是很满意。

make it clear phr. 澄清　　satisfied adj. 满意的

10. I don't mean to offend you by this observation.

我的这些评论，并没有要冒犯你的意思。

offend v. 冒犯　　observation n. 观察；评论

11. I apologize for being straightforward.

说得这么直接，我很抱歉。

straightforward adj. 直接的

12. I hope this doesn't sound too blunt.

我希望这听起来不会太直接。

blunt adj. 直率的；直言不讳的

13. I regret to tell you that your performance could have been better, but I don't mean this as an insult.

我很抱歉地告诉你，你本可以表现得更好，但我这样说并没有侮辱你的意思。

regret v. 后悔；感到遗憾　　performance n. 表现
insult n. 侮辱

14. I'm afraid that I have to say that I'm not too happy with this.

我恐怕得说我对此不是太高兴。

I'm afraid that... phr. 我恐怕……

15. I understand that you have been working very hard, but there are areas in which you can improve.

我知道你工作很努力，但有些地方你还可以改进。

area n. 区域　　improve v. 进步

16. By no means think that I mean to be rude.

千万不要认为我有冒犯的意思。

by no means phr. 千万不要……　　rude adj. 粗鲁无礼的

17. I'm sorry, but I think it's best that I be frank.

我很抱歉，但我想我还是直接说最好。

18. Please understand that I still hold you in high regard.

请理解，我仍然非常尊敬你。

hold sb in high regard phr. 十分尊敬某人

19. I'm sorry, but I beg to differ.

我很抱歉，但我有不同的意见。

beg to differ phr. 有不同的意见

20. I'm really sorry for being blunt, but you don't have the faintest idea about how this is done, and I don't blame you.

恕我直言，但你根本搞不清楚这该怎么做，而我并不怪你。

faint adj. 微弱的；模糊的　　blame v. 责怪

21. This isn't really your fault, but I don't think we should continue with this project.

这真的不是你的错，但我觉得我们不该继续进行这个计划。

fault *n.* 错误；过失　　continue *v.* 继续

22. I'm very sorry to say that this wasn't the kind of performance I was expecting from you.

我很抱歉，但我必须说你这种表现并不是我所期望的。

23. I regret to say that you can still improve in many aspects.

我很抱歉地要说你在很多方面仍然亟须提高。

aspect *n.* 方面；方向

24. Forgive me for being blunt, but this is for the good of the company.

恕我直言，但这对公司比较有益。

for the good of... *phr.* 为了……好

25. I hope you don't mind if I'm direct.

我说得很直接，希望你不要介意。

direct *adj.* 直接的

表达感激的心情

1. I know I speak for everyone on this when I say I really appreciate your kindness.

当我说我很感激你的好意时，我知道我是代表每个人说的。

kindness *n.* 好意；善意

2. Words cannot convey my gratitude.

我的感激之情无法用言语表达。

convey v. 表达；传达　　gratitude n. 感激

3. We hope to return your favor soon.

我们希望很快能报答你的恩惠。

favor n. 恩惠

4. Thank you for your patience.

感谢你的耐心。

patience n. 耐心

5. Thank you for understanding.

感谢你的理解。

6. I'm very much indebted to you.

我欠你太多了。

indebted adj. 亏欠的

7. Your help is very much appreciated.

非常感谢你的帮忙。

8. I would repay you if I could.

可以的话，我一定回报你。

repay v. 回报

9. I want you to know that I'm really grateful for your help.

我想让你知道，我非常感谢你的帮忙。

grateful adj. 感激的

10. Thank you so much for your time.

谢谢你抽出时间。

11. I'm so glad that amazing people like you exist.

世界上有像你这么棒的人，真是太令人高兴了。

amazing adj. 非常棒的　　exist v. 存在

12. I'm really thankful for your assistance.

我真的很感谢你的帮助。

assistance n. 帮助；协助

13. I can't ever thank you enough.

我感激不尽。

14. Thank you for what you did for me.

谢谢你为我做的事。

15. I don't know how to express my gratitude, so here's a little something for you.

我不知道如何表达我的感激之情，所以这是我给你的一点心意。

express v. 表达　　a little something phr. 一点小东西；小心意

16. People like you make my life so much better. Thank you!

像你这样的人让我的人生更精彩。谢谢你！

17. I could say thank you a thousand times and it wouldn't be enough.

说一千遍“谢谢你”都不够。

thousand n. 一千

18. Thank you for all that you've done.

谢谢你所做的一切。

19. I would never have managed it if not for your help.

如果没有你帮忙，我根本没法应付。

manage *v.* 管理；没法应付

20. If it weren't for you, I would never have been successful.

要不是有你，我根本不会成功。

successful *adj.* 成功的

21. Thank you for being there for me.

谢谢你总是在支持我。

be there for sb *phr.* 陪伴、支持某人

22. Thank you for your kindness.

谢谢你对我这么好。

23. I'm so lucky to have you.

我有你真是太幸运了。

24. Thank you for what you did for me, and please don't hesitate to let me know if there's anything that I can do for you.

谢谢你为我做的一切，如果有什么我可以为你做的，也请你不要犹豫，马上让我知道。

25. I really appreciate what you did.

我真的很感激你所做的。

觉得自己要求太多或造成麻烦时，表达歉意

1. Please accept my apology for any inconvenience this has caused.

对于这件事造成的不便，请接受我的道歉。

apology n. 道歉　　inconvenience n. 不便　　cause v. 造成

2. I'm sorry about asking for a favor from you out of the blue.

我很抱歉这么突然请你帮忙。

out of the blue phr. 突然

3. I'm really sorry about asking you to put some time aside for me.

要你为我抽出些时间，我真的很抱歉。

put some time aside phr. 抽一点时间

4. I'm really sorry for bothering you when you're so busy.

我真的很抱歉在你这么忙的时候打扰你。

bother v. 打扰

5. I really don't want to trouble you, but you're the only one who can help me.

我真的不想给你造成麻烦，但你是唯一可以帮助我的人。

6. Sorry for the trouble, and thank you so much for the help!

造成麻烦我很抱歉，非常谢谢你的帮忙！

7. I apologize for causing so much trouble.

造成这么多麻烦，我很抱歉。

8. I'll be really, really grateful if you can help me, but please don't hesitate to let me know if it's too much trouble.

如果你能帮我，我会很感激，但如果太麻烦了，也尽请告诉我。

9. I apologize for making you clean up after the mess I made.

我很抱歉让你帮我收拾残局。

clean up *phr.* 清理　　mess *n.* 混乱

10. I hope this doesn't cause you too much inconvenience.

我希望这不会给你造成很大的不便。

11. Please let me know if this is too much trouble for you.

如果这对你来说太麻烦的话，请让我知道。

12. I'll be really grateful for the favor, and will try my best to repay your kindness.

我会很感激你的帮助，也会尽全力报答你的善意。

13. I know that you are very busy, and will be very grateful if you can spare some time to help me.

我知道你非常忙，如果你能抽出一些时间帮助我，我会非常感激。

spare some time *phr.* 抽出一点时间

14. I'll be forever grateful if you can help me.

如果你能帮我，我会永远感激你。

15. I'm so sorry for being a bother, but you're the only person I can turn to.

我这样打扰你真的很抱歉，但你是唯一一个我可以求助的人。

turn to sb *phr.* 转向某人求助

16. I know that I might be asking too much, and you have every right to refuse.

我知道我可能要求太多了，你有足够的理由拒绝我。

ask too much *phr.* 要求太多　　refuse *v.* 拒绝

17. Is there anything I can do to repay your kindness?

我可以做些什么来回报你吗?

18. I know that this is terrible timing, but may I ask a favor of you?

我知道这是个糟糕的时机，但我可以请你帮忙吗?

timing *n.* 时机

19. I know that you're really busy, but I would appreciate your thoughts on the subject.

我知道你真的很忙，但如果你告诉我对这件事的意见，我会很感激。

subject *n.* 主题

20. Would you please fill me in if it's not too much trouble?

如果不是太麻烦，你可以向我说明一下状况吗?

fill sb in *phr.* 向某人说明状况

21. If it's really inconvenient for you, please tell me so.

如果对你来说真的很不方便，也请告诉我。

22. I'm sorry about asking such a huge favor of you.

请你帮我这么大的忙，我很抱歉。

23. I don't mean to trouble you. Please let me know if it's really too much to ask.

我不想麻烦你，如果我真的问太多了，请让我知道。

24. I hate to trouble you, but you're the best person for this.

我不想给你造成麻烦，但你真的是最能帮助我的人。

25. My work is still ridden with flaws, and I would really appreciate it if you could help me make it better.

我的作品还是充满瑕疵，如果你能帮我把它变得更好，我会非常感激。

ridden *adj.* 满是……的；受……困扰的 flaw *n.* 瑕疵

委婉地通知坏消息

1. Thank you so much for your interest, but there's something I need to tell you.

谢谢你对这件事的关注，但有件事我得告诉你。

2. Please understand that what I'm going to say does not reflect my own feelings.

请理解，我接下来要说的并不代表我个人的立场。

reflect *v.* 反映

3. I am regretful that I have to bring you this bad news.

带来这个坏消息，我很抱歉。

regretful *adj.* 抱歉的；遗憾的

4. Unfortunately, things have taken a turn for the worse.

很不幸地，事情越来越糟了。

unfortunately adv. 不幸地
take a turn for the worse phr. 恶化；变得更糟

5. Sadly, things don't look too bright at the moment.

很不幸地，事情现在看起来不太明朗。

bright adj. 光明的　　at the moment phr. 现在

6. I am sorry that I have to inform you of some bad news.

很抱歉，我不得不通知你一些坏消息。

7. I'm sorry about bringing you this bad news, but I promise to try my hardest to assist you through this crisis.

带来这个坏消息我很抱歉，但我答应会努力帮助你渡过这个危机。

promise v. 答应　　assist v. 协助　　crisis n. 危机

8. I was informed of some terrible news and have to let you know.

我被告知了一些糟糕的消息，而且必须让你知道。

9. I was informed of some bad news and want to put you on guard.

我被告知了一些坏消息，我要让你先有个心理准备。

put sb on guard phr. 让某人有心理准备；让某人不掉以轻心

10. There is something that I have to tell you.

有件事情我必须告诉你。

11. It's unfortunate that this has to happen.

很不幸发生了这种事。

12. Please be forewarned that this letter is not going to be pleasant.

请有心理准备，这不会是一封很愉快的信。

forewarn v. 预先警告　　pleasant adj. 愉快的

13. I thought you would probably need to know that things didn't work out too well.

我想你可能需要知道，事情进行得不太顺利。

work out phr. 发展；解决

14. I'm sorry that I have to notify you of this problem.

很抱歉我必须告知你这个问题。

notify v. 告知

15. You might be already aware that we are facing some problems.

你可能已经察觉到了，我们遇到了一些问题。

aware adj. 知道的；察觉到的

16. As you may have noticed, things aren't working out so well.

你可能已经注意到了，事情进展得不是太顺利。

17. Please understand that my thoughts are with you.

请了解，我会一直为你祈福。

18. I'm very sorry, but I have to bring this to your attention.

我非常抱歉，但我得让你注意到这件事。

bring sth to sb's attention phr. 让某人注意到某事

19. Please accept my most heartfelt sympathies.

请接受我诚挚的慰问。

accept v. 接受　　heartfelt adj. 衷心的；真诚的
sympathy n. 同情；慰问

20. It's difficult to say anything to make this easier.

我实在很难说些什么让这个状况变得更简单。

21. I offer my deepest sympathies.

向你致以最深切的慰问。

22. I understand that this must be a difficult experience for you.

我了解这对你来说一定是段困难的经历。

23. I'm saddened by what happened.

我对于发生的事感到非常难过。

sadden v. 使伤心

24. I'm sorry to inform you of this, but I know everything is going to turn out fine.

我很抱歉告知你这件事，但我知道一切都会好转的。

25. Should you require any guidance, please let me know.

如果你需要引导，请让我知道。

require v. 需要　　guidance n. 指导

想要纠正对方，但不能表达太直接，只好把错揽到自己身上的状况

1. I'm sorry, but I might have misheard it. Could you repeat what you said?

我很抱歉，但我可能听错了。可以请你重复一次吗?

mishear v. 听错；误听　　repeat v. 重复

2. I thought it was the other way around, but I must have been mistaken.

我以为是反过来的，但一定是我搞错了。

the other way around phr. 相反；反过来　　mistaken adj. 搞错的

3. I'm very sorry, but I must have forgotten about it.

我很抱歉，但我一定是忘记了。

4. I've never been good at this. Could you explain a bit more in detail?

我从来就不擅长这个。你可以再说得详细点吗?

in detail phr. 详细地

5. We appear to have different views on this. May I ask you to explain your reasoning further to me? I might need to rethink.

看来我们对这事有不同的看法。我可以请你进一步跟我解释一下你的推论吗? 我可能得重新思考。

reasoning n. 推论；论证　　further adv. 更进一步地

rethink v. 重新考虑

6. I'm so sorry, but it must have slipped my mind. Please remind me of what you said.

我很抱歉，但我大概是一时忘了。请再提醒我你讲了什么。

slip sb's mind *phr.* 被某人遗忘　　remind *v.* 提醒

7. I might have made an error in my calculations. Please let me know what you think.

我可能算错了，请告诉我你的想法。

error *n.* 错误；失误　　calculation *n.* 计算

8. I apologize if our miscommunication was due to any error on my part.

如果是因为我方的失误致使我们沟通不畅，我道歉。

miscommunication *n.* 沟通不畅　　due to *phr.* 因为

9. It must have been a mistake on my part. Let me recheck.

一定是我搞错了，请让我再检查一次。

recheck *v.* 再检查

10. I'm sorry, it must be my fault for not double-checking with you.

我很抱歉，这一定是我没跟你复核才造成的过错。

double-check *v.* 复核

11. I'm sorry for not notifying you of this issue sooner.

我很抱歉没有提早通知你这件事。

issue *n.* 议题；事件

12. I'm very sorry. That was not what I meant to imply.

我很抱歉，那不是我想表达的。

13. It must be due to my carelessness that this has happened.

会发生这件事一定是因为我的疏忽。

carelessness *n.* 疏忽

14. I get the impression that you don't really value our relationship, and I would like to apologize for any problems on my part.

我感觉你不是很重视我们的关系，而我要为我这边的问题道歉。

get the impression that *phr.* 得到……印象　　value *v.* 重视

15. I get the feeling that you're dissatisfied. I really value our partnership, and will try hard to improve and meet your expectations.

我觉得你好像不满意。我很重视我们之间的合作关系，会努力提高以达到你的期望。

dissatisfied *adj.* 不满意的
partnership *n.* 合作关系；伙伴关系
meet sb's expectations *phr.* 达成某人的期望

16. I sense that you don't see your experience with us as positive. Please let me know if there's anything we can improve on.

我感觉你对与我们的合作经历不太满意。请让我知道我们有哪些方面可以改进。

sense *v.* 感觉　　positive *adj.* 积极的；正面的

17. I apologize for being unclear and causing this misunderstanding.

我没讲清楚，造成这场误会，我很抱歉。

unclear *adj.* 不清楚的　　misunderstanding *n.* 误会

18. I'm sorry about the miscommunication. It is no fault of yours.

对于我们之间的沟通错误我很抱歉。这不是你的错。

19. I'm very sorry for rushing you. It is my fault for letting you know so late in the first place.

我很抱歉要催你。当初那么晚才让你知道是我的错。

rush *v.* 催促

20. Please accept my apology for the inconvenience I caused you. I will make sure this doesn't happen again.

对于我所造成的不便，请接受我的道歉。我会保证这种事不会再发生。

21. Please be assured that this is not your fault.

请安心，这不是你的错。

assure *v.* 保证；使放心

22. I'm sorry for the confusion. I will be pleased to supply any information you need.

我很抱歉造成困扰。我很乐意提供任何你需要的信息。

confusion *n.* 困惑　　supply *v.* 提供

23. I apologize for not being in contact with you sooner.

没有早点和你联系我很抱歉。

be in contact with *phr.* 与……联系

24. I apologize in advance for any inconvenience this might cause.

如果这可能造成任何不便，我先道歉。

in advance *phr.* 事先

25. Thank you for bringing this to my attention. We'll deal with this immediately.

谢谢你让我注意到这件事。我们会立刻处理。

deal with sth *phr.* 处理某事

表达恭喜、祝贺

1. My warmest congratulations!

衷心地祝贺你！

warm *adj.* 温暖的　　congratulation *n.* 恭喜

2. Please accept my sincere congratulations.

请接受我诚挚的祝贺。

sincere *adj.* 诚挚的

3. I wish you the best of luck in the future too.

我祝福你未来也一样好运连连。

in the future *phr.* 未来

4. Congrats on your great accomplishment!

恭喜你取得这么大的成就！

congrats n. 恭喜（congratulations 的简略说法）
accomplishment n. 成就

5. Don't you know that this is an incredible achievement? Give yourself a pat on the back!

你难道不知道这是难以置信的成就吗？好好奖赏自己吧！

incredible adj. 难以置信的
give sb a pat on the back phr. 好好奖赏、鼓励某人

6. I wish you happiness in the years to come.

我祝你未来几年都开开心心。

happiness n. 快乐　　the years to come phr. 未来几年

7. I wish you a lifetime of happiness!

我祝你开开心心一辈子！

lifetime n. 人生

8. Congratulations on the big announcement!

恭喜你宣布了这个大消息！

announcement n. 宣布；宣告

9. I'm so honored to know someone as successful as you.

我很荣幸认识像你这么成功的人。

honored adj. 荣幸的

10. We need to throw a party for this.

我们得为这个开一场派对。

throw a party phr. 开派对

11. I'm sure that this will be very memorable for you.

我相信这对你来说会是永生难忘的。

memorable adj. 难忘的

12. I'm so happy to hear of your achievement.

听到你的成就，我很开心。

achievement n. 成就

13. I'm looking forward to talking to you about this great news.

我很期待和你谈论这个好消息。

14. I'm so happy for you.

我为你感到非常高兴。

15. I'm so glad for you, you don't even know!

我有多为你高兴，你都不知道!

16. Congratulations on the amazing news.

恭喜你宣布了这么棒的消息。

17. I'm so excited for you!

真为你感到兴奋!

18. My warmest congratulations on your success.

我诚挚地恭喜你取得成功。

19. This will be really rewarding for you.

这对你来说会很值得。

20. This counts as a special occasion.

这算是一个特别的场合。

count as phr. 算是　　occasion n. 场合；事件

21. This is such exciting news for all of us.

这对大家来说都是很兴奋的消息。

22. I hope to follow your example.

我希望能以你为榜样。

follow sb's example *phr.* 以某人为榜样

23. I hope to achieve success like you in the future.

我希望以后也能像你一样成功。

24. My heartiest congratulations.

我衷心祝贺你！

hearty *adj.* 诚心的；热诚的

25. I wish you continued success and happiness.

我祝你一如既往地成功、快乐。

continued *adj.* 未完的；持续的

表达对未来的期待与祝福

1. We wish you all the best.

我们祝您一切顺利。

2. I hope everything goes according to plan.

我希望一切都按计划进行。

according to *phr.* 根据；按照

3. I look forward to maintaining our relationship for years to come.

我期待未来几年我们也能维持我们的关系。

maintain v. 维持　　relationship n. 关系

4. Please take care of yourself and your family.

请照顾好你自己和家人。

5. Hope everything goes smoothly!

希望一切顺利!

6. We look forward to working with you.

我们期待和你们的合作。

7. I'm sure we'll make a great team together.

我相信我们一定会合作愉快。

8. I hope the days ahead of you are full of fun and laughter.

我希望你未来的生活充满欢乐与笑声。

9. I hope you have a great year.

我希望你今年过得好。

10. I hope we can remain friends forever.

我希望我们可以一直是朋友。

11. I foresee that we will be very successful in the future.

我预期我们以后会很成功。

foresee v. 预见

12. Good luck on all your endeavors in the future!

祝你未来所有的努力都有好运!

endeavor n. 努力尝试

13. I hope we can always keep in touch.

我希望我们可以一直保持联系。

keep in touch *phr.* 保持联系

14. No matter what happens, I hope we can always remain in touch.

无论发生了什么事，我希望我们都能保持联系。

15. I wish you the greatest success.

我希望你取得巨大的成功。

16. Best wishes to you!

祝福你！

17. I hope we can meet soon!

我希望我们可以尽快见面！

18. I wish you the best of luck.

祝你好运！

19. I look forward to meeting you!

我很期待见到你！

20. I'll keep my fingers crossed for you.

我会为你祈福的。

keep sb's fingers crossed *phr.* 祈福

21. I know you have a bright future ahead of you.

我相信你的未来一片光明。

22. I know that you will be a valuable asset to us.

我相信你是我们的一笔宝贵财富。

valuable *adj.* 有价值的；珍贵的　　asset *n.* 有价值的人或物；资产

23. I'm confident that you will be a valuable contributor to our team.

我相信你会对我们的团队做出宝贵的贡献。

contributor *n.* 贡献者

24. I believe that we will work very well together.

我相信我们会合作顺利。

25. Thank you for your good wishes, and the same to you!

谢谢你的祝福，也祝福你！